DAVID SCHIMMELPFENNIG

CONFLICT AND COOPERATION ON TRANS-BOUNDARY WATER RESOURCES

NATURAL RESOURCE MANAGEMENT AND POLICY

Editors:

Ariel Dinar
Agricultural and Natural Resources Dept.
The World Bank
1818 H Street, NW
Washington, DC 20433

David Zilberman
Dept. of Agricultural and
Resource Economics
Univ. of California, Berkeley
Berkeley, CA 94720

EDITORIAL STATEMENT

There is a growing awareness to the role that natural resources such as water, land, forests and environmental amenities play in our lives. There are many competing uses for natural resources, and society is challenged to manage them for improving social well being . Furthermore, there may be dire consequences to natural resources mismanagement. Renewable resources such as water, land and the environment are linked, and decisions made with regard to one may affect the others. Policy and management of natural resources now require interdisciplinary approach including natural and social sciences to correctly address our society preferences.

This series provides a collection of works containing most recent findings on economics, management and policy of renewable biological resources such as water, land, crop protection, sustainable agriculture, technology, and environmental health. It incorporates modern thinking and techniques of economics and management. Books in this series will incorporate knowledge and models of natural phenomena with economics and managerial decision frameworks to assess alternative options for managing natural resources and environment.

Disputes over water-rights allocation and over responsibility to protecting trans-boundary water quality have become major concerns at both domestic and international forums. This book explores possibilities for practical applications of the economics of bargaining and game theory approaches in water-related conflicts, in order to evaluate the potential from cooperation in management of trans-boundary water resources.

The Series Editors

Recently Published Books in the Series

Schmitz, A., Moulton, K., Buckwell, A. and Davidova, S.:
Privatization of Agriculture in New Market Economies: Lessons from Bulgaria
Pingali, P. and Roger, P.:
Impact of Pesticides on Farmer Health and the Rice Environment
Bovenberg, L. and Cnossen, S.:
Public Economics and the Environment in an Imperfect World
Martin, Wade E. and McDonald, Lisa A.:
Modeling Environmental Policy
Parker, Douglas D. and Tsur, Yacov:
Decentralization and Coordination of Water Resource Management

CONFLICT AND COOPERATION ON TRANS-BOUNDARY WATER RESOURCES

edited by

Richard E. Just and Sinaia Netanyahu
University of Maryland, College Park, USA

KLUWER ACADEMIC PUBLISHERS
Boston / Dordrecht / London

Distributors for North America:
Kluwer Academic Publishers
101 Philip Drive
Assinippi Park
Norwell, Massachusetts 02061 USA

Distributors for all other countries:
Kluwer Academic Publishers Group
Distribution Centre
Post Office Box 322
3300 AH Dordrecht, THE NETHERLANDS

Library of Congress Cataloging-in-Publication Data

A C.I.P. Catalogue record for this book is available
from the Library of Congress.

Printed on acid-free paper.

Printed in the United States of America

Contents

PART III
THE EXPANDING SCOPE OF TRANS-BOUNDARY WATER PROBLEMS

PART IV
ECONOMIC ISSUES IN TRANS-BOUNDARY WATER ALLOCATION

PART V
POTENTIAL CONTRIBUTIONS OF ECONOMIC ANALYSIS TO TRANS-BOUNDARY WATER COOPERATION

List of Figures

List of Tables

Preface

This book demonstrates what the discipline of economics has to offer as support for analyzing cooperation on management of trans-boundary water resources. It also considers what the discipline of economics has to acquire to become a more effective contributor to trans-boundary water resource management given political, legal, social, physical, scientific, and ecological realities. This book has its genesis in a symposium of the International Water and Resource Economics Consortium held at Annapolis, Maryland, April 13-16, 1997. The symposium was organized by the editors and the book contains papers presented at the symposium with subsequent revisions. The symposium brought together both economists and agency management personnel for the purpose of discussing not only how economic tools apply to trans-boundary water management, but also of identifying the obstacles to making such tools useful and informative to politicians and negotiators in public decision making roles.

INTERNATIONAL VERSUS DOMESTIC TRANS-BOUNDARY PROBLEMS

Trans-boundary water problems arise in many dimensions. The two most important types of problems emphasized in this book are international and domestic interstate or interregional problems. Cooperation on international problems is especially difficult because enforcement must be voluntary given the sovereignty of nations and the absence of an effective legal enforcement mechanism. Agreements must be sustainable and self-enforced if they are to have lasting benefits. Every negotiating country must be convinced it will receive benefits before it gives its consent to cooperation. In the absence of enforceable agreements, trans-boundary (i.e., open) markets may not evolve. In this framework of negotiation and self-enforcement, traditional tools of economics may have little to contribute to the process of finding politically feasible conflict resolution. On the other hand, recently developed economic tools based on game theory seem to be a natural approach for supporting and facilitating cooperation because it is crafted to fit exactly these circumstances. Nevertheless, game theory has been exploited only recently with respect to international water cooperation. Furthermore, the potential of game theory as a tool of economics appears to be little understood or appreciated by negotiators and public agency personnel who are involved in crafting trans-boundary agreements. This book attempts to demonstrate some of this potential.

Domestic interstate/interregional water problems are somewhat different in character from international water problems because courts exist to enforce agreements and, in many cases, federal laws govern allocation and management of water resources. Historically, when water allocation—rather than water quality—was an exclusive issue of concern, water rights were acquired largely on a first-come-first-served basis and then perpetuated by the doctrine of prior appropriation. With growing scarcity of water resources, such use patterns are coming under increasing scrutiny. While revising allocation rights across political boundaries can become hotly contested when reassigned by political action, establishment of water markets by clearly defining property rights and compensation mechanisms may offer a more politically feasible alternative. Economists have long propounded assignment of property rights and establishment of water markets as a means of efficient water allocation. Increasingly such solutions are finding application. This book attempts to demonstrate the potential of economic analysis in evaluating the benefits in cases where markets have been developed and in identifying the steps that must be taken to achieve the benefits.

Interstate/interregional water quality issues, however, are a different matter. The right to pollute versus the right to clean water is not easily solved by appealing to the doctrine of prior appropriation, which often implies the undesirable "victim pays" outcome. Furthermore, environmental interests and awareness are growing stronger and likely dictate changes politically. An important characteristic of water quality problems is that costs and benefits of both pollution and clean up tend to depend heavily on local characteristics of water resources, surrounding land and land use patterns, population density, industrial activity, etc. As these realities have been realized, federal governments have become reluctant to pass uniform laws governing water resource quality or have found that such laws do not work effectively. Alternatively, federal governments in some cases have shifted the responsibility for regulating pollution to states or have begun to work in cooperation with states on such issues. As a result, interstate water quality problems—for example, the Chesapeake Bay—have begun to take on characteristics of international water problems where participation is voluntary and agreements have to be crafted to entice membership. This book attempts to define some of the most important problems and potential solutions related to such trans-boundary water quality problems.

OVERVIEW OF THE BOOK

This book is divided into five parts. Part I gives an overview of international trans-boundary water problems and illustrates the economic tools that can be used to facilitate cooperation and assess its benefits. Part II gives an overview of domestic trans-boundary water problems that exist among states and the approaches that have been used to manage them. Part III considers the expanding scope of trans-boundary water problems that has emerged with increased environmental and political awareness and improved information. Part IV of the book considers the economics of water allocation in trans-boundary problems including water market implementation with inefficient existing resource allocation, and water allocation issues in the presence of uncertainty which pervade most intertemporal water allocation problems. Part V assesses the potential of economic analysis to adapt to the expanding scope of trans-boundary water issues at both international and domestic levels and to contribute as a discipline to understanding the process of sustainable cooperation on trans-boundary water problems.

Part I. International Trans-Boundary Cooperation

Multiple users of common natural resources often face conflicts. Because economic growth of water-scarce countries depends on the sustainability and development of water resouces and because many major water resources span international boundaries, water has great potential as a basis for cooperation. In this book, Just and Netanyahu survey sources of international water conflicts and patterns of cooperative agreements. The success and failure of bilateral and multilateral agreements for exploitation of international water resources are discussed. They observe that most water resources agreements are bilateral even in basins where more than two countries claim rights to use of the resource and present a potential economic explanation for why attempts to expand coalitions beyond two-party coalitions have been unsuccessful. The explanation which is based on the theory of coalition formation suggests that efforts of international agencies to foster grand coalitions might be better redirected.

Frisvold and Schimmelpfennig focus on the structure of agreements which facilitate sustainability and self-enforcement. They demonstrate that much of the potential benefit of international agreements may be lost, depending on circumstances, by having to accommodate sustainability and self-enforcement. However, the alternative of having no agreement is inferior relative to such restricted agreements.

Netanyahu, Just and Horowitz investigate potential cooperation on common groundwater. They discuss the difficulty in defining property rights for a mobile resource with stock uncertainty and the effect of topographical differences in elevation on inequitable exploitation of water resources due to non-uniform pumping costs across aquifers. While equitable use may not result from cooperation in such cases, substantial benefits beyond existing cases with no cooperation are possible.

Bennett, Ragland and Yolles demonstrate the potential of game theory for improving cooperation when multiple bargaining issues are considered. A typical problem with international water resources is that most of the potential benefits may flow to one country which is well situated to take advantage of geography, topography, demography, and infrastructure. A neighboring country may have little inclination to facilitate such a lopsided outcome. However, if another issue, possibly non-water and perhaps non-resource related, can be identified where the benefits of cooperation tend to run in the opposite direction, then much better potential for cooperation exists. This chapter demonstrates the benefits of cooperation involving multiple issues when no satisfactory outcome is possible from negotiation over individual issues.

By and large, the economic tools demonstrated in this part of the book facilitate cooperation and mutual enhancement of social benefits in problems where markets may not exist and where political concerns of equity and other matters override concerns of efficiency.

Part II. Domestic Trans-Boundary Cooperation

Zilberman and Parker provide the overview chapter for domestic trans-boundary water conflict and cooperation. They argue that water markets provide a major tool for solving domestic trans-boundary water problems among states or districts—particularly with respect to water allocation. However, lawmakers need to be convinced of the effectiveness of water markets in order to let markets operate. Presently, many states do not allow water trading or have not granted rights that facilitate trading. Rather, most water rights are only for use.

Wayland and Lewicki give the institutional history and setting of the U.S. water quality experience. In particular, they give a summary of past and current directions of of the U.S. Environmental Protection Agency (EPA) in regulating water quality and pollution. They give the justification for the current EPA approach of encouraging watershed-based control leaving implementation in the hands of local authorities.

Cairo gives an overview of the alternative U.S. experience where the federal government remains a player in management of interstate water resources but does so in cooperation with individual states rather than as a unilateral governing authority. This approach, called the federal interstate compact experience, provides the framework for cooperation among states (as sovereigns) but leaves the mechanism for optimal cooperation among non-sovereigns (local jurisdictions) unaddressed.

Pigram and Musgrave discuss a similar approach, called cooperative federalism, used in Australia as an approach to management of the Murray-Darling Basin. They describe how this approach has led to (i) defining rights to facilitate water markets, (ii) protecting the environment by establishing diversion caps and environmental entitlements to water, and (iii) increasing efficiency through raising water prices (i.e., reducing subsidies to water associated with public capital investment financing).

The institutional setting and experience described in this part of the book suggests that establishment of water markets is an important approach for many allocation problems and identifies the major obstacles to their establishment. On the other hand, water quality problems appear to be couched in a morass of local concerns and political constraints that make federal solution difficult. When the federal government transfers control to states and local jurisdictions or becomes only a cooperative participant, the institutional setting becomes more akin to that of international cooperation. In this context, the economic tools illustrated in Part I of this book may find further use although they have been rarely applied to date.

Part III. Expanding Scope of Trans-Boundary Problems

With increasing demands on water resources, heightened environmental awareness, progress in scientific knowledge, advances in satellite imaging and information processing ability, and increased understanding of the political environment surrounding water policy, the scope of trans-boundary water problems has been expanding significantly. McConnell and Strand begin this section by considering the popular current approach of tributary level policy. They argue that balancing demands on water resources is achieved only by equating marginal costs and benefits across tributaries. When policy is formulated at the tributary level, no mechanism serves to achieve such efficiency. They estimate gains that are possible through coordination across tributaries.

Bockstael and Bell investigate the use of Geographic Information System (GIS) data to determine how differential development density policies among counties cause trans-boundary effects on a common water resource. They demonstrate gains in information possible through modeling spatial autocorrelation and assess the need for trans-boundary coordination of development policy.

Burke, Adams, and Draper demonstrate the importance of GIS data for basing water markets on return flows to streams. Heterogeneous characteristics of land and land use play a major role in determining how much irrigation returns to streams through runoff, etc. If an upstream water user sells his water rights to a downstream user, the upstream user's full water use is not added to the stream flow because return flows would normally reach the downstream user anyway. Only by perfecting return flow coefficients can the incremental water available for sale be measured accurately. Otherwise, water markets will likely fail or function inefficiently.

Austin, Krupnick, and McConnell consider the added dimension that air quality plays in determining water quality. They recognize that air quality significantly affects the Chesapeake Bay and that the relevant airshed is much larger than the watershed. Optimal balancing of quality standards or pollution pricing across different jurisdictions affecting the Bay depends on which pollutant is most critical. Interestingly, in some cases it is appropriate for water quality concerns to drive pricing of air pollution outside of the relevant watershed.

Milon, Kiker and Lee expand the scope of economic modeling of trans-boundary water resource problems by integrating ecology with economic considerations. They demonstrate the importance of endogenizing ecology in economic policy considerations because of two-way implications of political action. Significant ecological-economic interactions are found empirically in the Florida everglades.

Brill, Hochman and Hochman emphasize that care must be taken in crafting water policy when policy makers are uncertain about economic structure. They demonstrate how uncertainty regarding the coefficient of return across jurisdictions can cause taxes to be preferred in some cases and standards to be preferred in others as a trans-boundary allocation mechanism. Elasticities of demand are the key structural parameters that determine which economic instrument is preferred. They also demonstrate that the choice of taxes or standards matters in following the optimal trajectory and that even a correct steady state policy, if enacted without regard to trajectory, may not lead to convergence to the optimal steady state.

Lichtenberg and Olson examine the case where pollutants are cumulative but degrade at some natural rate over time. They demonstrate how critical the degradation rate is for problems of managing trans-boundary water resources in the case of accumulative pollutants. They find that the intensity of restrictions on agriculture for phosphorus loadings vary greatly among Chesapeake Bay jurisdictions.

Tsur and Zemel consider the problem of political uncertainty for trans-boundary water resource management. When water projects take longer to complete, political uncertainty is higher and the likelihood that an international project is abandoned increases. Therefore, political uncertainty makes investment less attractive. They consider the problem of whether, and under what circumstances, speeding up investment is an optimal strategy to reduce political uncertainty as opposed to slowing down investment because of reduced attractiveness of projects under political uncertainty.

This section of the book suggests that trans-boundary water problems are highly complex and multi-dimensional. The scope of issues has expanded greatly and is likely to continue to expand. It also illustrates that the tools of economics can be adapted in many ways to address the expanding concerns that can be quantified.

Part IV. Economic Issues in Trans-Boundary Water Allocation

This section begins by investigating the potential of water markets as a mechanism facilitating efficient trans-boundary water allocation and the steps that must be taken to implement them. Easter, Dinar and Rosegrant investigate the importance of formalizing water markets depending on the jurisdictional setting of water transfer. They find that if water allocation by markets is within jurisdictions, informal markets may be sufficient to facilitate efficient allocation. On the other hand, if trans-boundary trading of water is optimal across jurisdictions, then formal markets should be developed. They discuss the steps that must be taken to do so and the importance of considering transaction costs in evaluating whether such steps are appropriate.

Bjornland and McKay investigate the political inertia of converting water entitlement policies to tradeable water policies. In particular, they consider the problem of sleepers and dozers who, though not using water allocated to them under entitlement policy, will begin selling water once markets are enabled. Aggregate water resources have been overallocated successfully without aggregate overuse under entitlement policy because sleepers and dozers have not fully utilized their allocations. Making a successful transition to market allocation will require finding policies that adapt total use to availability as sleepers and dozers begin to trade.

Moncur, Roumasset and Smith address the problem where future users need to be considered in current allocation problems. Considering the allocation of present water supply to the windward and leeward sides of Oahu in Hawaii, they demonstrate the importance of allocating current water heavily toward enhancing groundwater which can support future users and postpone desalination needs.

Barrett considers the problem of establishing market architecture where aggregate water availability is uncertain and the benefits to individual parties (possibly countries) is uncertain. He examines the implications of various contract structures to determine how well various pricing and control strategies fare in the Lesotho Highlands project. Results show that nonlinear pricing (the case where a single price does not prevail) can greatly enhance efficiency and that control should be given to the party with more variable marginal benefits or to the party with greater marginal benefits (depending on which varies relatively more).

Netanyahu discusses the joint management of common natural resources by economically interdependent political entities. This chapter develops the thesis that uncertainty creates various incentives for governments to coordinate trans-boundary water policies depending on the source of the uncertainty.

Brill, Chakravorty and Hochman consider the problem of establishing water trade when a mix of water products are available—namely, desalinated, recycled and fresh water. They identify conditions in the Israeli-Palestinian water-scarce region of the Gaza Strip where fresh water trade in one direction and recycled water trade in the opposite direction is preferred because of geographic features of water source location and use and because of locational features of agricultural activity.

This section of the book demonstrates that water markets have large potential benefits but that many institutional and management features must be provided through government action before markets can be expected to function well. For example, tradeable property rights must be clearly established, infrastructure supporting trade must be provided, contract structure for trade must be determined appropriately, possibly by legal action, and tradeable rights must be defined in a way that accounts adequately for unrepresented parties such as future generations.

Part V. Contributions of Economic Analysis to Cooperation

The concluding section of the book turns to examining what economics has to offer negotiation and cooperation efforts regarding trans-boundary water problems. Sunding first considers the past and future potential role of economists in domestic problems in the United States. Then Just, Frisvold, Harrison, Oppenheimer, and Zilberman consider the role of economists and their interaction with agencies more generally in both domestic and international trans-boundary water problems. The final chapter is a summary of points made as part of a panel discussion held at the conclusion of the symposium which generated this book. The panel discussion was intended to bring to a head the interchange between economists and agency management personnel about whether and how economic models support cooperation on trans-boundary water problems.

We submit this book as a contribution to practical application and use of economics and as a support for cooperation on trans-boundary water issues. Although the book contains mostly articles written by economists, we hope it finds interest in much broader circles because of its outward orientation toward potential users of economics and because it demonstrates potential applications of economics for many problems where the classical tools of economics fail and noneconomists have little understanding of what the discipline of economics has to offer.

Richard E. Just and Sinaia Netanyahu
University of Maryland, College Park

Acknowledgments

The editors would like to thank the International Water and Resource Economics Consortium and its members for participating in the conference on Conflict and Cooperation on Trans-Boundary Water Resources held in Annapolis, Maryland, April 13–16, 1997, from which this book is derived. The International Water and Resource Economics Consortium's members are the Department of Agricultural and Resource Economics, University of California, Berkeley, USA; the Department of Food and Resource Economics, University of Florida, Gainesville, USA; the Department of Agricultural Economics and Management, Hebrew University of Jerusalem, Rehovot, Israel; the Centre for Water Policy Research, University of New England, Armidale, Australia; the Department of Applied Economics, University of Minnesota, St. Paul, USA; and the Department of Agricultural and Resource Economics, University of Maryland, College Park, USA. The editors also thank the reviewers for discussing individual chapters during the conference. The editors especially thank Liesl Koch for her excellent technical assistance in formatting and organizing this book and without whom the quality and timeliness of its publication would have been lost.

The editors would also like to acknowledge the financial support provided by the International Water and Resource Economics Consortium; the Center for Agricultural and Natural Resource Policy, University of Maryland, USA; the University of Maryland Agricultural Experiment Station; the United States-Israeli Binational Agricultural and Research Development Fund; and the Research and Scientific Exchange Division, Foreign Agricultural Service, United States Department of Agriculture.

PART I

INTERNATIONAL TRANS-BOUNDARY WATER: COOPERATION WITH LIMITED ENFORCEMENT

1

INTERNATIONAL WATER RESOURCE CONFLICTS: EXPERIENCE AND POTENTIAL*

Richard E. Just
Sinaia Netanyahu
University of Maryland, College Park, USA

In recent years, significant discussion has focused on the scope and scale of management of international water resources. Some have advocated that since countries' actions within a basin are interdependent, the management of trans-boundary water basins should consider the basin as a unit and cooperation should include all riparian countries. However, in practice, multilateral agreements on management of trans-boundary water resources with participation of all riparian countries are rare. This chapter discusses the sources of difficulties in forming international river basin agreements involving all riparian countries. Some countries may prefer a small blocking coalition that does not include all riparian countries. Moreover, coalitions can be formed, dissolved, and re-formed by various groups of countries within a basin. This raises the question of how sensitive the sustainability of agreements is to the size and membership of the coalition? We conclude by looking at past and present experience in the management of international water resources, where cooperation is still limited in scope and scale. The implications of our discussion are that a new concept of cooperative management less ambitious than basin-wide management may be desirable because of practicality considerations.

1. INTRODUCTION

In recent years, the scope of issues and the scale to be considered in management of trans-boundary water basins have been ever expanding. Apparently, much of this expansion has been due to identification of additional issues of scope and suggestions of broader management scale. For example, cooperation in the Danube basin was initially over navigation, but the scope has been expanded recently to

* This research was funded in part by the Binational Agricultural Research and Development Fund, BARD Grant No. US-2637-95R, and by Cooperative Agreement 58-319R-4-046 with the Research and Scientific Exchanges Division, Foreign Agricultural Service, United States Department of Agriculture.

include issues such as flood and pollution control (Linnerooth 1990). The Nile is an example of expanding management. Various partial agreements have been signed among riparian countries of the Nile since 1891, but in the last 30 years the Hydromet program and the TECCONILE initiative have promoted a comprehensive cooperative framework culminating in 1994 with twenty-one proposed projects, at a cost of $100 million, to benefit all riparian countries (Alemu et al. 1996; World Bank 1996).[1] Seemingly, a more general definition for trans-boundary water basin management is critical in eliminating most ambiguities in negotiations. Such definitions must identify the participating players, the sources of conflict, and the potential domain for cooperation. Once these elements are identified, it is still questionable whether any particular trans-boundary water resource should be managed as a unit by cooperation among all its riparian states (countries). This chapter focuses on the difficulties in forming such grand coalitions.

In two recent documents, the "Draft Articles of the Law of Non-navigation Uses of International Watercourses" adopted by the International Law Commission in 1994 and the 1997 United Nations report, "Comprehensive Assessment of the Fresh Water Resources of the World," water basins are identified as the logical unit for water management.[2] The "unit" argument is supported by the claim that actions taken by riparian states are interdependent. Action taken by any individual country is likely to impact water resource availability to all or most parties linked to the unit. Lack of coordination among riparian countries likely yields non-optimality from a basin-wide point of view (Biswas 1993). For example, the United Nations report states that "there is a clear need for cooperation in the management of international and trans-boundary watercourses to maximize mutual benefits for all riparian countries." Similarly, Chapter 18 of Agenda 21 of the 1992 Rio de Janeiro Conference suggests that because countries' actions are interdependent, riparian countries should multilaterally participate in forming agreements over water systems.[3]

There are more than two hundred international river systems worldwide.[4] Approximately one hundred and fifty of these water basins are shared by two countries where non-comprehensive bilateral agreements are common. The remaining fifty-three water basins are shared by three or more countries. Of them, eleven basins are shared by more than half a dozen countries (Table 1). Of the fifty-three trans-boundary water basins involving three parties or more, it is difficult

[1] See Alemu et al. (1996), for a detailed chronological discussion of Nile treaties.

[2] Among the first United Nations documents to discuss this concept is the "Integrated River Basin Development" document (1958). This report stated, "It is now widely recognized that individual water projects—whether single or multiple purpose—cannot as a rule be undertaken with optimum benefit for the people affected before there is at least the broad outline of a plan for the entire drainage area."

[3] In particular, see section 18.4 on "Protection of the Quality and Supply of Freshwater Resources: Application of Integrated Approaches to the Development, Management and Use of Water Resources."

[4] This number is given by a United Nations (1975) publication. It does not include rivers that became international following political disintegration of the former Soviet bloc nor does it include trans-boundary aquifers.

to find comprehensive basin-wide treaties. Most water basins shared by three or more nations are managed according to bilateral or trilateral treaties or initiatives which are subject to a limited scope of cooperation.

Table 1. Trans-Boundary River Basins Shared by Three or More Riparian Countries

Number of Riparian Countries	Number of Rivers	Names of Rivers
3	31	
4	9	
5	2	La Plata, Elbe
6	4	Chad, Volta, Ganges-Brahmaputra, Mekong
7	2	Amazon, Rhine
8	1	Zambezi
9	2	Niger, Congo
10	1	Nile
17[a]	1	Danube
Total	53[b]	

[a] Only 11 countries have a significant share in the Danube basin while 6 additional countries with a minor role are also included (The Task Force for the Danube River Basin Environmental Program 1995).
[b] Out of the 53 international river basins of 3 or more riparian countries, 32 basins cover areas greater than 100,000 square kilometers. Out of these 32 basins, 9 cover areas greater than 1 million square kilometers (The United Nations 1975).

Thus, in practice, multilateral agreements on the management of non-trivial trans-boundary watercourses including all riparian countries are a rare exception rather than the rule (World Bank 1993). In fact, empirical work on multilateral organizations shows that large regional, especially international, organizations are less successful than small ones. In particular, Russett and Sullivan (1971) found that the performance of twenty multilateral organizations correlates negatively with the number of states being members. Given this reality, an obvious question is whether international agencies should continue their four-decade-long promotion of the establishment of comprehensive river basin management institutions. Or should policy makers examine the reasons for the failure of grand coalition formation and begin forming less ambitious goals in river basin management?

Using the literature on coalition formation, this chapter discusses difficulties in forming water agreements in water basins with multiple (three or more) riparian states. Despite policy makers' calls for basin-wide treaties we suggest, in contrast to other types of international environmental agreements,[5] that negotiations over international water resources tend to yield bilateral arrangements in higher frequency than multilateral agreements.

The organization of the chapter is as follows. Section 2 describes the evolution of definitions of trans-boundary water resource management. Section 3 discusses common doctrines for allocation of international water resources, recent guidelines for the management of a river basin by its riparian states, and factors to be considered in achieving an equitable water apportionment. Section 4 outlines common obstacles to cooperation that ultimately affect the outcome of negotiations over international water treaties and the size of coalitions. Section 5 reviews trans-boundary water coalitions in practice with a comparison to implications of coalition theory. Section 6 discusses the theory of coalition formation. Special attention is given to explaining countries' motivations to act unilaterally or to join blocking coalitions rather than to forming a basin-wide grand coalition. Section 7 demonstrates theoretically a situation where a group of riparian states in a river basin are better off forming a partial rather than a grand coalition in the presence of externalities. Section 8 summarizes conclusions regarding the formation of treaties on international water resources.

2. THE SCOPE OF TRANS-BOUNDARY WATER RESOURCE MANAGEMENT

The scope of international water resource management has evolved substantially over the past century in order to address increasing conflicts of demand for trans-boundary water resources. In the nineteenth century, agreements over the development and protection of international water resources were limited mainly to navigation.[6] Trans-boundary river basins were called "water of international concern." In this century, increasing competing demands have accelerated conflicts among users and the scope of cooperation has expanded to include allocation of use. In 1966, the Helsinki Rules expanded this scope further, introducing the term "drainage basin." A drainage basin potentially consists of rivers, lakes, channels, groundwater, canals, small and seasonal streams, estuaries, floods, erosion, flora, and fauna. More recently, increasing economic activity has contributed to an increase in pollution of water resources. As a result, pollution issues were formally integrated into the definition of a drainage basin beginning in the late 1960s.[7]

---

[5] For discussions of various economic aspects of international environmental agreements see for example the special issue of *Environmental and Resource Economics* (2) 1992 and Barrett (1995).

[6] Caponera (1987) lists the following examples: the 1915 Final Act of the Congress of Vienna on European Rivers, the 1856 Treaty of Paris on the Danube, and the 1885 Act of Berlin on the Congo and Niger Rivers.

[7] The integration of pollution issues into water treaties was suggested in the following documents: the 1967 Helsinki Rules on "The Uses of Waters in International Rivers," the 1982 Montreal conference report on "The Rules on Water Pollution in an International Drainage Basin," and the 1994 "Draft Articles on the Law of Non-Navigational Uses of International Water Courses," all of which were adopted by the International Law

As the scope of river basin management has expanded from navigation or water diversion to multipurpose environmental issues, another term was introduced: "shared resources." "Shared resources" refer to the resources within a given geographic boundary. However, as pointed out by Szekely (1987), the definition used for this term is not consistent with legal rights and sovereignty because it does not explicitly reflect the presence of political boundaries within geographic boundaries (e.g., the Mountain Aquifer which is utilized by both Israelis and Palestinians). Alternatively, Szekely suggests that "trans-boundary resources" is a more appropriate term.

In 1983, the International Law Commission of the United Nations adopted yet another term: "watercourse system." The intention was to indicate that resources shared by nations have a scope larger than the physical and geographical dimensions suggested by the term "drainage basin." The term "watercourse system" suggests that nations' activities in international water are interdependent (International Law Commission Report 1983). For example, diversion of water from the Ganges-Brahmaputra River by India or from the Yarmouk River by Syria reduces availability for other riparian countries, e.g., Bangledesh/Nepal and Jordan/Israel, respectively.

In contrast, Teclaff and Teclaff (1987) suggest that the term "watercourse system" can potentially misrepresent the concept of a "drainage basin" because it ignores the ecosystem in which the water exists. They suggest that a more useful concept should include the natural land boundaries and the overlaying atmosphere. That is, in addition to water activities in a river basin, interdependence occurs because of economic activities that create externalities and damage the water resources in the basin. For example, coal-burning in neighboring regions and the resulting acid rain near the headwaters of the Danube causes substantial loss of water quality downstream. Similar problems for the Chesapeake Bay from energy choices in the Ohio River Valley are discussed in this volume by Austin et al.

Although trans-boundary water resources have different characteristics, it is only by considering all interactive components that benefit from or cause damage to the resource that optimal use from a social standpoint can be achieved. Thus, for example, all economic activities that interact with a given water resource should be considered. That is, not only economic activities that impose demands on the water resource and therefore create allocation conflicts among users should be included, but also economic activities that affect water quality. When such activities as toxic spills occur, it is clearly in the riparian countries' best interest to minimize damage when the benefits exceed the cost of protection. However, issues become less clear when economic activities related to air quality and land use are involved. Economic activities such as those that reduce air quality and thereby contribute to increased water pollution and/or activities that influence land uses that affect water resources (adversely or beneficially) should not be ignored.[8] With these

---

Association and the 1971 United Nations Economic Commission for Europe report, "Recommendation to ECE Governments Concerning River Basin Management."

[8] The significance of land use and air quality in trans-boundary domestic water resources is discussed in chapters by Bockstael and Austin et al., in this volume.

considerations in mind, it is useful to define water resource boundaries not just according to the physical and geographical boundaries of the resources themselves, but also according to the boundaries of the economic activities that affect the water resources.

Given this discussion, one may conclude that not only is the concept of management of an international river basin as a "unit" logical, but that the definition of the unit should be expanded to include relevant economic boundaries. The problem is that as the scope of management has been successively broadened, the scale of management also tends to increase. For example, only transportation flows are important in determining scope when management is limited to navigational interests (e.g., La Plata). As appropriation is added to the scope, more riparian countries tend to become part of the management problem, but not all riparian countries necessarily participate (e.g., the Nile). Finally, as pollution is added, all riparian countries tend to become part of the equation, both affecting and being affected by water quality, but experience shows that often only bilateral agreements are formed (e.g., the Danube). As atmospheric conditions and land use in the watershed are considered, even non-riparian countries may be involved, causing joint management to become more difficult to achieve (consider, for example, the polluting states both upriver and upwind from the Chesapeake Bay). Despite all intentions, these various complexities and approaches taken to them clearly demonstrate that the unit concept to management of trans-boundary water resources has not been adopted in practice. Moreover, evidence is accumulating that suggests a less ambitious concept of management may have more practical relevance.

3. DOCTRINES AND GUIDELINES FOR WATER APPORTIONMENT

In the last sixty years, a significant increase in water consumption for multiple competing purposes has occurred worldwide. The resulting increase in competition over the use of trans-boundary water resources has instigated an increasing number of conflicts among riparian countries. In order to resolve conflicts, governments can potentially agree to follow certain rules specified in a treaty. The task of writing treaties, however, has become complex because actions must be contingent on various states of natures. For example, agreement on the allocation of a water resource is more difficult when one does not know in advance how much water supply or demand will be generated by future weather conditions (variable monsoons in the Ganges-Brahmaputra basin represent a case in point). In the past few decades, the legal profession has begun to develop principles and guidelines for management of trans-boundary surface and groundwater resources. The most notable documents relating to surface and groundwater have been the Helsinki Rules and the Bellagio draft, respectively. Although not binding, such principles provide a basis for legal rules.

In 1966, the International Law Association adopted the Helsinki Rules. The Helsinki Rules consist of guidelines for a "reasonable and equitable" water allocation in international drainage basins. They suggest vague criteria for water allocation but do not offer a concrete framework for basin-wide cooperation. The Rules imply that reasonable and equitable apportionment should be determined according to the following non-exclusive set of factors: geography, hydrology, size of drainage area, climate, past utilization, economic and social needs of each riparian state, population, availability of alternative sources, comparative cost of such alternatives, pollution avoidance, prevention of harm (waste and damage) to other riparian countries, and compensation as a means of adjusting conflicts. In 1983, a more comprehensive document that included pollution issues was also introduced.

In 1994, the International Law Commission approved the "Draft Articles on the Law of Non-Navigational Uses of International Water Courses." These Articles offer a more comprehensive discussion than the Helsinki Rules on the meanings and implications of principles such as "equitable use" and "no-harm."

Together, LeMarquand (1977) and Rogers (1991) list six main theories which govern international rivers: prior appropriation, the Harmon Doctrine, riparian rights, mutual development, linkage, and reasonable and equitable utilization.

- *Prior Appropriation.* The prior appropriation concept basically advocates historical rights—"first in time, first in rights." An example is the position held by Egypt on its right to use Nile water (Alemu et al. 1996).
- *Harmon Doctrine.* The Harmon Doctrine advocates absolute sovereignty. It implies that the most powerful riparian country—whether determined by the relative strength of economies, militaries, or by who controls the water source—can shape the way the international basin is managed. For example, Turkey is the upstream riparian country and thus the dominant power on the Euphrates River (Lowi 1993).
- *Riparian Rights.* The riparian rights approach adopts the idea of absolute territorial integrity, which guarantees downstream riparian states the use of the river in an unaltered state. For example, recent clean-up efforts in the Danube reflect this principle.
- *Mutual Development Theory.* Mutual development theory is based on the premise that drainage basin development should be consistent with mutual development of a river's waters by all riparian states. For example, the Mekong River Committee was established to deal with such issues.
- *The Linkage Principle.* The linkage principle argues for tying water agreements to other (not necessarily related) issues that are of concern and can benefit one or more countries. For example, it can be argued that in the recent Israeli-Jordanian agreement on the Jordan River, the agreement over the Arava Aquifer and land was based on the linkage principle. The chapter by Bennett et al. (1997) on interconnected games gives a detailed discussion of the implications of the linkage principal for international river basin agreements.
- *The Helsinki Rules.* The doctrine promoted by the Helsinki Rules is that of reasonable and equitable utilization. These guidelines are consistent with reasonable and equitable utilization theory, which is based on limited territorial sovereignty. Limits are determined to ensure that a river's water can be used beneficially without doing harm to other riparian countries. For example, this

position is taken by the upstream riparian states of the Nile that claim to be disadvantaged by the current agreement between Egypt and Sudan.

As the examples indicate, all six doctrines are practiced to some extent in international river basins around the world. Where basins are shared by three or more parties and are not managed by a grand coalition, it is seemingly impossible to determine which of these doctrines dominate in basin-wide agreements. With little doubt, countries determine bargaining objectives on river basin use according to perceptions of ability to gain from various scales of regional cooperation relative to unilateral action. Thus, countries are naturally inclined to adopt doctrines that best defend their position and which give them the highest share of the resources.

Consider, for example, the notable example of the Nile River. The Nile has ten riparian countries. While Egypt and Sudan are downstream countries and contribute essentially no water to the river, they are undoubtedly the river's biggest consumers. The other eight riparian countries obviously dispute Egypt's and Sudan's claims. Egypt's bargaining position on the Nile consists of a combination of appeals to the principle of prior appropriation (i.e., historical rights) and the principle of riparian rights because no alternative resources are available and, hence, Egypt is completely dependent on Nile waters. In addition, it has the economic and military ability to keep other riparian countries from blocking the water before reaching Egypt (the Harmon doctrine). Ethiopia and Sudan base their bargaining position on the principle of reasonable and equitable use (as the Helsinki Rules suggest). In addition, Sudan claims water with the argument that it can benefit relatively more than the other riparian countries from use of Nile water for agricultural purposes due to its excess land availability—a position also supported by mutual development theory. Ethiopia defends its claims by applying the equitable utilization doctrine to argue for potential use of water in the future. Obviously, the various riparian countries have priorities on river basin issues that are in direct conflict with one another. These priorities motivate preferences for different guiding principles with respect to management of the international river basin. While Egypt promotes regional cooperation consistent with status quo allocation, Ethiopia would like to resolve water allocation issues first.

4. OBSTACLES TO TRANS-BOUNDARY WATER RESOURCE COOPERATION

The multiple scope and scale of choices defining international river basin management as well as the existence of competing doctrines that offer guidelines for international water resource management are evidence of numerous conflicts over international water resources. The lack of confinement of hydrological systems within the boundaries of respective political systems (Kneese 1971), the absence of a common goal among riparian countries, and the increasing scarcity of water arising from multiple uses are the core of conflicts in international water basins. Even when conflicts are resolved and treaties are signed, agreements on trans-boundary water resource use can still be easily broken. In particular, agreements are susceptible to change when new information becomes available. Such information can cause incentives for individual nations to exercise the principle of sovereignty strategically. This problem, in turn, raises the possibility that agreements will need to be renegotiated following every major change in the information set. These considerations underscore the importance of achieving broader understanding among riparian countries regarding the exploitation, administration, and protection of such resources.

While the problems in trans-boundary cooperation can be grouped into the two broad categories of asymmetry and sovereignty, they can be usefully identified more explicitly according to the specific obstacles that hinder cooperation. These obstacles originate from asymmetric information among riparian countries, existence of scientific gaps and technical uncertainties, lack of effective enforcement mechanisms and institutions, natural claims for sovereignty, unavoidable conflicting national and international interests, obvious asymmetric country characteristics, and geographical upstream/downstream considerations. The presence of these elements and their consequences ought to be considered simultaneously when an international treaty is designed. We discuss each element in turn.

• *Asymmetric Information.* A fundamental barrier to negotiation arises from asymmetric information among players (Wilson 1995). Riparian countries generally have asymmetric access to data and information because of differing data accessibility and differing abilities to process data. Beyond the social value of data for planning purposes, data has a bargaining value to the country that owns it because it can be used strategically in negotiation with other riparian countries. Having information that is unavailable to other riparian countries allows a country to better calculate its potential payoff from cooperation and to better assess risks. Even when information is available, sharing it openly may not be advantageous to individual countries. A country may decide strategically to share the information fully, partially, or abusively (mislead other parties). This strategic possibility raises uncertainty around the reliability of shared data. Therefore, effective agreements need to be as robust as possible to information asymmetries.

• *Scientific Gaps and Technical Uncertainties.* While the relevance of asymmetric information is well understood in domestic water projects, the implications of asymmetric information intensify when water arrangements are expanded to the international arena. The source of missing information can be scientific and/or strategic. Scientific gaps often originate from inability to fully understand and measure physical processes or from nations' inability to agree on standards (Heal 1991). Scientific gaps create uncertainty with respect to the availability or quality of water. Examples are how acid rain affects water quality in lakes, how much water is available in a particular aquifer, how cyclical are climatic changes, how irreversible are water pollution and over-exploitation, etc. Quantifying water resources can be very difficult, especially in the case of an aquifer. Basin countries often dispute evaporation rates, flow rates (seasonal or annual), number of aquifers and the interactions among them, etc., which makes the estimation of water quantities very difficult. Reaching understanding over water quality is even harder because standards vary among countries. In addition, water quality is often determined in a pragmatic way depending on the purpose of the water use.

• *Enforcement Limitations.* Agreements are not likely to be effective if they cannot be enforced. Enforcement often requires some degree of monitoring. Monitoring has three major limitations: (i) high cost, (ii) technical feasibility, and (iii) sovereignty. When monitoring is feasible and acceptable to all parties, it provides the parties with valuable information. The likelihood of noncompliance should, of course, be small if effective monitoring exists. However, for international issues, monitoring by itself does not provide an enforcement tool. Existing international legal rules offer guidelines for water allocation principles but have no effective enforcement power. In addition, sovereignty principles give nations the option of either not exposing themselves to international court or not accepting a third party ruling.

• *Sovereignty.* Typically, protection of absolute sovereignty is in a country's best interest. Therefore, a country is not likely to accept a decision made by a third party from outside a basin or dictated by a basin-wide planning commission whenever the decision would decreases that country's rights in the river basin.

• *Conflicting National and International Interests.* In defining their interests in trans-boundary water resources, countries are affected by various water-related national priorities. Such national priorities are likely to be influenced by, among other things, administrative and water law procedures, alternative resources for their solution, stage of industrial development, and geographic location (e.g., upstream/downstream considerations). While negotiating a basin-related treaty, a country faces competing domestic and neighboring country pressures simultaneously. For example, giving up water quantities to another country may affect the country's various domestic economic sectors differently, likely affecting the agricultural sector most. Accepting stringent pollution standards (i.e., giving up the "right" to pollute) as part of an international treaty may have a dramatic effect, for example, on specific domestic industries and/or agricultural practices. Domestic interest groups associated with these sectors are likely to oppose a "soft" international position, which complicates the work of policy makers and negotiators.[9]

• *Asymmetric Country Characteristics.* Various countries have different levels of population densities, national income per capita, military power, access to

[9] The seventeen riparian countries in the Danube Basin give a classical example of where entities have conflicting interests and therefore different priorities. Upstream countries use the water mainly for producing electricity, while downstream countries use the water for drinking, irrigation, fisheries, and tourism. While water quality is of interest to downstream countries, the upstream countries have no incentive to invest in pollution prevention (Linnerooth 1990).

natural resources, etc. These differences affect each county's ability to bargain over its share in a basin and to set effective threats to be executed if either agreements or, alternatively, the status quo are broken. Part of the debate concentrates on questions such as the following. Should rich countries that can offer side-payments (or other forms of compensation such as technical know-how) have greater rights over water than highly populated countries? Should income per capita or water availability per capita determine water allocation? Should rich countries have a greater liability to clean up water than poor polluting countries?

• *Upstream/Downstream Considerations.* Although the overall power of a country determines its impact on the size of the ruling coalition in a river basin, its geographical location may be an important factor as well. A country that controls the source of water can potentially hold a major position in the negotiation process (e.g., Syria controls four major heads of the Yarmouk River and therefore has dominant power in its negotiations over water with Jordan). An upstream country can divert water at its source or pollute water carelessly, decreasing both water quantity and quality available to downstream countries. Thus, in certain situations, an upstream country can be an essential player in basin-wide negotiations, without which no agreement can be reached (e.g., when upstream pollution or diversion is a major problem). The Nile River, however, serves as a counter example. While Ethiopia contributes about 86 percent of the Nile water as an upstream country, Egypt contributes essentially no water and uses 66 percent of the annual 84 billion cubic meters flow as a downstream country (Alemu et al. 1996). Although an upstream country can form a credible threat with respect to polluting water, being an upstream country by itself does not give a country the ability to form a credible treat with respect to water diversion. A credible threat must be backed up with, among other things, financial and technical abilities.

Clearly, asymmetric abilities existing among riparian countries provide incentives to basin partners to join a grand coalition or a blocking one or even to keep the status quo in which countries act unilaterally. The major sources of asymmetry among countries that serve as obstacles to cooperation on trans-boundary water resources include, among other things, asymmetry in power (economic or military power), asymmetry regarding information (lack of reliable data and inability to predict an opponent's move), and asymmetry in location (upstream/downstream geographic location). With this perspective, Lowi (1993) suggests that agreements over cooperation can be achieved only if the dominant power in the basin accepts it. We conclude that the formation of a coalition and its size are influenced by the various countries' abilities to provide incentives or disincentives to other riparian countries in bringing them closer to or further from the negotiation table.

5. TRANS-BOUNDARY WATER COALITIONS IN PRACTICE

Currently, there are more than 300 international water resource treaties (Rogers 1992). Such international river basin treaties often deal with multipurpose development projects that increase water supply, prevent wide-scale flooding, produce electric energy, and improve navigation. In recent years, there has also been an increasing awareness with respect to water pollution prevention, environmental quality, and the ecosystem, but few such treaties have been developed.

Given the potential scope of cooperation, an important issue is determination of the optimal scale of cooperation in international river basins. Given the unique characteristics of each river basin, an unambiguous answer to this question is likely not attainable. Three typical approaches to the scale of international river basin management are observed: unilateral action, bilateral treaty formation, and multilateral treaty formation. A fourth approach of basin-wide treaty formation is rarely observed.

Experience shows that many riparian entities act unilaterally on issues that have a direct impact on other riparian water resources. Unilateral acts often correlate with strong domestic political pressure that urges protection of a country's water rights in international river basins (e.g., Turkey's recent irrigation projects in the Euphrates Basin). By promoting its own agenda, a country can increase tension in a basin. Unilateral action tends to escalate international disputes. For example, the dispute among Turkey, Syria, and Iraq over the Euphrates escalated following unilateral moves by Turkey.

The majority of international river basin treaties are between two countries that share resources in a two-riparian-country basin. Examples include cooperation between the U.S. and Canada and between the U.S. and Mexico. The Indus River provides an interesting example of a two-riparian-country basin where the two countries agreed to facilitate independent action. The two riparian countries simply divided the water resources to eliminate interdependence and interaction in allocation between the two countries (Mehta 1986). Even though this agreement facilitates independent allocation, it is expected that environmental problems that can only be solved jointly will induce the two parties to cooperate in the future.

Interestingly, bilateral agreements on water resource management are also observed in basins with three or more riparian countries. Just as unilateral acts can escalate international disputes, a partial coalition consisting of a subset of riparian countries can raise tensions and be an obstacle to formation of a grand basin-wide coalition.

Few treaties involve multilateral cooperation. The common denominator of such treaties is that they are limited in scope of cooperation. The issues in such agreements are simply too complex to consider all possible state contingencies or to get agreement on all such contingencies among countries. As a result, agreements are incomplete and suffer deficiencies of incomplete contacts. State contingent agreements lose their usefulness when countries face states which are not specified in the agreement. As a result, conflicts inevitably arise and treaties are abandoned or must be renegotiated. The situation is more complex when a country is located within a multiple number of basins. In Sub-Saharan Africa, ten major basins have more than four riparian countries each. Twelve countries are co-riparian on between four and twelve basins. For example, Guinea, Mozambique, Cameroon, and Liberia are co-riparian on 12, 9, 6, and 6 basins, respectively (World Bank 1996). Nevertheless, multilateral agreements exist, for example, for environmental management of Lake Victoria and for a hydroelectric project on the Ruzizi River in the Congo Basin (World Bank 1996). Also, multilateral agreements are clearly envisioned for other cases. For example, a basin organization was developed as an initial step for cooperation on the Senegal River, Lake Chad, and the Gambia River Basin (Caponera 1987). Also, potential cooperation among the six riparian countries in the Volta River Basin was recently identified by the World Bank (World Bank 1996).

Treaties formed by grand coalitions are rare in river basins that consist of more than three states. Perhaps such coalitions will only be obtained in an advanced state of multilateral coordination that must first be preceded by bilateral agreements. On the other hand, other factors may explain why such a grand level of cooperation is not optimal in the near term and possibly in the long run. In the remainder of this chapter, we turn to considering such factors. Inefficiencies associated with grand coalitions, for example, may require bilateral treaties in the case of La Plata. In the case of the recent Israeli-Jordanian treaty over the Jordan River, bargaining motivation and strategic positioning may have induced a bilateral treaty when a multilateral treaty over the Jordan Basin might otherwise have been expected. In the Danube case, moral hazard problems may explain why basin leaders favored piece-wise bilateral implementation of the environmental and monitoring program envisioned by a long-term multilateral agreement.

6. COALITION FORMATION IN THEORY

Having reviewed the various practical problems encountered in international river basin cooperation, we now turn to the implications of game theory for addressing such problems. We focus on two main topics: (i) elements that contribute to the rise of partial coalitions rather than formation of grand coalitions, and (ii) the relevance of cooperative game theory for modeling coalition formation in the presence of externalities.

In general, a set of countries will form a coalition when each country realizes potential gains in forming the coalition that could not be obtained if it were to engage in any alternative coalition possibility. In principle, in the absence of externalities the interaction among many players can be modeled as a series of two party games (Harsanyi 1977). However, when one party's action in a bilateral negotiation affects a third party, the game can no longer be modeled as a series of two-party games (Parson and Zeckhauser 1995). Clearly, in the case of a multiparty international water basin, some party's action is likely to have an effect on more than one riparian country (e.g., as in the Danube). Thus, externalities are created. Therefore, the multiparty interaction cannot be disaggregated into series of bilateral games. Moreover, due to the presence of externalities and potential strategic behavior by individual countries in river basin problems, the ability of cooperative game theory to accurately model coalition formation is questionable. We suggest that, for such games with more than two players, it is sensible to examine not only unilateral actions and formation of grand coalitions that comprise all game participants, but also to consider formation of coalitions that include only some of the players. Since most current ruling coalitions in international river basins with more than two riparian countries represent only some of the riparian countries, understanding the reason for this partial-coalition choice is interesting.

6.1 Why Do Coalitions Form?

First, consider why individual countries form partial coalitions (a subadditive game) rather than a grand coalition consisting of all riparian countries (a superadditive game).[10] Both cooperative and noncooperative behavior in trans-boundary resource management has been shown to generate partial coalitions in various circumstances (see e.g., Mäler 1989, and Hoel 1992). The obstacles to cooperation in international river basins noted in Section 4 are affirmed by Auman and Dreze's (1974) discussion on the sources of incentives that give rise to non-superadditive games. They note the following three sources: inefficiency of the grand coalition, moral hazard, and free riding. Owen (1977) and Hart and Kurz (1983) offer a bargaining-driven argument as an additional reason for existence of subadditive games. Each source is discussed in turn.

- *Inefficiency of the Grand Coalition.* Forming a grand coalition forces all members to act together. Acting together may be difficult and costly. Even when full cooperation is economically optimal, players may have subjective reasons not to act together (Aumann and Dreze 1974). India and Pakistan's Indus River serves as an example where the grand coalition was considered to be economically optimal among solutions advocated by the World Bank. However, the sub-optimal solution was adopted by which each country was allocated three streams in the river to be managed independently (Mehta 1986). With increased scope and scale of cooperation, productivity of basin-wide projects can possibly become negative at the margin. The consumption of water by individual countries may decline, for example, if there is congestion in consumption. In these situations, individual countries are likely to face diminishing utility as the size of the coalition increases. In other words, the value of the grand coalition is not necessarily an increasing function of the size of the coalition.

[10] A game (N, V) is superadditive (relative to N) if for any two disjoint coalitions S and T (satisfying $S \cap T = N$) $x \in V(S) \cap V(T)$ implies $x \in V(S \cup T)$. In particular, a transferable utility game is superadditive (relative to N) if for any two disjoint coalitions S and T (satisfying $S \cup T = N$), $V(S \cup T) \geq V(S) + V(T)$. See Greenberg (1994) for details.

• *Moral Hazard Problems*. The inability of countries to monitor compliance of other riparian countries can create an incentive for some countries to take actions that lead to sub-optimal outcomes. Moral hazard problems are readily apparent in the management of international river basins where countries cannot perfectly monitor the clean-up effort invested by polluting riparian countries. Extensive water-pumping from an aquifer beyond agreed quantities is another example. Although action can be detected eventually, it likely takes several years, a period longer than most riparian countries would be willing to wait, especially if the damage is irreversible. Monitoring is less costly and more feasible when the coalition is small and can operate successfully based on mutual trust.[11] As coalition size increases, more costly monitoring is required. In the absence of perfect monitoring, the chances of being caught can potentially decrease with coalition size. As a result, the number of incidents of deviation may increase. Thus, the coalition must impose a more costly threat, suggesting that the cost involved in executing the threat is an increasing function of the size of a coalition.

• *Free Riders Problems*. Subadditive games tend to form when individual countries have (1) different preferences for goods from which they cannot be excluded or (2) benefits from improved management that cannot be limited according to their participation in management (Greenberg 1994). Obviously, in international river basin problems, preferences for the resource are driven, in part, by geographic location (upstream versus downstream) and availability of alternative water resources. Thus, permitting some countries to free-ride and/or rely on alternative water resources is possible during the period when trans-boundary waters are being cleaned up by other co-riparian states. When preferences are sufficiently different, the work of Guesnerie and Oddou (1988) on taxation games related to public goods and of Farrell and Scotchmer (1988) on labor-managed economies suggest that individuals who share the same preferences will form their own institutional arrangements, i.e., form a coalition. This is one argument that can explain the existence of bilateral limited-scope agreements in multiparty international river basins. Parties who share similar preferences have an incentive to form a coalition that addresses their countries' specific concerns. Experimental work in coalition formation predicts the same result. Note also that players tend to form coalitions consisting of members who share close "ideologies" (see references in Section 10 of Greenberg 1994). Even when a superadditive game is played, it is likely that it will not result in the formation of a grand coalition. Medlin (1976) supports this assertion by using empirical data in the case of a three-player superadditive game.

---

[11] The Ruzizi II Regional Hydroelectric Project between Rwanda, Burundi, and Zaire is an example to an unsuccessful performance of trilateral cooperation. The project failed mainly due to the presence of asymmetric information and lack of trust among the countries (World Bank 1996).

• *Bargaining Motivation.* In contrast to Auman and Dreze (1974), Owen (1977) and Hart and Kurz (1983) assume that society operates efficiently and that a grand coalition will eventually be formed. However, prior to the formation of the grand coalition, other coalitions will be formed. They justify the formation of these coalitions as strategic steps to bargaining for a larger share of the pie when the grand coalition is formed. In Hart and Kurz's words (p. 1048), "Coalitions do not form in order to obtain their 'worth' and then leave the game. But rather, they 'stay' in the game and bargain as a unit with all other players." The piece-wise solution of bilateral agreements for Danube environmental problems suggested by the river's heads of states serves as an example for this bargaining motivation.

6.2 Suitability of Cooperative Game Theory for Modeling International River Basin Problems

Cooperative game theory offers a framework for modeling the structure of coalitions. An N-person cooperative game consists of (1) a set of N players and the possible coalitions that can be formed as nonempty subsets of the N players, (2) a set of feasible actions associated with each possible coalition, and (3) a utility function for each individual measuring benefits as a function of the chosen coalition and the action taken. Let A(S) denote the set of feasible actions for each coalition S and let the utility level of individual j who joins coalition S follow $U^j(a, S)$ when action a is taken by the coalition, $a \in A(S)$. Then there is a set of vectors of utility levels V(S) (known also as the characteristic function) that the members of S can attain. Note that (1) V(S) is not dependent on the payoff of nonmembers of the coalition, and (2) the utility level, $U^j(a, S)$, depends on the individual action and the membership and size of coalition S. Clearly, this framework suggests that the selection of actions is made independently by members and non-members and that the coalition's utility level is not affected by actions taken by non-members. According to Greenberg (1994):

(1) Actions A(S) available to coalition S are assumed to be independent of the actions chosen by nonmembers of the coalition.

(2) Utility levels V(S) attained by members of S are also assumed to be independent of the actions chosen by nonmembers of the coalition.

These two assumptions are unlikely to hold in reality for international river basins. Externalities are likely present because actions taken by members are dependent on actions taken by non-members and visa versa, and because members' utilities are likely to be affected by non-members' actions. Greenberg concludes, "in the presence of externalities the characteristic function is no longer unambiguous" (p. 1308). An important issue is thus sensitivity of solutions of cooperative games to externalities.

Recent papers (e.g., Dinar et al. 1992; Dinar and Wolf 1994; Rogers 1993; Tijs and Driessen 1986; Peyton et al. 1982) apply the approach of N-person cooperative game theory to problems of allocation of benefits and costs among riparian countries in international river basin problems. Given the apparent significance of externalities in international river basin management, however, we suggest a closer look at the applicability of cooperative game theory. The shortcomings of cooperative game theory might not be a problem if only grand coalitions are considered as in many previous studies. However, problems may be significant when partial coalitions are considered because actions taken and utility gained by individual countries belonging to particular coalitions are clearly dependent on actions taken by riparian nonmembers of the coalition.

Alternatively, noncooperative game theory appears to be better suited to modeling coalition formation in the presence of strategically driven externalities. For this purpose, Auman (1959, 1967) has introduced the concepts of "strong Nash equilibrium" and "the α and the β core." Models that relax the assumption of binding contracts and stress self-enforcing agreements yield the notion of "coalition proof Nash Equilibrium" (e.g., Bernheim et al. 1987). Coalition formation as an extensive form game that captures sequential bargaining is modeled, for example, by Moldovanu (1990).

The above examples from the literature on coalition formation suggest that coalition formation may be modeled appropriately in the framework of cooperative games, strategic games, or sequential games, depending on the environment in which the game is played.

7. FORMATION OF COALITIONS IN A RIVER BASIN WITH ECONOMIC EXTERNALITIES

In this section, we introduce a simple game in which coalition formation takes place in an environment with economic externalities. A stylized example of coalition structures is introduced to illustrate two results. First, the example shows how the choice of partial rather than grand coalitions may be optimal either when the transaction costs of coordination are considered or when water conservation encounters decreasing returns to scale. Second, the example shows that once a coalition is formed, optimal behavior may allow members of the coalition to carry different environmental responsibilities or receive different allocation opportunities, even when countries are identical, with inequities compensated by side payments.

Consider N identical riparian countries in a basin, all of which consume water from a given body of water. Each country can either treat its wastewater or not and must purify the water it consumes. Let

A	=	cost to an individual country of treating wastewater before discharge
B(n)	=	cost to an individual country of purifying water for consumption when n countries do not treat waste

T(n) = transaction costs to an individual country to cooperate with n-1 other countries

C(n,m) = total waste treatment and water purification cost per member country incurred by a coalition of size n if exactly m members of the coalition treat waste.

This framework can also represent the essential elements of a water allocation problem if, for example, B(n) represents the cost of acquiring water from a stream when, say, n other countries are using water. The cost A can represent the cost of building water storage facilities which facilitate off-peak withdrawals from the river or the cost of building facilities that route recycled water back to the river. Transaction costs of cooperation and the overall cost of water can be defined essentially as in the water quality problem.

Assume that the cost of purifying/obtaining water is increasing in the number of countries that pollute/use the water, i.e., $B_n(n) > 0$, and that transaction costs are increasing in the number of countries cooperating, i.e., $T_n(n) > 0$. Subscripts are used to represent first differences so that $B_{n+}(n) = B(n+1) - B(n)$ and $B_{n-}(n) = B(n-1) - B(n) = -B_{n+}(n-1)$. For simplicity and without loss of generality, assume B(0) = 0 and T(1) = 0. Also, assume no coalition forms if members are exactly as well off with a coalition as without it.

For brevity, we discuss only the water quality interpretation of the example. With unilateral action, each country can treat or not. If n is the number of other countries that do not treat waste, then an individual country's cost is

$$C(1,1) = B(n) + A$$

if it treats waste, or

$$C(1,0) = B(n+1)$$

if it does not treat. Two possibilities arise. First, an individual country acting unilaterally will choose to treat waste if

$$B_{n+}(n) > A, \tag{1}$$

in which case C(1,0) > C(1,1). If (1) holds for all n then all countries will treat waste without cooperation, or if $B_{nn+}(n) > 0$ and (1) holds for some resulting coalition where all n members treat waste, then all nonmember countries will treat waste as well. We dismiss these cases as uninteresting because the typical problem is how to achieve water conservation through cooperation when it does not occur otherwise. The second possibility is where an individual country acting unilaterally will choose not to treat,

$$B_{n+}(n) < A, \tag{2}$$

in which case $C(1,0) < C(1,1)$. In this case, countries acting individually become free riders. If (2) holds for all n, then coalition nonmembers will never treat waste, or if $B_{nn+}(n) < 0$ and (2) holds for some resulting coalition of size n, then nonmembers will not treat waste no matter how large the coalition grows.[12]

Assuming (2) applies with a coalition of size n, the coalition can choose the number m of its members that treat waste such that $0 < m < n \leq N$, where N is the total number of riparian countries. In other words, a coalition may choose to have some of its members forgo treatment as long as costs are shared appropriately. The cost per member country is thus

$$C(n,m) = B(N-m) + T(n) + mA/n. \tag{3}$$

The incremental per member cost in having one additional member treat waste is

$$C_{m+}(n,m) = B_{n-}(N-m) + A/n. \tag{4}$$

We treat this problem in two cases: $B_{nn+} < 0$ and $B_{nn+} > 0$ (i.e., cost to an individual country of purifying water for consumption is increasing at a decreasing rate, and increasing at an increasing rate in the number of countries not treating waste, respectively).

Case 1. If $B_{nn+} < 0$, then the problem of choosing m for a given n has a corner solution so that either m = n or m = 0. Either all coalition members will treat waste or none will treat. If no members treat, then m = 0 and $C(n,0) > C(1,0) = B(N)$ so cooperation is dominated by noncooperation. Thus, the only possible cooperation is a coalition where all members treat waste yielding per country cost $C(n,n)$. This cooperation would occur only if $B(N) > B(N-n) + T(n) + A$. The intuition of this case is perhaps more transparent and leads most casual observers to conclude that all identical members of a coalition should agree to participate in conservation when conservation is desirable.

Assuming $B_{nn+} < 0$ so that all coalition members treat waste, the cost per country for coalition members is

---

[12] If $B_{nn+}(n) > 0$, an individual country could choose not to treat waste when the number of countries treating waste is small but to treat if the number of countries treating is large. This would be the case if $B_{n+}(n) < A$ for small $n \geq 1$ and $B_{n+}(n) > A$ for large $n < N$. In this case, all countries acting unilaterally will follow the lead of countries forming a coalition (in which all members treat waste) if the coalition is sufficiently large. In this case, trans-boundary treaties can only be effective in the presence of a minimal coalition, much as in the case analyzed by Black et al. (1992). However, if a country is better off being a free rider when the treating coalition is small, then it will be better off being a free rider if the size of a treating coalition becomes large. Alternatively, if $B_{nn+}(n) < 0$ and a country is better off being a free rider with a small treating coalition, treatment could become advantageous with a sufficiently large treating coalition, i.e., $B_{n+}(n) > A$ for small $n \geq 1$, and $B_{n+}(n) < A$ for large $n < N$. While the results below can be generalized to consider these possibilities, for simplicity we consider only the case where $B_{n+}(n) < A$ for all n after dismissing the case with $B_{n+}(n) > A$.

$$\overline{C}(n) / C(n,n) = B(N-n) + T(n) + A.$$

The incremental per member cost from increasing the coalition size assuming all members treat is

$$\overline{C}_{n+}(n) = B_{n-}(N-n) + T_{n+}(n).$$

If $\overline{C}_{nn+}(n) = B_{nn+}(N-n) + T_{nn+}(n) < 0$, then a corner solution occurs for coalition size n. Since $\overline{C}(1)$ is dominated by no cooperation [(2) implies $\overline{C}(1) = B(N-1) + A > B(N)$], the result is that either no cooperation occurs with per country cost B(N) or a grand coalition forms where all countries cooperate and treat waste with per country cost $\overline{C}(N) = T(N) + A$.

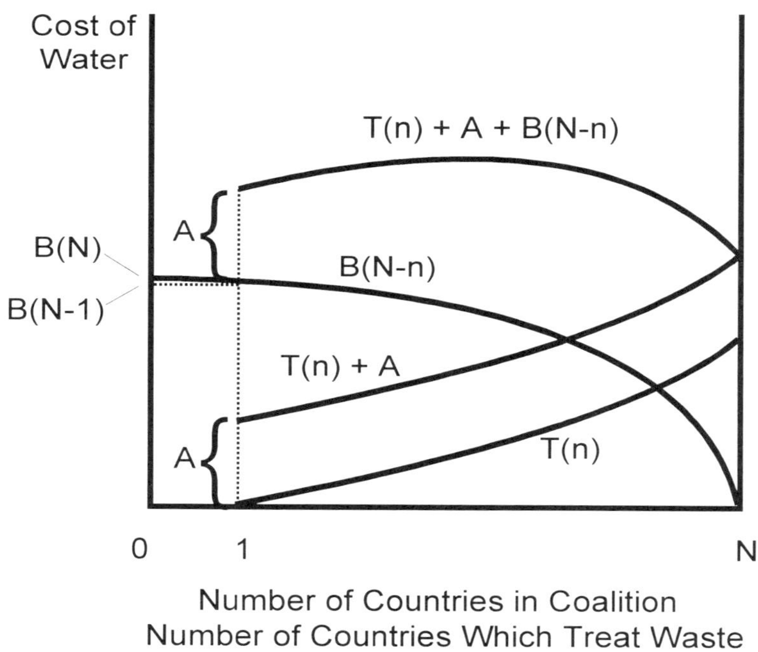

Figure 1. Either No Coalition or a Grand Coalition Dominates

The intuition of this case follows Figure 1 and agrees with the line of casual thought whereby, if some conservation is good, then complete cooperation in a grand conservation coalition is better. Transaction costs are represented by an upward sloping curve T(n) in the number of countries cooperating and treating, which is assumed to increase at an increasing rate. The cost of treatment B(N-n) is a decreasing curve in the number of countries cooperating and treating, which is assumed to decrease at an increasing rate (the cost of treatment is increasing at a decreasing rate in the number of countries not treating waste). Adding these two curves to A obtains the total per country cost of B(N-n) + T(n) + A. The concave curvature of B(N-n) + T(n) + A occurs in the case $B_{nn+}(N\text{-}n) + T_{nn+}(n) < 0$ and implies minimum cost is obtained either with no cooperation (n = 1), in which case no treatment dominates, or a grand coalition (n = N). If T(N) + A < B(N), then cost is less for the grand coalition (n = N) so all countries treat waste. If B(N) < T(N) + A, then cost is minimized when no country treats waste (n = 0).

An alternative case occurs when $B_{nn+}(N\text{-}n) + T_{nn+}(n) > 0$ because an internal solution occurs for n if $B_{n-}(N\text{-}1) + T_{n+}(1) < 0$ and $B_{n-}(0) + T_{n+}(N) > 0$. This case is represented in Figure 2 where B(N-n) + T(n) + A is convex and has an interior minimum. This curve is sufficiently tilted so that minimum cost occurs with a grand coalition at n = N if $B_{n-}(0) + T_{n+}(N) < 0$ or with no coalition at n = 0 if $B_{n-}(0) + T_{n+}(N) > 0$ (these cases are not shown in Figure 2). Otherwise, a grand coalition is inferior to a smaller coalition, such as size n*, but some coalition is preferable to no coalition.

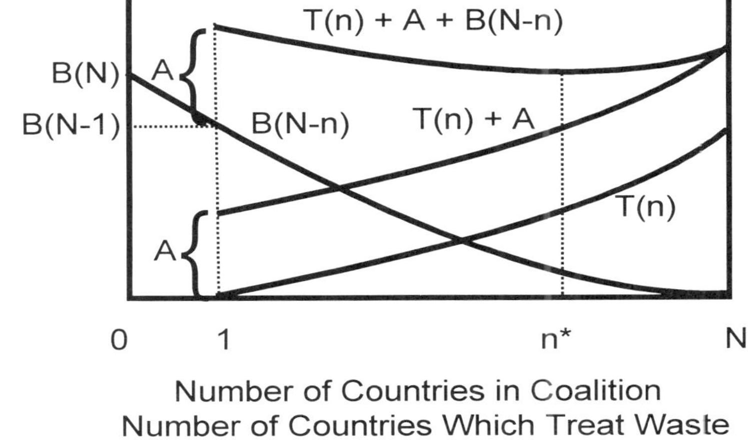

Figure 2. A Partial Coalition Dominates

This case occurs if either purifying costs or transaction costs are increasing at a sufficiently increasing rate. For example, this case occurs when transaction costs increase at a faster rate than the rate of increase in purifying cost decreases (if purifying cost is increasing at a decreasing rate). Furthermore, if purifying cost is increasing at an increasing rate, $B_{nn+} > 0$, then this result can hold (a partial coalition with all members treating can be preferred to a grand coalition with all members treating), but allowing some members of the coalition to forgo treatment may rank even higher.

Case 2. Where $B_{nn+} > 0$, an internal solution occurs for m if $A/n < B_n(N-1)$ and $A/n > B_n(N-n)$. To see this, suppose the coalition size is fixed at n and consider varying the number of coalition members who treat, in which case per country costs follow (3). The marginal effect of changing the number of coalition members who treat is given by (4). Starting from a situation where m members treat, the coalition is better off to have an additional member treat if $B_{n-}(N-m) + A/n > 0$ or, equivalently, if $A/n > B_{n+}(N-m-1)$. If $A/n < B_{n+}(N-n)$, then the coalition will choose to have all its members treat because $B_{nn+} > 0$ then implies that A/n will be less than $B_{n+}(N-m)$ for all m < n. However, if $A/n < B_{n+}(N-1)$ and $A/n > B_{n+}(N-n)$, then there will be some optimal number of coalition members m < n which should treat assuming $T(n) + A + B(N-n) < B(N)$.

The intuition of this case is presented in Figure 3. If the cost of treatment curve is sufficiently convex, then A/n for a fixed coalition size n* can lie between the slopes of the B curve at 1 firm treating and n* firms treating. This implies that the per member benefits of spreading costs over non-treating members (A/n*) are more beneficial than having additional member firms treating waste [which has benefit $B_{n+}(N-m)$].[13] In other words, member countries that treat waste are better off than if the member non-treaters are excluded from the coalition because the member non-treaters share in the cost of treatment. On the other hand, the member countries who do not treat are better off as cost-sharing members if $B(N-m) + T(n) + mA/n < B(N)$, where B(N) is the cost faced by such members if the coalition breaks down.[14] In practicality, this sharing may take place through various forms of side payments of either a pecuniary or non-pecuniary nature. For example, a country with superior

[13] With slight manipulation, the case in Figure 3 can also imply that partial treatment in a grand coalition can be optimal. This case would be obtained, for example, if the transaction cost curve is insignificant, so that the $T(n) + A + B(N-n)$ curve has a minimum at n = N.

[14] Depending on the rate of increase in B(n), a coalition of this type could result in an instability where non-treating members would rather be nonmembers if the coalition could be preserved in such a move because $B(N-m) < B(N-m) + T(n) + mA/n$. If an attempt to achieve this result causes the coalition to break down, then the cost of free riding increases from B(N-m) to B(N) where the latter may be higher than C(n,m). This outcome would represent a prisoner's dilemma. For further discussion of prisoner's dilemma problems in the management of trans-boundary resources, see Ostrom (1990) for the one-shot prisoner's dilemma problem and Mäler (1989) for the repeated prisoner's dilemma problem.

water technology such as Israel may be able to make a greater claim to water by sharing its technology.[15]

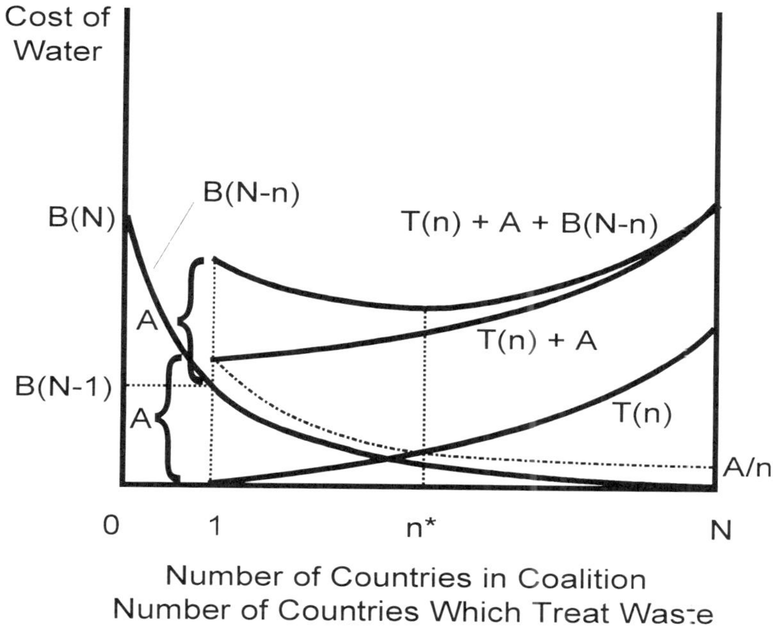

Figure 3. Partial Treatment as Grand Coalition Behavior

As this example shows, broad conditions exist under which partial coalitions can dominate permanently, i.e., in a static sense, rather than simply during the transition to a grand coalition. Furthermore, coalitions that permit inequitable behavior (partial treatment among their membership) can be optimal where inequities are offset by side payments that may or may not be related to water. Interestingly, the conditions under which these arrangements become optimal are those where costs are increasing at an increasing rate—an imminently plausible condition by economic standards. Typical economic reasoning suggests (i) that the expense of purifying water should increase at an increasing rate as pollution increases and as the limits of technology are reached, (ii) that the cost of pumping water from a river should increase at an increasing rate as water demands on the river increase and the limits of the river flow are exploited, and (iii) that coordination with other countries should become increasingly difficult as the number of countries gets large and the limits to cooperation and diplomacy are exhausted.

---

[15] The presence of transfers in coalition formation are explained as a bargaining strategy by Stranlund (1994). Situations that give rise to formation of partial coalitions rather than to establishment of full cooperation and expansion of existing coalitions by means of self-financed welfare transfer are analyzed by Carraro and Sinscalco (1993).

8. CONCLUSIONS

Managing water effectively requires consideration of all the interacting players. Decisions made unilaterally or by a partial coalition likely will not serve the interests of the non-participating parties due to conflicting priorities of nations. Nevertheless, water basins consisting of more than two countries are currently dominated by bilateral rather than multilateral water treaties. In this context, free riding becomes a major concern.

This chapter suggests and demonstrates by example that grand coalitions often envisioned by international agencies and championed by environmental interests may not be optimal arrangements. Either increasing costs due to physical and technological limitations associated with trans-boundary water resources may explain incomplete conservation actions among the members of water basin coalitions, or increasing transaction costs of monitoring and coordinating efforts among many nations with conflicting interests may make complete contracts and comprehensive treaties less desirable. Results show that inequitable water arrangements can become optimal even among identical countries when creative transfer arrangements are considered. While the empirical relevance of these explanations has yet to be tested, the general framework appears to be consistent with casual analysis of the evidence.

REFERENCES

Alemu, S., A. Dinar, and G. LeMoigne. 1996. The Nile Basin: Data Review and Riparian Issues. Center for International Development and Conflict Management, University of Maryland.

Aumann, R. 1959. Acceptable Points in General Cooperative N-Person Games, *Annals of Mathematics Studies* 40:287-324.

Aumann, R. 1967. A Survey of Cooperative Games without Side Payments. In M. Shubik, ed., *Essays in Mathematical Economic*. Princeton: Princeton University Press.

Aumann, R., and J. Dreze. 1974. Cooperative Games with Coalition Structures, *International Journal of Game Theory* 3:217-237.

Austin, D., Krupnick, A., and V. McConnell. 1997. Efficiency and Political Economy of Pollution Control with Ancillary Benefits: An Application to NO_X Control in the Chesapeake Bay. In R. Just and S. Netanyahu, eds., *Conflict and Cooperation on Trans-boundary Water Issues*. Boston, MA: Kluwer Academic Publishers.

Barrett, S. 1995. Toward a Theory of International Environmental Cooperation. Paper No. 60.95, Economics, Energy, Environment, Fondazione Eni Enrico Mattei.

Bennett, L., S. Ragland, and P. Yolles. 1997. Unidirectional Problems with International River Basins: An Interconnected Game Approach. In R. Just and S. Netanyahu, eds., *Conflict and Cooperation on Trans-boundary Water Issues*. Boston, MA: Kluwer Academic Publishers.

Bernheim, B.D., B. Peleg, and M. Whinston. 1987. Coalition Proof Nash Equilibria, I: Concepts, *Journal of Economic Theory* 42:1-12.

Biswas, A.K. 1993. Management of International Waters: Problems and Perspective, *Water Resources Development* 9:167-188.

Black, J., M.D. Levi, and D. De Meza. 1992. Creating a Good Atmosphere: Minimum Participation for Tackling the 'Greenhouse Effect', *Economica* 60:281-294.

Bockstael, N., and K. Bell. 1997. Land-Use Patterns and Water Quality: The Effect of Differential Land Management Controls. In R. Just and S. Netanyahu, eds., *Conflict and Cooperation on Trans-boundary Water Issues*. Boston, MA: Kluwer Academic Publishers.

Caponera, D.A. 1987. Patterns in International Water Law: Principles and Institutions. In A.E. Utton and L.A. Teclaff, eds., *Transboundary Resources Law*. Boulder: Westview Press.

Carraro, C., and D. Siniscalco. 1993. Strategies for the International Protection of the Environment, *Journal of Public Economics* 52:309-328.

Dinar, A., A. Ratner, and D. Yaron. 1992. Evaluating Cooperative Game Theory in Water Resources, *Theory and Decision* 32:1-20.

Dinar, A., and A. Wolf. 1994. International Markets for Water and the Potential for Regional Cooperation: Economic and Political Perspectives in the Western Middle East, *Economic Development and Cultural Change* 43:43-66.

Farrell, J., and S. Scotchmer. 1988. Partnership, *Quarterly Journal of Economics* 103:279-97.

Greenberg, J. 1994. Coalition Structures. In R.J. Aumann and S. Hart, eds., *Handbook of Game Theory*. Vol. 2. Amsterdam: Elsevier Science.

Guesnerie, R., and C. Oddou. 1988. Increasing Returns to Size and their Limits, *Scandinavian Journal of Economics* 90:259-72.

Harsanyi, J.C. 1977. *Rational Behavior and Bargaining Equilibrium in Games and Social Situations*. Cambridge: Cambridge University Press.

Hart, S., and M. Kurz. 1983. Endogenous Coalition Formation, *Econometrica* 51:1047-64.

Heal, G. 1991. Economy of Climate: A Preliminary Framework for Microeconomic Analysis. In R.E. Just and N. Bockstael, eds., *Commodity and Resource Policies in Agricultural System*. Berlin: Springer-Verlag.

Hoel, M. 1992. Emission Taxes in a Dynamic International Game of CO_2 Emissions. In R. Pethig, ed., *Conflict and Cooperation in Managing Environmental Resources*. Berlin: Springer-Verlag.

International Law Association. 1967. The Helsinki Rules on The Uses of Waters in International Rivers.

International Law Association. 1982. Montreal Conference Report on The Rules on Water Pollution in an International Drainage Basin.

International Law Commission. 1983. Draft Report on International Drainage Basin.

International Law Association. 1994. Draft Articles of the Law of Non-Navigational Uses of International Watercourses.

Kneese, A.V. 1971. Background for Economic Analysis of Environmental Pollution. In P. Bohm and A.V. Kneese, eds., *The Economics of Environment*. London: Macmillan.

LeMarquand, D. 1977. *International Rivers: The Politics of Cooperation*. Vancouver: Westwater Research Center.

Linnerooth, J. 1990. The Danube River Basin: Negotiating Settlements to Trans-boundary Environmental Issues, *Natural Resources Journal* 30:629-60.

Lowi, M.R. 1993. *Water and Power: The Politics of a Scarce Resource in the Jordan River Basin*. Cambridge: Cambridge University Press.

Mäler, K.G. 1989. The ACID Rain Game. In H. Siebert, ed., *Valuation Methods and Policy Making in Environmental Economics*. Amsterdam: Elsevier.

Mehta, J.S. 1986. The Indus Water Treaty. Presented at the Conference on the Management of International River Basins, International Institute for Applied Systems Analysis.

Medlin, S.M. 1976. Effects of Grand Coalition Payoffs on Coalition Formation in Three-Person Games, *Behavioral Science* 21:48-61.

Moldovanu, B. 1990. Sequential bargaining, cooperative games, and coalition-proofness. Mimeo, Bonn University.

Ostrom, E. 1990. *Governing the Commons: The Evolution of Institutions for Collective Action*. New York: Cambridge University Press.

Owen, G. 1977. Values of Games with a Priori Unions. In R. Hein and O. Moeschlin, eds., *Essays in Mathematical Economics and Game Theory*. New York: Springer-Verlag.

Parson, E.A., and R.J. Zeckhauser. 1995. Cooperation under Unbalanced Commons. In K. Arrow, R.H. Mnookin, L. Ross, A. Tversky, and R. Wilson, eds., *Barriers to Conflict Resolution*. New York: W.W. Norton & Company.

Peyton Young, H., N. Okida, and T. Hashimoto. 1982. Cost Allocation in Water Resources, *Water Resources Research* 18:463-475.

Rogers, P. 1991. International River Basins: Pervasive Unidirectional Externalities. Presented at conference on The Economics of Transnational Commons, Universita di Siena, Italy.

Rogers, P. 1992. Economic and Institutional Issues in International River Basins. In Le Moigne, G., S. Barghouti, G. Feder, L. Garbus, and M. Xie, eds. *Country Experience with Water Resources Management*. Washington, D.C.: The World Bank.

Rogers, P. 1993. The Value of Cooperation in Resolving International River Basin Disputes, *Natural Resources Forum* 17:117-131.

Russett, B., and J.D. Sullivan. 1971. Collective Goods and International Organizations, *International Organization* 25:845-865.

Stranlund, J.K. 1994. Bargaining and Transfers in International Environmental Relations. Department of Resource Economics, University of Massachusetts, Amherst.

Szekely, A. 1987. Trans-Boundary Resources: A View from Mexico. In A.E. Utton and L.A. Teclaff, eds., *Trans-Boundary Resources Law*, Boulder: Westview Press.

Task Force for the Danube River Basin Environmental Program. 1995. Strategic Action Plan for the Danube River Basin 1995-2005.

Teclaff, L.A., and E. Teclaff. 1987. Trans-Boundary Toxic Pollution and the Drainage Basin. In A.E. Utton and L.A. Teclaff, eds., *Trans-Boundary Resources Law*. Boulder: Westview Press.

Tijs, S.H., and T.S.H Driessen. 1986. Game Theory and Cost Allocation Problems, *Management Science* 32:1015-28.

United Nations. 1958. Integrated River Basin Development, Department of Economics and Social Affairs, U.N. Doc. No. E/3066.

United Nations. 1971. Recommendation to ECE Governments Concerning River Basin Management.

United Nations. 1975. Management of International Water Resources: Institutional and Legal Aspects, Natural Resources/Water Series No. 1. New York.

United Nations. 1992. Protection of the Quality and Supply of Freshwater Resources: Application of Integrated Approached to the Development, Management and Use of Water Resources, Rio de Janeiro Conference, Brazil, Agenda 21, Chapter 18.

United Nations. 1997. Comprehensive Assessment of the Fresh Water Resources of the World.

Wilson, R.B. 1995. Strategic and Informational Barriers to Negotiation. In K. Arrow, R.H. Mnookin, L. Ross, A. Tversky, and R. Wilson, eds., *Barriers to Conflict Resolution*. New York: W.W. Norton & Company.

World Bank. 1993. Water Resources Management, Policy Paper, p. 12.

World Bank. 1996. African Water Resources: Challenges and Opportunities for Sustainable Development, World Bank Technical Paper No. 31, African Technical Department Series.

2

POTENTIAL FOR SUSTAINABILITY AND SELF-ENFORCEMENT OF TRANS-BOUNDARY WATER AGREEMENTS

George Frisvold
David Schimmelpfennig
Economic Research Service, USDA, USA

Recent theoretical research on self-enforcing international environmental agreements (IEAs) has yielded rather pessimistic results: IEAs can be sustained only in cases where the outcome of cooperation is only a minor improvement over noncooperation. However, trans-boundary water pollution problems and negotiations have salient differences from previous theoretical representations of IEAs. This chapter discusses these differences and presents some counter-examples where trans-boundary water agreements can achieve significant increases in pollution abatement and abatement benefits over noncooperation.

1. INTRODUCTION

Many environmental and natural resource management problems are trans-boundary in nature. Examples include climate change, stratospheric ozone depletion, acid rain, and loss of biological diversity. Water resource development and pollution abatement are often trans-boundary problems as well. Water bodies either form or flow across borders. Oceans, lakes, rivers, and inland seas are environmental resources often shared by multiple national, state, or provincial entities.

Trans-boundary environmental management is particularly challenging because agreements must be reached between sovereign entities. One cannot appeal to a central authority to impose Pigouvian taxes or set emission standards. Consequently, many international environmental agreements are made up of non-binding commitments. Whether countries adhere to these commitments is often a matter of choice. So, international environmental agreements must be self-enforcing. The fundamental problem countries face is to overcome free-riding. Each country has an incentive to allow others to abate pollution while withholding abatement effort.

There is a growing economic literature on international environmental agreements

(IEAs) and on the capacity for such agreements to (a) be self-enforcing, and (b) achieve meaningful improvements relative to countries' unilateral pollution abatement (Barrett 1994a, 1994b, forthcoming; Bauer 1992; Black et al. 1992; Carraro and Siniscalco 1993; Hoel 1992). Results from this literature have rather pessimistic implications for IEAs. Given the assumptions of their theoretical models, Hoel finds that no more than three countries will enter into a self-enforcing IEA, while Bauer finds that no more than two will. Barrett (1994a, 1994b) has shown that many countries may accede to an IEA, but only under circumstances when pollution abatement under full cooperation is not much greater than abatement under noncooperation. In other words, agreements with a large number of signatories do little more than codify countries' noncooperative strategies. In cases where potential gains from full cooperation are large, a self-enforcing IEA involving many countries cannot be sustained. Barrett obtains these basic results in both static and repeated game frameworks.

The theoretical results obtained by Barrett and by others are particularly pessimistic considering all the complications of international negotiations *left out* of these models. These simple formulations abstract from broader political conflicts between nations sharing watersheds. There is no uncertainty about the political stability of bargaining nations or entities. Neither is there uncertainty about the environmental benefits or economic costs of pollution abatement. Moreover, negotiators follow the precepts of economic rationality. Countries choose to accede or not to accede to an agreement and influence the conditions of agreements in order to maximize their net benefits. In other words, economists (or at least economic concepts) are not excluded from the bargaining table. Even though many daunting real-life complications are ignored, these models illustrate how difficult it can be to overcome simple free-rider problems.

These pessimistic results, however, are based on highly stylized models. For example, Barrett (1994a, 1994b) assumes that (i) all countries have identical pollution abatement benefit functions, (ii) all countries have identical abatement cost functions, (iii) country abatement costs are independent of other countries' abatement efforts, and (iv) a large number of countries pollute a common resource and benefit directly from abatement. One would rarely expect (i) and (ii) to hold in reality, while (iii) and (iv) may be particularly inappropriate in the case of water resources.

This chapter examines trans-boundary water agreements as self-enforcing IEAs. Our goals are twofold. First, we would like to see what insights this approach provides for understanding water agreements. Second, we wish to draw on salient features of water conflicts and agreements to see if these features alter the pessimistic results of previous literature on IEAs.

The chapter is planned as follows. Section 2 develops a model of abatement of pollution of a common environmental resource. The model extends the work of Barrett (1994a, 1994b, and forthcoming) in that benefits to countries are heterogeneous and costs are both heterogeneous and interdependent. Barrett's earlier models are shown to be special cases of this model. Although we refer to the agents who share the resource as countries, these agents could represent any jurisdictional body or even individual farmers in a watershed. Central here is the notion of sovereignty. We are concerned with cases where agents sharing the resource cannot compel other agents to accede to the terms of any agreement. Environmental problems arise in an inter-jurisdictional setting with no central authority to impose Pigouvian taxes or set emission or loading standards.

While problems of self-enforcement apply most obviously to international watersheds, the model in Section 2 could also represent firms considering collective pollution abatement as a means of avoiding regulation. In the United States, a common approach to control agricultural non-point water pollution has been to encourage voluntary programs with regulatory measures used as a backstop. Under this approach, taxes or regulation are imposed on producers if a predetermined ambient water quality standard is not met in particular water bodies.[1] Producers may collectively seek out abatement measures on their own and then meet the standard and avoid more costly regulation or taxes that would be imposed by regulatory agencies. However, individual producers still have incentives to free-ride and allow others to incur abatement costs. Here, the collective benefit to polluters of abatement may not be direct benefits from improved quality of a common water resource. Rather, the common benefit is the avoidance of taxes or regulatory costs.

Section 3 examines the circumstances where there are significant *potential* gains from full cooperation of all countries sharing the resource. It also identifies complementary cases where countries can pursue unilateral abatement strategies that closely approximate the full cooperative solution. In these latter cases, countries need only worry about domestic pollution abatement and need not devote resources to negotiating international agreements.

Section 4 uses the basic model to characterize a self-enforcing IEA. Following Barrett, a self-enforcing IEA can be sustained if no signatory to the agreement can be made better off by unilaterally withdrawing from it and no non-signatory can be made better off by unilaterally acceding to the agreement. The model simultaneously solves for each country's decision to sign or not sign the agreement, the terms of the agreement, and the actions of non-signatories. Non-signatories maximize their individual benefits, taking the behavior of all other countries as given. Signatories act strategically to maximize their own collective net benefits. When a country accedes to the agreement, signatory countries increase their abatement levels, rewarding the joining country. If a country withdraws from the agreement, signatory countries reduce abatement levels. Because each country benefits from aggregate abatement, this amounts to punishment for pulling out of the agreement. It is this mechanism that sustains the self-enforcing IEA. Section 4 identifies two special cases where a self-enforcing IEA can increase pollution abatement levels significantly above the noncooperative case. In particular, we illustrate this for a case where the number of countries sharing the water resource is relatively small and a case where the number of countries is large but heterogeneous. Section 5 summarizes major results and raises questions for future research.

---

[1]One example of this approach is control of nutrient pollution under the Everglades Forever Act. Other cases are discussed below.

2. A MODEL OF A SHARED WATER RESOURCE

In this model, N countries emit pollution that damages a shared water resource. Each country's effort to abate this particular pollution is q_i. This represents *incremental* abatement specifically affecting the shared resource in addition to a country's purely domestic pollution abatement. Each country's environmental benefit from abatement, β_i, depends on total abatement, $Q = \Sigma_i q_i$:

$$\beta_i = b_i[aQ - 1/2Q^2] \tag{1}$$

where a and the b_i terms are positive constants. The b_i terms are the slopes of each country's marginal abatement benefit function. Each country's abatement costs Γ_i depend on both its own abatement as well as abatement by other countries:

$$\Gamma_i = q_i[c_{i0} - \mathbf{c_{ij}}\mathbf{q_{ij}}] + 1/2\, c_{ii}\, q_i^2 \tag{2}$$

where the c_{ii} and c_{i0} terms are positive parameters relating the ith country's costs to its own abatement, $\mathbf{c_{ij}}$ is an N -1 dimensional vector of positive parameters, and $\mathbf{q_{ij}}$ is a vector of abatement levels chosen by the N - 1 other countries.

Equations (1) and (2) show that countries may differ with respect to (i) how much they benefit from abatement of pollution of the shared resource, (ii) how their abatement costs depend on their own abatement, and (iii) how their abatement costs depend on abatement by other countries. Abatement benefits can be asymmetric, while abatement costs can be both asymmetric and interdependent. This abatement cost interdependence, itself, can be asymmetric. For example, an upstream country's abatement effort can reduce the cost of a downstream country's abatement of pollution flowing into a bay, while the converse would not be true. Barrett's (1994a) earlier model is a special case of (1) and (2) where c_{i0} and $\mathbf{c_{ij}}$ terms all equal zero and where the b_i and c_{ii} terms are equal for all countries. Barrett's later model (forthcoming) allows the b_i and c_{ii} terms to vary across countries but assumes the c_{i0} and $\mathbf{c_{ij}}$ terms all equal zero. This implies that the marginal abatement cost function has no intercept and that there is no abatement cost interdependence.

3. GAINS FROM COOPERATION

Total benefits are $B = \Sigma_i\beta_i$ while total costs are $C = \Sigma_i\Gamma_i$. In the Nash noncooperative solution each country maximizes $\beta_i - \Gamma_i$ with respect to its own q_i, taking the abatement effort of other countries as given. The full cooperative solution maximizes total net benefit, B - C, with respect to each q_i for i = 1, . . . , N. Country i's net benefits under the noncooperative and cooperative solutions are π^n and π^c, while total net benefits are Π^n and Π^c. The total gains from full cooperation are $G = \Pi^c - \Pi^n$.

First, we are interested in cases where there is *potential* for significant gains from cooperation. This depends on the size of G. When G is large, there are significant gains from cooperation. When G is small, achieving full cooperation is less important.

Countries can focus on domestic pollution abatement policies without worrying about the actions of other countries. For the case where countries are heterogeneous, it is difficult to obtain tractable analytical solutions that characterize the effects of changing parameter values on G. Consequently, we rely on numerical simulations. The potential gains from cooperation are sensitive to assumptions about the slope of the marginal abatement benefit function, b_i, and the slope of the marginal abatement cost function, c_{ii}:

A. When c_{ii} is small relative to b_i, unilateral abatement Q_n is almost as large as cooperative abatement Q_c. Gains from cooperation, G, are small.

B. When c_{ii} is large relative to b_i, unilateral abatement is relatively low and G is relatively large.

C. When $c_{ii} \approx b_i$ and both rise, G rises.

The above results were derived assuming that all countries were identical and where there was no abatement cost interdependence. The following simulation results summarize what happens when those assumptions are relaxed:

D. A mean-preserving spread in the distribution of b_i raises G.

E. A mean-preserving spread in the distribution of c_{ii} lowers G.

F. Increasing the values of the cost interdependence parameters raises Π^c by a greater amount than it raises Π^n, so that G rises.

4. SELF-ENFORCING IEAs

We now consider a case where countries can negotiate an IEA to abate pollution of the shared water resource. The outcome of this agreement determines (i) which countries are signatories and which are not, (ii) the abatement levels chosen by both signatories and non-signatories, and (iii) net benefits to each country sharing the resource. Non-signatory countries maximize $\Pi^n = \beta_i - \Gamma_i$ with respect to their own q_i taking the abatement effort of other countries as given. This is the same as the noncooperative strategy and yields $q_{in} = q_{in}(\mathbf{q_{jn}}, \mathbf{q_s})$ where the first term is a vector of abatement levels for other non-signatories, and the second term is a vector of abatement levels for signatories. The M signatories maximize their mutual net benefit $\Pi^s = \Sigma_k \beta_k - \Gamma_k$ with respect to their own abatement levels, q_k for k = 1, . . ., M. Following Barrett (1994a, 1994b), we assume signatories take into account the effect of their abatement choices on the abatement choices of non-signatories. Signatories use the N - M equations, $q_{in} = q_{in}(\mathbf{q}_{jn}, \mathbf{q}_s)$, as reaction functions.[2]

In what sense is this IEA self-enforcing? Consider a country that chooses to accede to the agreement. Current signatories reward the new entrant by increasing their own abatement levels. This raises aggregate abatement to the benefit of the new signatory (benefiting free-riders as well). What happens if a member of this coalition decides to drop out and violate the agreement? The remaining coalition of signatories' only mechanism to punish this behavior is to reduce their own abatement levels. This lowers aggregate abatement and punishes the exiting coalition member.

The next step is to determine which coalition of M countries (if any) will form the self-enforcing IEA. Following Barrett (1994a, 1994b, forthcoming), a self-enforcing IEA is sustainable if no signatory can make itself better off by unilaterally withdrawing from the agreement and no non-signatory can make itself better off by unilaterally acceding to the agreement.[3] This condition is similar to the one employed by d'Aspremont et al. (1983) to characterize cartel stability. To determine if an IEA might exist, the first step is to generate a payoff matrix for each possible combination of signatories and non-signatories. The next step is to examine which agreements fulfill this stability condition. With heterogeneous agents there could well be multiple sustainable IEAs. In addition, the solution does not guarantee which IEA would be chosen or even if the chosen agreement is the one that yields the largest net benefits to signatories.

Table 1 shows some results under conditions when N is large, countries are identical, and where there are no cost interdependence effects. The conditions were first examined in Barrett (1994a, 1994b). Under these circumstances, the self-enforcing IEA can improve upon unilateral action only to a negligible extent. The increase in benefits is small as a percentage of noncooperative benefits. When benefits from full cooperation are substantial, the IEA does little to close the gap between the noncooperative and cooperative outcome. In one case (where $b_i = 100$ and $c_{ii} = 0.01$), abatement under the IEA equals that under full cooperation. However, both are essentially equal to the noncooperative solution.

Table 2 shows country abatement levels for the same nine cases considered in Table 1. The IEAs yield minor percentage increases in *total* abatement, even though abatement by signatory countries increases 50-100 percent in some cases. Evaluations of actual IEAs often place emphasis on percentage reductions in pollution by signatory countries. Tables 1 and 2 suggest that this may not accurately measure total pollution reduction or the actual environmental and economic benefits of the IEA. In cases where signatories increase their abatements significantly above noncooperative levels, there are few signatories. In cases where there are large numbers of signatories, their

---

[2] Barrett's assumption, followed here, is that non-signatories form Cournot conjectures about the behavior of signatories, while signatories behave as Stackelberg leaders. One could well assume different conjectural variations in the static case considered here (e.g., all countries form Cournot conjectures) or treat the problem in a repeated game framework as in Barrett (1994a, 1994b).

[3] At present we ignore the possibility of side payments. Barrett (forthcoming) addresses this. We also ignore the possibility that countries could form sub-coalitions of two or more to accede or

percentage increase in abatement above the noncooperative solution is low. Table 2 also illustrates the free-rider effect. When signatories increase their abatement above the noncooperative solution, non-signatories abate *less* than under the noncooperative solution.

Table 1. Percent Increase in Net Benefits of Water Pollution Abatement Under IEA and Full Cooperative Solutions*

		c_{ii}		
		0.1	1	100
b_i	0.01	0.5 32.5 1.5	2.0 451 0.1	0 4900 0
	1	0 0 2	0.5 32.5 1.5	2 2451 0.1
	100	0 0 100	0 0 2	0.5 32.5 1.5

* The first number is the percent increase under IEA, the second number is the percent increase under full cooperation, and the third number is the first number as a percent of the second. Calculations assume N = 100, a = 1,000, countries are identical, and there are no abatement cost interaction effects.

Table 3 shows a more favorable case, where N is small (N = 5). Trans-boundary water conflicts often involve a small number of countries. Of more than 200 international river systems worldwide, only 53 basins are shared by three or more countries (Just and Netanyahu, 1997). Of these, 46 river basins are shared by three to six countries. Here, at least in theory, free-riding can be controlled more effectively because withdrawal or accession to the agreement has a greater effect on signatories. Table 3 illustrates that in these cases there may be scope for significant improvements over the noncooperative solution. The IEA increases total abatement 26 percent and net benefits 10.5 percent. The IEA makes up more than half the difference between the noncooperative and the cooperative solutions.

Row 1 of Table 3 shows that the percentage increase (above the noncooperative case) of the *physical measure* of abatement is more than double the percentage increase in abatement *benefits*. Row 5 shows that non-signatory abatement declines substantially (over 25 percent) under the IEA, while Row 6 shows that the major beneficiaries of the IEA are the non-signatories.

Table 2. Percent Increase in Signatory Abatement, Non-signatory Abatement, Total Water Pollution Abatement, and Number of Signatories Under IEA Solutions*

		c_{ii}		
		0.01	1	100
b_i	0.01	51 -0.8 0.8 3	100 0 2 2	0 0 0 2
	1	1 -1 0 51	51 -0.8 0.8 3	100 0 2 2
	100	0 0 0 100	1 -1 0 51	51 -.08 0.8 3

* The first number is the percent increase in signatory abatement under IEA, the second number is the percent increase in non-signatory abatement under IEA, the third number is the percent increase in total abatement, and the fourth number is number of signatories. Calculations assume N = 100, a = 1,000, countries are identical, and there are no abatement cost interaction effects.

In the United States, a common approach to control agricultural non-point water pollution has been to encourage voluntary programs with regulatory measures used as a backstop. Under this approach, taxes or regulations are imposed on producers if a predetermined ambient water quality standard is not met in particular water bodies. Producers may collectively seek out abatement measures that are less costly than submitting to regulation or taxes. However, individual producers still have incentives to free-ride and allow others to incur the costs of meeting the ambient water quality standard. The collective benefit to polluters of abatement may not be direct benefits from improved quality of the common water resource. Rather, the common benefit is the avoidance of taxes or regulatory costs.

For example, under the Everglades Forever Act, agricultural producers can avoid land taxes by reducing nutrient concentrations in downstream water bodies. Agricultural groups in New York City's watershed have agreed to work with the City of New York to control water pollution and avoid land use and production regulations that could be imposed under state law to protect municipal drinking water. In Nebraska, the Central Platte Natural Resource District has imposed restrictions on such fertilizer use to control nitrate pollution of groundwater. Restrictions are imposed on beneficiaries of the IEA are the non-signatories. an area-wide basis, becoming more stringent as average nitrate concentrations in an area increase. Producers can avoid more stringent controls if average nitrate concentrations in an area fall below pre-specified levels.

Table 3. Percent Changes in Total, Signatory, and Non-signatory Pollution Abatement and Abatement Benefits Under IEA and Full Cooperative Solutions: An Example When N Is Small*

	Total Abatement	Total Net Benefits
IEA solution (percent increase from noncooperative case)	25.5	10.5
Full cooperative solution (percent increase from noncooperative case)	66.7	19.1
Improvement from IEA solution as a percent of improvement with full cooperative solution	38.3	55.0
	Country Abatement	Country Net Benefits
Individual signatory country (percent change from noncooperative case)	59.6	4.9
Individual non-signatory country (percent change from noncooperative case)	-25.5	19.1

* Calculations assume N = 5, a = 100, countries are identical, and there are no abatement cost interaction effects, $c_{ii} = b_i = 1$.

For water pollution abatement agreements involving individual farms or local jurisdictional bodies in a watershed, N may be relatively large. The results from Table 1 suggest that if agents are identical, then self-enforcing agreements may have little scope for improving water quality above the noncooperative solution. But how much does this result depend on the assumption that all countries are identical?

Table 4 considers a case where N = 100, but where 3 large firms account for half of the loadings to the water body and 97 firms each account for about 0.5 percent of total loadings. Further, we assume that the marginal cost of abatement is the same for all countries for the same level of percentage abatement.[4] Thus, it would cost more for smaller-scale producers to increase total abatement, Q, by a given absolute amount,

[4] This case is adapted from Barrett (forthcoming), who uses it to characterize greenhouse gas emissions. In his case, the three large "firms" were the United States, the European Union, and the former Soviet Union.

because that would require a much greater percentage increase in their individual abatement.

Table 4 shows that the self-sustaining agreement can yield significant percentage increases in abatement benefits. In two cases, total net benefits under the self-enforcing agreement increase about 12 percent above the noncooperative solution. The self-enforcing agreement also makes up nearly 26 percent of the difference between the noncooperative and cooperative outcome. If one focuses on percentage improvements, then the self-enforcing agreements can be both sustainable and significant. Barrett (forthcoming), who first considered this example, dismissed this result, arguing that the "absolute gain to cooperation is small." We feel, however, that this assessment is too pessimistic. This is because the values used in the simulations have no units of measure attached to them. The absolute gain is only small relative to other arbitrarily chosen unitless values. It could be quite large in terms of some money metric.

Table 4. Percent Increase in Net Benefits of Water Pollution Abatement Under IEA and Full Cooperative Solutions with Heterogeneous Agents*

		c_{ii} (large polluters)		
		0.1	1	100
b_i	0.01	0.0 0.2 0.0	11.8 45.9 25.7	0.0 300.0 0.0
	1	0.0 0.0 ---	0.0 0.2 0.0	11.8 45.9 25.7
	100	0.0 0.0 ---	0.0 0.0 ---	0.0 0.2 0.0

* The first number is the percent increase under IEA, the second number is the percent increase in net benefits under full cooperation, and the third number is the first number as a percent of the second. Calculations assume N = 100 with no abatement cost interaction effects. The c_{ii} values for small polluters are 3 percent of those for large polluters, while the b_i values for small polluters are 1.5 percent of those for large polluters.

5. CONCLUSIONS

In this chapter, we presented a simple static model where countries negotiate an international environmental agreement (IEA) to abate pollution of a shared water resource. The model determines (i) which countries are signatories to the agreement and which are not, (ii) pollution abatement undertaken by each country , and (iii) net benefits of abatement accruing to each country. The problem is akin to cartel formation and stability. In our case, however, agents attempt to restrict production of pollution (rather than a commodity) to increase their net benefits.

While previous research suggests that IEAs can be sustained only in cases where the outcome of cooperation is a minor improvement over noncooperation, we have identified some cases where self-enforcing IEAs can be sustained and can achieve substantive improvements over unilateral pollution abatement.

This was demonstrated for a case where the number of countries sharing the water resource is small (5 in number). This case is particularly relevant for international water agreements where the number of relevant negotiating countries is often very small, about 5 or less. Significant increases in abatement and abatement benefits were also demonstrated for a case with many agents (100 in number) but where pollutant loadings were concentrated among a small number of agents. The example applies more readily to domestic watershed-based agricultural water pollution programs such as those in Florida, Nebraska, and New York. Such programs rely on collective action by producers to improve ambient water quality measures and to avoid costly taxation or regulation. Both these results parallel those from studies of cartel behavior. Enforcing agreements is easier if there are a small number of countries (firms) involved, or if production (in our case, production of pollution) is highly concentrated among a few large agents.

While previous literature has focused on absolute gains obtainable from IEAs, this chapter has focused on percentage improvements from the noncooperative solution. Because the model used here and in earlier studies generates unit-free measures of abatement levels and benefit, we feel that percentage improvements are a more appropriate yardstick to evaluate IEAs.

Evaluations of actual IEAs often place emphasis on percentage reductions in pollution by *signatory countries*. The results presented here illustrate that percentage increases in signatory abatement can be much larger than percentage increases in total abatement. This occurs because non-signatory abatement declines as signatory abatement increases through the free-rider effect. Percentage increases in total abatement, in turn, can also be much larger than percentage increases in the economic benefit of abatement. Physical measures of signatory abatement may not be a particularly accurate measure of the economic benefits of IEAs.

To conclude, we note some issues not fully addressed in this chapter and suggest some areas of future research. First, our main purpose was to present examples where trans-boundary water agreements could have scope for success. Consequently, we have focused on a small number of simulations. Further sensitivity analysis is warranted to see how robust results are to different assumptions about parameter values and functional forms.

Second, although we have shown how one may account for interdependence of country abatement costs, we have yet to formally examine how this interdependence affects self-enforcing IEAs. All else given, marginal abatement benefits for a country decline as other countries abate more. This gives rise to a free-rider effect. However, cost interdependence acts in the opposite direction. A country's marginal abatement costs fall as other countries abate more. More formal simulations are needed to

Third, we have not formally examined the effects of side payments. Side payments may be particularly important in trans-boundary water agreements, however, because they may be the only way to bring "upstream" countries into comprehensive agreements. An upstream country may affect abatement costs of downstream countries, while the reverse might not be true. Also, upstream countries may derive little or no direct benefit from improved conditions of a water body shared by downstream countries. Bringing such countries into trans-boundary water agreements is often crucial, but would be very difficult without some form of side payment. Barrett (forthcoming) has demonstrated how the ability to make side payments can increase benefits under IEAs.

Fourth, we followed Barrett's assumption that non-signatory countries form Cournot conjectures while signatories act as Stackelberg leaders. This raises the question of why a country would form one type of conjecture if it signed but another if it did not. Future research could examine alternative conjectural variation assumptions or repeated game solutions.[5]

Finally, the strict definition of self-enforcement used here does not allow for the possibility that countries can make legally binding pre-commitments to abatement levels, future side payments, or sanctions. Nor does this definition allow for the possibility of voluntary agreements whose terms can be enforced *ex post* by some central authority. These enforcement mechanisms could be more readily applied to state-level watershed-based agreements among individual producers than they could be applied to countries in international agreements. Such mechanisms would require countries to give up a measure of pure sovereignty and to be willing to submit to some form of *ex post* arbitration. However, precedent exists in international trade agreements (such as the GATT and NAFTA) for this type of arbitration process. Multi-lateral development agencies could also require countries to make binding pre-commitments as a condition of aid or loans (Zilberman 1992; Frisvold and Caswell 1997). Work on the role of arbitration and third-party enforcement would be a fruitful area of future research.

REFERENCES

d'Aspremont, C.A., J. Jacquemin, J. Gabszeweiz, and J.A. Weymark. 1983. On the Stability of Collusive Price Leadership, *Canadian Journal of Economics* 16:17-25.

Barrett, S. Forthcoming. Heterogeneous International Environmental Agreements. In C. Carraro, ed., *International Environmental Agreements*. Aldershot, United Kingdom: Edward Elgar.

Barrett, S. 1994a. Self-Enforcing International Environmental Agreements, *Oxford Economic Papers* 46:878-894.

Barrett, S. 1994b. The Biodiversity Supergame, *Environmental and Resource Economics* 4:111-122.

Bauer, A. 1992. International Cooperation over Environmental Goods. Volkswirtschaftliches Institute, University of Munich, mimeo.

[5]Barrett (1994a, 1994b) found that IEAs with a large number of identical agents could not improve substantially from the noncooperative outcome even in an infinitely repeated game framework. This occurs because punishments large enough to maintain cooperation between signatories would be subject to renegotiation and the ability of the agreement to be self-enforcing would break down.

Black, J., M. Levi, and D. de Meza. 1992. Creating a Good Atmosphere: Minimum Participation for Tackling the Greenhouse Effect, *Economica* 60:281-293.

Carraro, C., and D. Siniscalco. 1993. Strategies for the International Protection of the Environment, *Journal of Public Economics* 52:309-328.

Frisvold, G., and M. Caswell. 1997. Transboundary Water Agreements and Development Assistance. In D. Parker and Y. Tsur, eds., *Decentralization and Coordination of Water Resource Management*. Boston: Kluwer Academic Press.

Hoel, M. 1992. International Environment Conventions: The Case of Uniform Reductions of Emissions, *Environmental and Resource Economics* 2:141-159.

Just, R.E., and S. Netanyahu. 1997. International Water Resource Conflicts: Experience and Potential. In R.E. Just and S. Netanyahu, eds., *Conflict and Cooperation on Trans-Boundary Water Resources*. Boston: Kluwer Academic Publishers.

Zilberman, D. Environmental Aspects of Economic Relations Between Nations, *American Journal of Agricultural Economics* 74:1144-1149.

3 BARGAINING OVER SHARED AQUIFERS: THE CASE OF ISRAEL AND THE PALESTINIANS*

Sinaia Netanyahu
Richard E. Just
John K. Horowitz
University of Maryland, College Park, USA

This chapter considers the application of game theory to international cooperation on shared groundwater resources. Sharing of the Mountain Aquifer between Israel and the Palestinian Autonomy is considered as an example. Since 1967, the management of the Mountain Aquifer has been under the unilateral control of Israel. The new peace accords, however, have begun a shift in control of parts of Judea and Samaria to the Palestinian authority. This may divide control of the Mountain Aquifer between Israel and the Palestinians. Without cooperation, over-pumping and degradation of water quality are likely. Both cooperative and noncooperative bargaining approaches for Israeli-Palestinian agreements are explored. The results show that the bargaining solutions implied by game theory are robust with respect to demand elasticity for water, user cost, and cost of pumping. Further considerations of potential strategic behavior and defection show that risk of breakdown in negotiations and impatience in realizing benefits of the agreement are also important. Because of limited empirical information, the analysis at this point is illustrative and demonstrates the role of various factors and potential behaviors.

1. INTRODUCTION

Aquifers common to two or more political entities can be managed with various degrees of cooperation ranging from no cooperation, to joint cooperation on specific projects or goals, to full cooperation on all issues. The degree of cooperation is

---

* This research was supported in part by Grant 58-319R-4-046 from the Research and Scientific Exchange Division, Foreign Agricultural Service, United States Department of Agriculture and in part by the Bi-national Agricultural Research and Development Fund, BARD Grant No. US-2637-95R. However, the views expressed herein are solely those of the authors and should not be construed to represent those of the USDA, agencies therein, or BARD.

influenced by hydro-geological, technological, economic, and political factors. Incentives to cooperate and the magnitude of cooperation are enhanced by definition of property rights in the Aquifer, externalities experienced in the absence of joint management, and external water resource opportunities. Other factors such as political stability and international image can also encourage cooperation.

The problem of water scarcity has become more acute for Israelis and Palestinians because of economic and population growth and because of degradation of historic sources. New sources and more efficient use of old ones are needed, but these may require large investments and effective institutions for allocating water, monitoring its use, and ensuring its quality. Coordination and cooperation are required for implementing each of these measures. The purpose of this chapter is to develop and apply a framework to evaluate the potential for cooperation in aquifer sharing, with an application to the Israeli-Palestinian Mountain Aquifer (MA). Based on potential gains from Israeli-Palestinian cooperation in management of the MA, a framework to analyze the process of negotiation and the factors that affect its outcome is suggested.

We start with a description of Israeli and Palestinian water supply and use. An overview on the hydrology of the Aquifer is then presented.[1] A single stage cooperative game is offered as a tool to suggest and motivate possible agreements. Then we expand the analysis by recognizing three distinct phases in water exploitation decisions. First, entities, possibly with the help of a third party, must design a contractual agreement for joint aquifer use. The outcome of this stage possibly depends on players' impatience and on the risk that unrelated political factors will cause negotiations to break down. The end of the first stage is marked by an agreement over the use of the joint resource. If the agreement is advantageous to each party, they are expected to move to the second stage which is the cooperation stage. In the third stage, the sustainability of the agreement is questioned. Parties may choose to deviate from the cooperative agreement. However, if a credible punishment is incorporated into the agreement in the first phase, then no defection occurs. To ensure that cooperation over water agreements will be sustained, the punishment must be robust to changes in governments, economic variables, political forces, and water availability.

2. ISRAELI-PALESTINIAN WATER ISSUES

High population growth and lack of water resource planning are leading to long term water scarcity and serious conflicts over exploitation of the MA for both Jewish and Arab populations. Because per capita water availability for both Israelis and Palestinians is much below the "water stress zone" (Falkenmark et al 1990), relying solely on intertemporal reallocation of MA water is inadequate from a long run perspective. Forecasted water demand for Israel including the Judea and Samaria regions suggest a significant increase in water needs in the region (see Table 1).

---

[1] This overview is intended to draw attention to issues that motivated this chapter. It is by no means a complete description of the hydro-geology of the Mountain Aquifer.

Effective water supply can be increased by improvement of water quality of existing resources (e.g., the lower Jordan River), water desalination, waste water recycling, water conservation, and cloud seeding. Water reallocation with accompanying structural changes in the agricultural sector is also a likely part of any efficient solution (Frederick 1993).

Table 1. Forecasted Water Demands for Israel, Judea and Samaria, 2000-2040[a]

Year	Urban and Industry		Agriculture	
	Israel	Judea and Samaria	Israel	Judea and Samaria
2000	750	93	1,220	155
2010	929	132	1,400	190
2020	1,098	224	1,600	280
2040	1,535	414	2,040	415

a Note that assumptions include annual population growth of 1.69 percent in Israel, status quo for irrigation water supply in Israel, and an accelerated growth of domestic and irrigation water demand in Judea and Samaria.

Source: Israel Water Study for The World Bank, Tahal Consulting Engineers LTD, Israel, March 1993, Draft.

While water resources can be reallocated within individual jurisdictions by conservation, recycling, and local desalination, these alternatives alone are likely not sufficient in the long run and may not be economically attractive even in the short run. The creation of water markets is widely propounded as a means of achieving economic efficiency in water resource allocation. However, water markets require well defined property rights, institutions, laws, regulations, financial arrangements, large investment in specific capital, monitoring capability, and quality control. Currently, most of these institutional features are inadequate in the area. Until such institutions are better developed, water markets are unlikely to be effective.

However, due to the urgency of water problems, an Israeli-Palestinian agreement over exploitation of the MA is becoming increasingly critical. Externalities resulting from current use patterns and from Israeli and Palestinian urban development in the area call for coordination and cooperation. That is, because the Aquifer is a common property resource, parties do not bear the full cost their use imposes on the other due to lack of proper management. In the absence of cooperation, the Aquifer is being depleted by over-pumping and its quality degraded by lack of effluent management. A solution to these problems must be reached by negotiation and mutual agreement. Bargaining appears superior to pure market solutions not only because infrastructure is insufficient to make markets operate efficiently but also because common property problems cannot be resolved efficiently by markets without public intervention.

Moreover, bargaining may be the only means of initially facilitating the institutions needed for water transfer.

The sources of the dispute between Israel and the Palestinians over the rights to use the MA are related to the climatic-hydrologic characteristics of the region. Israel pumps water that emerges in the Yarkon-Taninnim (YT) Aquifer of the coastal plain but which originates in rainfall in the Judean and Samarian mountains (Lowi 1992).

Increasing opportunities for voluntary water cooperation between Israel and the Palestinians have surfaced as a result of recent peace talks and agreements. These opportunities are broadened by the activities of international organizations such as the World Bank, and by efforts of academic and research institutions. However, at this time no long term agreement between Israel and the Palestinians appears to be close.

3. THE MOUNTAIN AQUIFER—A HYDROLOGICAL OVERVIEW

The MA lies under the mountains of Judea and Samaria and constitutes about one third of Israeli water supply. It consists of two major basins, the YT Basin (or the Western Aquifer) and the Eastern Mountain (EM) Basin. A large portion of the water collected in the mountains of Judea and Samaria emerges in springs or is pumped from groundwater in the YT Basin within Israel. The maximum sustainable yield of the YT Basin is 350 million cubic meters per year (mcm/yr). In recent years, Israeli and Palestinian pumping in the YT Basin has exceeded the maximum sustainable yield (Hydrological Service 1994) and an average of 382 mcm/yr has been removed from the YT Basin including both spring water and pumping. Palestinians have consumed only 20 mcm/yr of this amount (Gvirtzman 1994).

The EM Basin consists of seven sub-aquifers. In the EM Basin, production (springs and pumping) in the years 1987/88 through 1992/93 averaged 378 mcm/yr. Water quality is not uniform. Excluding saline spring water, annual water production from the EM is about 228 mcm/yr. The EM water resources are exploited by Palestinians (110 mcm/yr), Jewish settlers (40-60 mcm/yr), and Israeli local water associations (Schwartz 1988; Benvenisti 1987).

Knowledge of connections among sub-aquifers is crucial for water negotiations because they affect each party's ability to pump, the quality of the water, and the cost of pumping. In order for Israelis and Palestinians to reach a sustainable agreement over allocation and management of the aquifer, an understanding of the hydro-geological nature of the Aquifer is needed. In the EM Basin, some rainfall flows north (Northern Aquifer) and is mostly consumed by Israeli regional water associations while some flows east (Eastern Aquifer) and is consumed by Palestinians and Jewish settlers. Rainfall on the western slope contributes to groundwater which emerges in the YT Basin. Hence, MA water can be pumped in Israel in the YT Basin, on the northern slope (mainly by the water associations), and on the eastern slope. With Palestinian autonomy in areas overlying the aquifer, water use in the YT Basin and in the Northern Aquifer of the EM Basin will come under possibly competing control of both Israel and the Palestinians.

4. A COOPERATIVE SOLUTION TO AQUIFER SHARING BY ISRAEL AND THE PALESTINIANS

Market solutions to international aquifer use problems are often not feasible because of lacking infrastructure, monitoring capabilities, trading relationships, and the common property difficulties that tend toward over-drafting of groundwater supplies. Market solutions are often unacceptable politically as well because they tend to allocate water to the country with the highest marginal value of water. Market outcomes are typically lopsided due to differing levels of development and differing elevations and pumping costs. Because of political concerns and uncertainty in demand growth, countries are understandably reluctant to enter into agreements that allocate a given water supply mostly to another country. Bargaining theory explores the opportunities for cooperation among all participating parties. In some cases, bargaining solutions are remarkably robust but can be quite different from market solutions.

Sharing of an aquifer across boundaries of jurisdictions is not an uncommon problem. The issue of domestic aquifer exploitation across state or regional boundaries has been raised in India, Germany, the US, and Switzerland. International conflicts over aquifer use are more common and more heated because no over-arching process or mechanism exists to resolve disputes. Examples of aquifer sharing conflicts are the Sahara Desert Aquifer shared by Egypt and Libya, the Sinai Aquifer shared by Israel and Egypt, and the Yarmouk Aquifer shared by Syria, Jordan, and Israel. International aquifer sharing problems also exist between Egypt and Sudan; Saudi Arabia and Jordan; Saudi Arabia, Qatar, and the Emirates; Iraq and Syria; and Syria, Iraq, and Turkey (Soffer 1992). Substantial progress has not been made in any of these problems so little precedent exists to guide solution of the MA sharing problem.

The Oslo peace accords, which give some authority to the Palestinians, will necessitate cooperation in the use of the MA. To explore the application of bargaining theory to the MA cooperation problem, note that the eastern and western parts of the MA are hydrologically disconnected basins (Lowi 1993, p.185). Thus, the implications of sharing and pumping in the YT Basin can be studied independently from over-pumping and leakage among the seven sub-aquifers of the EM Basin.

Ideally, the multi-dimensional nature of water sharing considerations should be addressed. If, for example, less water were withdrawn from any of the central or southeastern sub-aquifers in the EM Basin (mainly used by Palestinians), solutions should consider whether more water could be sustainably pumped from the neighboring aquifers in the northern part of the EM Basin (mainly used by the Jewish population residing outside the 1967 Israeli borders). Alternatively, if sewage is untreated by Palestinians on the western slope, solutions must consider impacts on water quality in the YT Basin. The springs in the YT Basin are gravity fed. Water can be pumped in the lower elevations at less cost than at higher elevations. In the higher elevations where Palestinian water supplies are drawn, extraction can be quite costly and these costs are an increasing function of the amount pumped in the entire Basin. In other words, pumping costs increase as the amount of water remaining in the Aquifer decreases.

4.1 Application of Bargaining to Cooperation over YT Aquifer Withdrawal

Bargaining to achieve cooperation can be modeled using a game theoretic approach. In this section, a cooperative game framework is used to demonstrate potential gains from cooperation. Our approach is to fit the parameters of the game to observed data where data are available. For other parameters we assume values based on intuition and then perform sensitivity analysis. The results identify which parameters are crucial in determining gains from cooperation and, thus, those on which further empirical work should focus.

Results from game theory show that the only solution that satisfies particular standard and reasonable assumptions is the Nash bargaining solution (Nash 1953).[2] The Nash bargaining solution maximizes $[u_i - \bar{u}_i][u_p - \bar{u}_p]$ with respect to the variables over which bargaining occurs where u_i and u_p are the utilities or benefits achieved for the respective parties (i denotes Israelis and p denotes Palestinians) and $\bar{u}_i$ and $\bar{u}_p$ are reservation utilities below which the respective parties prefer no agreement. The payoffs of the reservation outcome reflect the status quo.[3]

The parameters of the game are determined as follows. To represent the cost of extracting water from the aquifer, suppose the short-run marginal cost of Palestinian pumping follows a linear function of the quantity pumped and that the short-run marginal user cost reflecting the effect of Israeli pumping on Palestinian marginal cost is proportional to Israeli use. Thus, Palestinian short-run marginal cost follows $\alpha_p + \beta_p(q_p + \theta q_i)$ where the user cost is $\beta_p\theta q_i$. This function can represent the substantially increasing marginal cost found by Isaac et al. (1994) where pumping at 20 mcm/yr costs \$.16 per cubic meter (cm) but pumping at 135 mcm/yr requires a cost of \$.34 per cm. The parameters α_p and β_p are determined proportional to the values that reflect these two cost-quantity pairs when Israeli pumping is held at its current level of 362 mcm/yr.[4] While the spillover coefficient θ which determines user cost is not known, we assume a value of .2 (an additional unit of Israeli pumping raises Palestinian pumping cost 20 percent as much as Palestinian pumping) for illustrative purposes and then investigate sensitivity of results to assumptions on θ.

For Israelis, a constant short-run pumping cost of \$.15 per cm is assumed because this is a realistic current pumping cost and Palestinian pumping apparently has little effect on water availability at the lower-elevation YT springs at least at historical levels of Palestinian pumping. For both parties, however, we consider a future user cost that occurs if the present level of pumping beyond maximum sustainable yield continues. We assume the future user cost for party j follows

---

[2] These axioms require that the parties are rational, treated symmetrically, and that the result be independent of irrelevant alternatives and not affected by linear transformations of each parties preferences.

[3] Over-pumping has adverse long-run consequences on future pumping possibilities. Therefore, countries' utilities in the absence of cooperation may be substantially less than the utility they get from current water use.

[4] We inflate the parameters that actually fit Isaac's data by 50 percent because his data apply only to total pumping from one sub-aquifer. The resulting current marginal pumping cost for Palestinians is \$.24 per cm.

$$UC_j = \frac{p_j(q_i + q_p)q_j}{rS}$$

where p_j is pumping cost for player j, r is the interest rate, and S is the maxmimum storage capacity of the aquifer. This user cost is an approximation derived from dynamic programming (see Burt 1964; Feinerman and Knapp 1983). Because S is unknown, we simply summarize $p_j/(rS)$ by an unknown parameter c_f and estimate it by solving for the future user cost which, if internalized, drives aggregate pumping down from the current level of 382 mcm/yr to the maximum sustainable yield of 350 mcm/yr (at equilibria determined from demands specified below).

Integrating the current marginal cost functions including present user costs and then adding the future user cost obtains the Palestinian social cost function,

$$c_p(q_p, q_i) = \alpha_p q_p + \beta_p(q_p^2/2 + \theta q_p q_i) + c_f(q_i + q_p)q_p,$$

and the Israeli cost function,

$$c_i(q_i) = \alpha_i q_i + c_f(q_i + q_p)q_i.$$

Fitted parameters of the Palestinian cost function are $\alpha_p = .023$ and $\beta_p = .0023$. For Israelis, $\alpha_i = .15$. The fitted value of the future user cost parameter is $c_f = .007$.

To represent utility of the two parties, linear demands of the form

$$p_j = \gamma_j - \delta_j Q_j, \quad j = i,p \tag{1}$$

are assumed where p_j represents the marginal value of water and Q_j represents total water consumption from all sources including YT water. The associated benefits of water consumption (consumer surpluses) are

$$b_j(Q_j) = \gamma_j Q_j - \delta_j Q_j^2/2, \quad j = i,p.$$

The parameters of demand are fitted to observed YT water use data, $Q_i = 362$ mcm/yr and $Q_p = 20$ mcm/yr, and marginal values of water determined by marginal costs specified above at observed water use. Actual water prices paid by both Israelis and Palestinians are higher but are presumably explained by transportation costs, implicit taxes imposed on household use in municipalities, etc. The parameters of (1) are determined by fitting elasticities at current use levels to a weighted average of elasticities reported by Fisher (1994) for agricultural, industrial, and domestic demands. The resulting elasticity of demand is .407 for Israel and .440 for Palestinians and the fitted parameters are $\gamma_i = .519$, $\delta_i = .000211$, $\gamma_p = .785$, and $\delta_p = .00420$.

With these specifications, Palestinian's consumer benefits less the cost of pumping from the aquifer are given by

$$u_p = b_p(Q_p) - c_p(q_p, q_i)$$

and similarly for Israel. These net benefits ignore the cost of supplying non-aquifer water, Q_p - q_p, but such costs are assumed to be largely overhead costs and, in any case, unaffected by bargaining over aquifer water. To consider bargaining to determine the quantities used by each party, Israeli and Palestinian reservation utilities are assumed to be represented by the status quo,

$$\bar{u}_i = b_i(\hat{Q}_i) - c(\hat{q}_i, \hat{q}_p),$$

$$\bar{u}_p = b_p(\hat{Q}_p) - c(\hat{q}_p, \hat{q}_i),$$

$$\hat{Q}_i = 1745 \text{ mcm/yr}, \ \hat{q}_i = 362 \text{ mcm/yr},$$

$$\hat{Q}_p = 130 \text{ mcm/yr, and } \hat{q}_p = 20 \text{ mcm/yr.}$$

Using these reservation utilities, the Nash bargaining solution is for Palestinians to pump 20.3 mcm/yr and for Israelis to pump 329.7 mcm/yr (Table 2, Case 1). The resulting marginal cost for water extracted by Palestinians is $0.225 per cm. This solution entails an 8.4 percent reduction in total water use (from 382 mcm to 350 mcm). Palestinian use increases by 1.5 percent while Israeli use decreases by 8.9 percent from the current noncooperative outcome.

These results raise the question of why Israel would be willing to participate in an agreement that so seriously reduces its water use from the YT Basin. The answer lies in future user cost savings that could not be realized if water use continues to exceed the maximum sustainable yield. The major saving from cooperation is that overall use levels are brought into line with sustainable yield so that adverse future consequences of over-pumping are avoided. From a short-term perspective, Palestinians appear to be the major benefactor of cooperation because their pumping costs are otherwise higher with over-pumping. However, overall use by Palestinians is low compared to Israelis. Because Israelis are the large user, they benefit much more from reducing future user cost that would occur if over-pumping continues and damages the Aquifer.

4.2 Sensitivity of the Cooperative Solution

The water quantities in the Nash bargaining solution are highly robust to changes in the assumed parameters. Comparison of Cases 1 and 2 of Table 2 shows that water allocation is not highly sensitive to changes in the assumed Israeli demand elasticity. A higher elasticity from a given initial point implies that Israelis lose less benefits in reducing consumption so slightly more consumption is given up. A high elasticity implies a much lower reservation utility because the consumer surplus triangle becomes flat and thin. According to the Nash bargaining solution, the Israelis thus capture much less of the benefits from bargaining. On the other hand, a higher Palestinian demand elasticity as in Case 3 also increases Palestinian water

Table 2. Alternative Nash Bargaining Solutions for Israeli-Palestinian Sharing of the Yarkon-Taninnim Basin[a]

	Demand Elasticity		Average Pumping Cost		Short-Run Spillover	Future User	Water Use		Palestinian Marginal	Reservation Utility		Surplus Gain		Side Payment
	(I)	(P)	(I) ($/cm)	(P) ($/cm)	Coefficient ($/cm)	Cost ($/cm)	(I) (mcm/yr)	(P) (mcm/yr)	Cost ($/cm)	(I) ($)	(P) ($)	(I) ($)	(P) ($)	(I to P) ($/cm)
1.	.407	.440	.150	.202	.2	.0070	329.7	20.3	.225	12.38	1.16	.126	.318	0
2.	.814	.440	.150	.202	.2	.0036	329.2	20.8	.226	6.18	1.23	.063	.322	0
3.	.407	.880	.150	.202	.2	.0072	329.5	20.5	.226	12.35	0.74	.129	.321	0
4.	.407	.440	.300	.201	.2	.0136	330.8	19.2	.224	24.86	1.02	.249	.315	0
5.	.407	.440	.150	.404	.2	.0071	329.2	20.8	.453	12.36	2.47	.126	.644	0
6.	.407	.440	.150	.187	.4	.0076	328.3	21.7	.212	12.26	1.15	.135	.681	0
7.	.407	.440	.150	.202	.2	.0141	329.0	21.0	.227	10.92	1.01	.357	.334	0
8.	.407	.440	.150	.203	.2	.0070	328.0	22.0	.229	12.38	1.16	.226	.226	-.056
9.	.407	.440	.300	.201	.2	.0136	330.2	19.8	.225	24.86	1.02	.282	.282	.206
10.	.407	.440	.150	.203	.2	.0070	327.5	22.5	.230	11.76	1.10	.730	.398	0
11.	.407	.440	.150	.204	.2	.0070	327.0	23.0	.231	0.00	0.00	11.489	1.500	0

[a] Note that (I) denotes Israel and (P) denotes the Palestinians. The initial Case 1 is based on current use in the YT Basin of 362 mcm/yr for Israelis and 20 mcm/yr for Palestinians. Subsequently, Case 2 increases Israeli demand elasticity, Case 3 increases Palestinian demand elasticity, Case 4 increases the cost of Israeli pumping, Case 5 increases the cost of Palestinian pumping, Case 6 increases the short-run user cost coefficient, Case 7 increases the future user cost, Case 8 and 9 generalizes cooperation by allowing a side payment for pumping YT water., and Cases 10 and 11 consider noncooperative game theory results that correspond to reducing reservation utilities proportionally by 5 percent and 100 percent, respectively.

quantity because from a given initial point relatively more surplus is gained by increasing quantity above the initial point.

The effect of errors in representing pumping costs are represented in Cases 4 and 5. If the Israeli pumping cost is doubled, then it exceeds the Palestinian cost. At equilibrium, the pumping costs represent the marginal value of water for the respective parties. Thus, if the Israeli marginal value exceeds the Palestinian marginal value, then there is benefit to transferring consumption to Israel. Accordingly, this case, which is the only one where Israeli cost exceeds Palestinian cost, is the only case where Palestinian pumping declines from the present use of 20 mcm/yr. Note also that the Israeli reservation utility is quite high in this case because the inferred demand for water is greater if the observed consumption is based on a higher marginal value.

In Case 5, the Palestinian pumping cost function is doubled which causes more water consumption to be transferred to the Palestinians in accord with the higher valuation of water implied thereby (from the current observed equilibrium).[5] Accordingly, the Palestinian reservation utility is roughly doubled. The higher reservation utility permits the Palestinians to gain relatively more.

In Case 6, the spillover effect of Israeli pumping on Palestinian cost is doubled. As expected, the cooperative effect is to reduce Israeli pumping because it has a higher current user cost. Palestinian pumping climbs to 8.5 percent above the status quo. The Palestinian surplus gain in this case is high even though the reservation utility is low because of the strength of the high user cost with cooperation.

Case 7 roughly doubles the future user cost. The effect of this change comparing to Case 1 is to penalize Israel relatively more because it carries the bulk of the responsibility in contributing to future user cost, i.e., because about 95 percent of current pumping from the aquifer is by Israelis.

The latter two cases introduce the possibility of a side payment into the bargaining solution. That is, bargaining can occur not only with respect to how much each party should pump but also on whether one party should pay the other for the right to increase use from the current status quo. Case 8 simply introduces this possibility into Case 1. Because pumping costs and the implied marginal valuation of water is higher for Palestinians, they are willing to pay Israel for the right to increase pumping ($.056 per cm). Accordingly, this case obtains the highest Palestinian use—a full 10 percent above the status quo—with Israeli pumping 9.4 percent below the status quo. On the other hand, if Israeli pumping cost is higher than Palestinian pumping cost as in Case 4, then cooperation entails Israelis paying Palestinians as much as $.206 per cm to reduce Palestinian use below the status quo. This is Case 9. Interestingly, introducing a side payment permits each party to receive the same magnitude of gain from Nash bargaining.

In general, the total utility outcome (reservation utility plus surplus gain) for Israelis is quite stable with respect to parameters that affect Palestinians directly and

---

[5] It is important to bear in mind that estimates are presented in each case based on a starting point represented by current consumption. Normally one would expect raising cost to reduce consumption. In this case, however, the demands are inferred from observed quantities and assumed costs. If cost is higher, then the inferred demand is greater at every price.

total Israeli utility is quite stable with respect to parameters that affect Palestinians directly. Overall, Israeli gains from cooperation amount to only about 1 percent of initial surplus except in Cases 7 and 8 where gains reach 3.3 and 1.8 percent, respectively. The Israelis gain relatively more in Case 7 because they capture most of the gains from a greater reduction in future user costs by the fact that they use more water from the aquifer. Overall, Palestinian gains from cooperation amount to more absolutely (in most cases) and relatively. Their gains range from a minimum of 19 percent of initial surplus (Case 8) to as much as 59 percent of initial surplus (Case 6). In many cases, their gains are around 26-28 percent. A property of the Nash bargaining solution that leads to this outcome is that Palestinian welfare can be raised relatively easier than Israeli welfare because of the sheer magnitude of initial use and welfare received by Israelis.

Overall, the solutions in Table 2 are remarkably insensitive to assumed parameters. Water use for Israelis does not vary more than 1 percent among all cases while Palestinian use varies by less than 15 percent even though parameters are doubled. While reservation utilities are sensitive to some of the parameters, the relative outcomes are not. Thus, by and large, the implications of Nash bargaining are relatively clear even though data are sparse.

4.3 Stability of the Cooperative Solution

Once the possibility of agreement is realized, the possibility must be considered that one of the parties will choose to deviate from the agreement after it is in place. The sustainability of an agreement may depend on both sides possessing sufficient retaliatory actions to make continued cooperation by the opposite side advantageous. The problem with the cooperative approach is that it does not explicitly model the incentives to abide by the agreement in the long run even if an agreeable pumping limit can be determined. Both parties have an incentive to renege because monitoring is difficult. In addition, Israeli pumping is at low cost and they can apparently deviate with little consequence. Palestinian deviation is not as advantageous because its marginal pumping costs are increasing. On the other hand, its use is so small compared to Israelis that it may seem somewhat inconsequential. Strategies that each can use to penalize the other and thus force compliance with the cooperative solution are not obvious or easily modeled. Further understanding of these considerations requires a noncooperative bargaining framework. In particular, in order to sustain the second (cooperative) stage and avoid defection, the consequences of defection need to be clearly spelled out in the first (negotiation) stage. Consideration of noncooperative outcomes also gives further insights into the choice of reservation utilities that define the cooperative outcome.

5. NONCOOPERATIVE SOLUTIONS TO JOINT AQUIFER MANAGEMENT

A major weakness of cooperative games lies in the potential for not following the terms of agreement. The essence of the problem is commitment. Cooperative games assume that players can commit to a contract. With binding agreements, players can agree to limit ground water pumping. However, binding agreements may not be possible because (i) groundwater property rights are ambiguous, and (ii) enforcement is difficult or impossible due to lack of enforcement agencies, limitations of technology, inadequate punishment for deviations, and high costs of monitoring. This section considers models in which parties cannot make commitments. Such models are categorized as noncooperative games. These games model the negotiation process by focusing on the players' strategies.

Dixon (1991) outlines several possible solutions for joint management of groundwater resources. He notes that at one extreme is the "social-planner" solution in which a third-party social planner (such as the World Bank or the U.S. government) maximizes the net present value of aquifer-wide rents using a social discount rate. If each of the players is identical, then each receives an equal share of the total payoff. In the more realistic case where players are not identical, the problem of sharing is more complicated. Accordingly, the social-planner approach becomes complicated in the realistic case where aquifer-wide rents depend on how the benefits are shared. The discount rate also represents a critical choice for which an objective selection may not be clear. With these complications, the social-planner approach is not likely to generate solutions which all parties can acknowledge *ex post* as fair and agreeable.

Dixon places the "myopic" solution at the other extreme. The myopic solution occurs when parties ignore the impact of current ground water extraction on future pumping cost and water quality. In other words, the parties use a zero discount factor. The resulting solution equates the marginal benefit from additional water to the marginal cost of extraction. Future user cost is taken as zero. When water extraction is left to private individuals without government regulation, the myopic solution will almost certainly lead to depleting the resource at an excessive rate. Current government regulations and enforcement likely do better than the myopic solution.

Another solution noted by Dixon is characterized by the case where each party maximizes its net present discounted value given the withdrawal path of the other parties. This approach obtains an "open loop" solution where neither party attempts to influence or take advantage of the other parties' behavior. If each party simply optimizes its own behavior taking the other parties' behavior passively as given, then strategic externalities are avoided and noncooperative extraction results in no inefficiencies. The problem here is that each individual has an incentive to take advantage of the other players' behavior if the other player behaves passively.

If each of the players considers optimal reactions in response to the other parties' actions, then the conventional "closed-loop" solution is obtained. The result is equivalent to the game theoretic concept of sub-game perfection. Under sub-game perfection, each party responds to an action by picking the extraction path that maximizes its payoff from the remainder of the game. The solution is found by solving the maximization problem for each agent by backward induction through time, assuming a finite terminal point. This solution may be difficult to calculate particularly with uncertainty. Aside from these difficulties, this solution is likely a reasonable standard of comparison and may approximate behavior that will result in absence of an agreement. (It may not correspond to the current status quo in the

Israeli-Palestinian problem because Israel controlled pumping in Judea and Samaria unilaterally prior to the Palestinian peace agreement.)

Comparison of this outcome to the cooperative outcome reveals a classic prisoner's dilemma whereby strategic behavior to try to improve individually on the cooperative outcome leads both players to a worse result. Although each player would be better off if all the players follow the cooperative solution, any individual player can be even better off by strategically taking advantage of the players who follow the cooperative solution. In other words, the cooperative solution is not a Nash equilibrium because players have an incentive to defect.

If the horizon is infinite (or if the final period is unknown but potentially infinite), then a much richer array of outcomes can result, including the cooperative outcome. These outcomes can be achieved through "trigger strategies." They allow agreements to be self-enforcing in equilibrium, so that agents obtain the collusive solution in a noncooperative game. Trigger strategies impose penalties on players who defect. Therefore, when there are incentives to deviate from the agreement and when it is unrealistic or impossible to write binding agreements, noncooperative games may offer a better framework for modeling the process of negotiation.

6. MODELING THE BARGAINING PROCESS EXPLICITLY

In this section we consider noncooperative bargaining games that can be used to model the bargaining process explicitly. However, the assumption of Section 5 is maintained whereby any agreement will continue once reached.

Binmore, Rubinstein, and Wolinsky (BRW 1986) identify two important motives that may induce negotiating parties to reach an agreement rather than insist indefinitely on incompatible demands: (i) impatience for benefits from the outcomes of an agreement, and (ii) risk of an exogenous breakdown of negotiations. In a cooperative bargaining game, the more concave a player's utility function, the weaker is his bargaining position. This weakness can be attributed either to impatience or risk aversion.

Israelis and Palestinians (players i and p, respectively) must agree upon a set of issues including monthly water withdrawal rates, x_W, punishment in the event of defection, x_T, and duration of punishment, x_D, all of which may be vector valued. To define steps in the process of negotiation, suppose player i proposes $\{x_W, x_T, x_D\}$ in period one. Player p must either accept or reject the offer. If the proposal is accepted then the game is over. If the proposal is rejected the game moves to the second period where player p proposes an alternative $\{x_W, x_T, x_D\}$. The game continues with alternating proposals in each period until either an agreement is reached or the negotiation breaks down.

6.1 Why Would Israel Agree to Negotiate?

Israel's incentives to negotiate appear rather small because it has the physical location and the technology to extract a significant share of MA water regardless of

Palestinian actions (especially in the YT Basin). However, high pumping also implies high future user cost which may in turn create an incentive to reduce pumping. Israel may also have incentives to gain international good will and to keep prior commitments. As part of the Camp David Accord and the Oslo Agreement, Israel declared that governance issues of the Palestinians will be discussed in phases. Administration of water resources is one such important issue. Transferring administrative authority for water is subject, among other things, to the ability of the Palestinians to form an institutional infrastructure for water management. Nevertheless, a transfer of authority appears to be highly likely so establishment of a plan when Israel is more in the driver's seat may be advantageous to Israel. Transferring authority to the Palestinians without agreeing on some type of water management framework may leave Israel worse off because it will not be able to monitor and enforce water use as it does currently.

Moreover, in view of the high Palestinian population growth rate and the increasing standard of living, water scarcity will likely contribute to regional destabilization in the future if an adequate water solution is not reached. Therefore, holding to the status quo will not only adversely effect Israel's image but it will ultimately discourage foreign investment in various projects in the region. Finally, for water quality in most parts of the aquifer to remain at satisfactory levels, treatment of sewage originating from Palestinian residential areas in the MA is crucial. Israeli experts advocate immediate development of sewage treatment systems. Therefore, prior commitment, regional stability, foreign investment, international image, monitoring, and sewage effluent control present a host of reasons that motivate Israel to negotiate a water agreement with the Palestinians.

6.2 Exogenous Risk of Breakdown

Osborne and Rubinstein (1990) suggest two definitions of the exogenous risk of breakdown. Exogenous risk of breakdown is concerned with action that a third party may impose on the other parties who participate in the game and that causes breakdown of the negotiations. In addition, a breakdown is an event in which the "threat made by one of the parties to halt negotiations is actually realized" (Osborne and Rubinstein 1990, p.72). Typically, the outcome of negotiations is uncertain and players have incomplete information on a third party's plan for action that either sustains or breaks the negotiation process. For example, suppose financial assistance is offered for an Israeli-Palestinian water project by an international aid agency. The aid may be conditioned on sufficient progress in negotiations. However, the agency may re-prioritize its portfolio for other reasons and eliminate the commitment of assistance. Such an action will likely cause breakdown of negotiations (see, for example, Caswell and Frisvold 1994).

Suppose, following BRW and Osborne and Rubinstein, that the event of breakdown occurs with probability q, $0 < q < 1$. The payoff function for player j where j = i, p is $V_j(x) = qb_j + (1 - q)u_j(x)$ where player j receives utility b_j under negotiation breakdown with probability q and receives utility $u_j(x)$ when an agreement is reached with probability 1-q. Assume that players are indifferent about the period in which an agreement is reached and that the probability is constant over time. Then the probability that no breakdown occurs in the first t periods is $(1-q)^t$ and the probability of a breakdown in the first t periods is $1-(1-q)^t$. The expected utility of player j at period t is $V_J(x) = [1 - (1-q)^t] b_j + (1-q)^t u_j(x)$. Where the parameters q and b_j are fixed, the endogenous variables are the agreement x and the date of the agreement.

Suppose an alternating offer game is played. In the first period, player i offers player p a proposal. If the proposal is accepted, the game is over. Otherwise, player p makes a proposal. The players keep proposing in turns until an agreement is achieved. If an agreement x is more desirable than breakdown and if players are risk averse,[6] the game has a unique sub-game perfect equilibrium, $x^*(q)$. Osborne and Rubinstein show that this agreement is proposed by the first player at period 0 and accepted by the second player. They also show that as the probability of a breakdown goes to zero, $x^*(q)$ approaches the Nash bargaining solution associated with a zero threat point.

To model the exogenous risk of breakdown of Israeli and Palestinian negotiations, information is needed on beliefs of each that offers of assistance for joint water projects will be withdrawn. Given current broad world interest in the Palestinian peace agreement, the likelihood that interests will wane over time could be quite high.

6.3 Impatience

Time preferences can also affect the outcome of negotiations. The value of time is relevant in models of negotiations for the following two reasons. First, the actual amount of the good to be divided may shrink over time. For example, mismanagement of water withdrawal or effluent discharge can cause salt water intrusion or contamination, reducing future water availability or productivity. Second, future payoffs are likely to be discounted. The outcome of an agreement reached today may be valued higher than the same outcome would be valued today if it were to be received at a later time (Friedman 1986). A player with a low discount factor is viewed as an impatient player and is likely to want to reach the negotiation conclusion quickly. In the process, the impatient player may need to compensate the relatively patient player for reaching an agreement more quickly and, as a result, receive a lower benefit.

To illustrate, consider the Rubinstein (1982) and BRW strategic bargaining model with discounting where players are induced to reach an agreement by their impatience for outcomes. Assume an infinite horizon, conflict of interests, mutually beneficial agreements,[7] time-indifferent agreements,[8] stationarity, time monotonicity, continuity,

---

[6] It can be shown that these three assumptions satisfy six conditions on the player's preference ordering: continuity, stationarity, inferiority of disagreement, desirability of x, valuing of time, and increasing loss of delay (Osborne and Rubinstein 1990, pp. 33-35, 74-75).

[7] Mutually beneficial agreements are assured if disagreement is the worst outcome.

[8] Time indifferent agreements occur when players have constant discount rates. However, agreements are not time indifferent if the parties suffer bargaining costs that are increasing in time.

and concave compensation.[9] If the preference ordering of player 1 over all possible choices of x and T satisfies these assumptions then, for each positive delay between offers, the game has a unique subgame perfect equilibrium. There exists a single pair of outcomes $[x^*(q), y^*(q)]$ where the equilibrium strategies are such that player 1 always demands $x^*(q)$ and rejects any offer strictly below $y^*(q)$, and player 2 always offers $y^*(q)$ and rejects any offer strictly above $x^*(q)$.

Players' time preferences can be either symmetric or asymmetric. When players have asymmetric time preferences, one player can gain a larger share of the resource relative to the other. A player prefers negotiating with an impatient player rather than with a patient player because an impatient player must compensate the patient player in order to reach an agreement in an earlier stage. Two possible scenarios span the set of possibilities for asymmetric time preferences in the Israeli-Palestinian problem: (i) Israelis are patient relative to Palestinians, or (ii) Israelis are impatient relative to Palestinians. Perhaps casual observation suggests that Israel is content to maintain the status quo with relatively heavy Israeli use of the aquifer, i.e., that Israel is patient.

6.4 Why Would Israel be Impatient Relative to the Palestinians?

While it is crucial for the Palestinians to assure a sufficient water supply for the future, water use of the West Bank may not be the most urgent issue currently on the Palestinian agenda. The Palestinian economy and the current size of its population do not yet require significant additional quantities of water. Since the Palestinians cannot yet utilize new water efficiently, their time preferences may reflect patience.

On the other hand, Israel has the desire to improve its image by fulfilling its promises with respect to the Palestinian Authority. Transfer of authority began in November 1994. Presumably, enforcing water pumping rates and quotas (as is currently done by the Israeli Civil Administration) will become more difficult and less desirable. On the other hand, leaving water resources unmanaged will create both short term and long term water shortages. Over-pumping, saline intrusion, and contamination by untreated sewage may be common and possibly irreversible problems. Such problems are not likely to affect Palestinians in the immediate future but will in the long run. However, Israel has already expressed some concerns. Thus, Israel may tend to be more impatient to conclude the agreement and to enjoy the outcome.

6.5 Incorporating Time and Risk Preferences

Consider the situation where Israel is impatient relative to the Palestinians and the Palestinians are more risk averse than Israel. Palestinians are likely more risk averse than Israel with respect to MA negotiations for several reasons. First, Israel tends to have much greater wealth and therefore is not as vulnerable to breakdown of funding. Second, the Palestinians have only the MA for a water source whereas Israel has many

---

[9] Concave compensation is assured if the loss of delay is increasing and concave in time.

other sources. Third, the Palestinians are affected relatively more in terms of pumping costs by a decline in the water table. Thus, Palestinians stand to suffer relatively more from continued over-pumping.

By comparison, even after Israel withdraws from the administrated territories, about two-thirds of MA water will still be available to Israel in the YT aquifer. Palestinians currently do not have the technology to pump sufficient amounts of water from the Western Aquifer to prevent water from reaching Israel's coastal plain. Even if water pumping technology becomes available for the Palestinians, it likely will be very costly. Thus, Israel suffers little risk that much water supply will be lost.

When integrating time and risk preferences, three possible scenarios are of interest: (i) the impacts of Israeli impatience and Palestinian attitudes toward risk may balance one another so the outcome will simulate the case with linear utility functions, (ii) the impact of Israeli impatience on the outcome may be smaller than the impact of Palestinian attitudes toward risk so that the outcome is less advantageous for the Palestinians, and (iii) the impact of Israeli impatience may be greater than the impact of Palestinian attitudes toward risk so that the outcome will be less advantageous for Israel than it would for the Palestinians.

Osborne and Rubinstein consider a model with both time preferences and risk of breakdown. Suppose players have symmetric preferences with respect to time and risk. Then a game of alternating offers has a unique sub-game perfect equilibrium that converges to the Nash bargaining solution in which the reservation utilities are $[\lambda/(\lambda - r)]\,(\overline{u}_i,\overline{u}_p)$ where λ is a factor which when multiplied by the delay in making an offer gives the probability of breakdown, r is the logarithm of the discount factor, and $\overline{u}_j$ represents the reservation utilities. As r goes to zero (the discount factor goes to 1), the reservation utilities converge to $(\overline{u}_i,\overline{u}_p)$. These reservation utilities are motivated by the risk preference argument. However, if the probability of breakdown, λ, goes to zero, then the reservation utilities approach (0, 0) and time preferences explain the outcome rather than risk preferences. These various outcomes demonstrate the relevance of the Nash bargaining solution and provide guidance on the range of reservation utilities that should be considered.

To consider the potential implications of risk and impatience for the Israeli-Palestinian MA problem using the Nash bargaining model of Section 4, alternative solutions can be calculated by modifying the reservation utilities used in developing Table 2. Compared to Case 1 of Table 2, if both reservation utilities are reduced by 5 percent then a further Israeli transfer of water to Palestinians is required (see Case 10 of Table 2). If both reservation utilities are decreased to zero, even further reallocation from Israel to the Palestinians occurs (see Case 11 of Table 2). The latter result applies if the probability of breakdown goes to zero. While the loss of water to Israel may become so large that Israelis are better off with desalination technologies and letting negotiations break down, the results in Table 2 suggest that potential noncooperative outcomes are not too different from the cooperative outcomes in Cases 1-9. Furthermore, with no discounting (a unit discount factor), the result in Case 1 of Table 2 applies directly. Thus, cooperative analysis seems to offer a good indicator of noncooperative outcomes for these problems.

7. CONCLUDING REMARKS

Israelis and Palestinians face a very difficult problem of sharing the Mountain Aquifer. Little progress has been made worldwide in the development of sharing arrangements for aquifers that cross international borders. Yet several factors demand that progress be made. Expanding population and economic growth are placing increased demands on the Mountain Aquifer. Plans and promises by Israel to turn authority over to the Palestinians imply that unilateral management of the aquifer will no longer be possible. Also, new prospects for peace have attracted interest from international agencies willing to finance Israeli-Palestinian water projects if progress on cooperation can be made. This window of opportunity may fade if such progress cannot be made in a timely fashion. With time, other international issues will appear and may induce international agencies to place a lower priority on Israeli-Palestinian water projects or withdraw support from these projects all together.

This chapter has argued that common property problems associated with shared aquifers imply that market solutions alone will not resolve water conflicts. Mutually acceptable agreements are needed. Cooperation can be achieved only through an initial phase of bargaining. This chapter has used a simple model to illustrate the cooperative bargaining solution. Even with the simple cooperative solution, however, enforcement and monitoring problems remain. Both sides face incentives to deviate from the agreement. Although both sides could be better off by following the cooperative solution, neither side has an incentive to adhere to it without further safeguards.

Calculations associated with the cooperative solution reveal what information is needed to guide sharing arrangements. This information includes (i) estimates of water demand or marginal benefits of water for each party, (ii) cost functions that reflect how pumping costs depend on both current Palestinian and Israeli pumping and past pumping practices, and (iii) reservation utilities that represent the payoffs below which the parties prefer not to have an agreement. Improved estimates of water demand and pumping cost are needed and feasible. Determination of appropriate reservation utilities is more difficult and depends on considerations that lead to noncooperative bargaining analysis.

Noncooperative bargaining models illustrate what is likely to happen in the absence of agreement. These models are valuable for understanding strategic behavior with respect to common resources. "Punishments" must be devised that are feasible based on the behavior of the other party and they must be both credible (in the interest of the punishing party) and sufficient to make honoring the agreement attractive compared to the perceived gains of other strategic behavior. Obviously, monitoring is necessary to facilitate such a solution.

Two factors that complicate the cooperative bargaining model based on an underlying noncooperative model are (i) risk that negotiations will break down or fail to reach success before offers of international financing are withdrawn, and (ii) impatience in realizing the benefits of a successful agreement. The relative preferences of the two parties with respect to risk and impatience influence the determination of the sharing arrangement. While these preferences are difficult to measure, a brief discussion suggests that the Palestinians are likely more risk averse regarding negotiation breakdown and Israelis are likely more impatient in realizing the benefits of an agreement. These considerations offer some guidance for choice of parameters that determine appropriate sharing arrangements.

The work in this chapter is admittedly tentative. The water demands and costs of pumping are only estimates and are used primarily for illustration. Nevertheless, this information should be obtainable through further research. Risk preferences and impatience will be more difficult to measure. Some rough estimates may be inferred by observing behavior of each party with respect to related matters. But ultimately, guidance for negotiations based on analytic research may amount more to provision of a range of possibilities under alternative scenarios than identification of a clear solution. Fortunately, bargaining outcomes for these problems appear to be insensitive to several factors. Finally, applicability of game theory to international negotiations regarding aquifer sharing is as yet untested. Specifically, applicability of alternative approaches has not been demonstrated. Thus, further research along these lines is needed until applicability is either proved or disproved.

REFERENCES

Benvenisti, M. 1987. *Judea and Samaria Lexicon.* Jerusalem.

Binmore, K., A. Rubinstein, A. Wolinsky. 1986. The Nash Bargaining Solution in Economic Modeling, *Rand Journal of Economics* 17: 176-188.

Burt, O.R. 1964. Optimal Resource Use over Time with an Application to Ground Water, *Management Science,* 11: 80-93.

Caswell, M.A., and G.B. Frisvold. 1994. Transboundary Water Quality Trading: Implications for Overseas Development Assistance. Economic Research Service, USDA, October.

Dixon, L.S. 1991. Common Property Aspects of Ground Water Use and Drainage Generation. In A. Dinar and D. Zilberman, eds., *The Economics and Management of Water and Drainage in Agriculture.* Norwell, MA: Kluwer Academic Publishers.

Falkenmark, M., J. Lunkqvist, and A. Widstrand. 1990. Water Scarcity—An Ultimate Constraint in Third World Development. Tema V, Report 14, University of Linkoping, Department of Water and Environmental Studies, Linkoping, Sweden.

Feinerman, E., and K.C. Knapp. 1983. Benefits from Groundwater Management: Magnitude, Sensitivity, and Distribution, *American Journal of Agricultural Economics* 65: 704-710.

Fisher, F.M. 1994. The Harvard Middle East Water Project: Model Overview and Results So Far. Harvard University.

Frederick, K.D. 1993. Balancing Water Demands with Supplies: The Role of Management in a World of Increasing Scaracity. The World Bank, Technical Paper No 189, Washington, D.C.

Friedman, J.W. 1990. *Game Theory with Applications to Economics* (second edition). New York: Oxford University Press.

Greenold, Z. 1989. The Water in Israel: 1962--1989. Water Commission, Ministry of Agriculture, Israel.

Gvirtzman, H. 1994. Groundwater Allocation in Judea and Samaria. In J. Isaac and H Shuval, eds., Water and Peace in the Middle East. Amsterdam: Elsevier.

Isaac, J., O. Youssef, M. Ja'fari, I. Shawwa, J.R. Deshazo, and R. Newell. 1994. Water Supply and Demand in Palestine: 1990 Baseline Estimates and Projections for 2000, 2010, and 2020 Prepared for the Harvard Middle East Water Project, Harvard University.

Lowi, M.R. 1992. West Bank Water Resources and the Resolution of Conflict in the Middle East, Environmental Change and Acute Conflict. American Academy of Arts and Sciences, Occasional Paper No. 1, 29-60.

Lowi, M. 1993. *Water and Power: The Politics of Scarce Resources in the Jordan River Basin.* Cambridge: Cambridge University Press.

Nash, J.F. 1953. Two-Person Cooperative Games, *Econometrica* 21: 128-140.

Osborne, M.J., and A. Rubinstein. 1990. *Bargaining and Markets.* San Diego: Academic Press Inc.

Rubinstein, A. 1982. Perfect Equilibrium in a Bargaining Model, *Econometrica* 50: 97-110.

Schwartz, J. 1988. A Prototype to Exploitation of Water in Israel. Tahal, Publication No. 1.8/70, Tel-Aviv.

Soffer, A. 1992. *Rivers of Fire: The Conflict of Water in the Middle East.* Israel: Am Oved. In Hebrew.

Tahal Consulting Engineers LTD. 1993. Israel Water Study for the World Bank. Working Paper, Draft, Israel.

4 FACILITATING INTERNATIONAL AGREEMENTS THROUGH AN INTERCONNECTED GAME APPROACH: THE CASE OF RIVER BASINS

Lynne L. Bennett
Yale University, USA
Shannon E. Ragland
Hagler Bailly, Inc., USA
Peter Yolles
Yale University, USA

International rivers are a substantial source of the world's freshwater supply. The resulting interdependence between nations leads to resource conflicts as upstream nations impose water quantity and quality externalities on downstream countries. As examples of international externalities, these conflicts can be analyzed using game theory. We show, however, that the predominantly unidirectional nature of these externalities produces an unsatisfactory "victim pays" outcome under a traditional game theory approach wherein the downstream country may need to bribe the upstream country into sharing water or improving water quality. Recognizing that nations in weak negotiating positions often try to improve their leverage by linking issues, we recommend using the interconnected game modeling approach for international rivers. To this end, we present two prospective case study examples of how the interconnected game can model issue linkages. The first is a study of tensions over a shared basin in Central Asia. The second is a study of shared water in the Euphrates and Orontes River Basins in the Middle East. An important innovation of the interconnected game is that it can generate outcomes that cannot be obtained when issues are modeled independently.

1. INTRODUCTION

International rivers are a substantial source of the world's freshwater supply. McCaffrey (1993) notes that international river basins account for almost 47 percent of the earth's land area.[1] Of the 63 major rivers listed in Gleick (1993), 47 are in international basins. International rivers are also the primary freshwater source in some countries. For example, almost 80 percent of the Ukraine's water supply comes from the Dneiper River, which flows into the Ukraine from Russia (Trofimova et al. 1996).

The resulting interdependence between nations leads to resource conflicts as upstream nations impose water quantity and quality externalities on downstream countries. As examples of international externalities, these conflicts can be analyzed using game theory. We show, however, that the predominantly unidirectional nature of these externalities produces an unsatisfactory "victim pays" outcome under a traditional game theory approach wherein the downstream country may need to bribe the upstream country into sharing water or improving water quality. Recognizing that nations in weak negotiating positions often try to improve their leverage by linking issues, we recommend using the interconnected game modeling approach for international rivers. To this end, we present two prospective case study examples of how the interconnected game can model issue linkages. The first is a study of tensions over a shared basin in Central Asia. The second is a study of a shared water in the Euphrates and Orontes River Basins in the Middle East. An important innovation of the interconnected game is that it can generate outcomes that cannot be obtained when issues are modeled independently.

2. TRANS-BOUNDARY EXTERNALITIES AND GAME THEORY

Because international rivers flow through more than one country, they are subject to trans-boundary externalities. A trans-boundary externality occurs when actions in one country affect the welfare of residents in another country. Trans-boundary externalities that involve natural or environmental resources are legion. Perhaps at the forefront of global concern today are the international environmental externalities associated with climate change in which greenhouse gas emissions in all countries potentially affect changes in long-run climate conditions in all countries. Other examples include harvests in pelagic fisheries, acid deposition in international airsheds, and water extraction from international aquifers.

When an upstream country alters the water quantity or quality in an international river, it imposes external costs on downstream riparian countries. Similarly, if upstream development activities provide better streamflow regulation, then downstream countries derive external benefits. This particular externality issue was raised between Canada and the United States when improvements on the Canadian portion of the Columbia River Basin promised better flood control and increased hydropower opportunities downstream in the United States (Barrett 1994).

When selecting a modeling context for trans-boundary environmental externalities like international rivers, one must consider two important characteristics of the problem: the countries' welfares are interdependent, and all solutions to the problem must embody the principle of national sovereignty. The interdependence characteristic

[1]This figure excludes the land area of Antarctica. McCaffrey is citing Biswas (1991).

comes from the externality relationship, and it implies that joint or international river management is more likely to produce optimal watershed management than independent management. However, an international agreement to jointly manage the river cannot be enforced in the same way that a national management agreement might be enforced. The principle of national sovereignty means that a country's compliance with international agreements or commitments must be strictly voluntary.

Noncooperative game theory is an effective modeling context for trans-boundary environmental externalities because it accommodates these two characteristics. Game theory studies the strategic interactions of agents who have interdependent welfares. The focus on strategic behavior is due to the uncertainty that the interdependence imposes on decision-makers, which are generally assumed to be countries in the case of trans-boundary externalities. Furthermore, noncooperative game theory studies the behavior of players in games where binding agreements or commitments are not possible. Because this notion closely represents the national sovereignty characteristic, we focus on noncooperative game theory concepts and solutions. Examples of noncooperative game theoretical approaches to modeling trans-boundary environmental externalities include (Barrett 1991, 1994; Carraro and Siniscalco 1991; Hoel 1990, 1991; Mäler 1989, 1990; Markusen 1975; Martin et al. 1993; van der Ploeg and de Zeeuw 1990; Tahvonen et al. 1993; Tulkens 1979).

3. INTERNATIONAL RIVERS AND VICTIM PAY OUTCOMES

International rivers' unidirectional flow make them somewhat unique among trans-boundary resources because game theoretic solutions tend to embody victim pay principles. For international river basins, static games may generate outcomes in which the upstream country has a dominant strategy not to cooperate with the downstream country, e.g., not share or clean up water. A dominant strategy would arise if a cooperative action makes the upstream country worse off; an action such as reducing water use or increasing effluent treatment is certain to make the upstream country worse off than if it did not undertake the action. Consequently, the country's "do not share water" action will dominate its "share water" action. Under these circumstances, the externality will not be internalized by the upstream country, and the outcome will be inefficient overall.

It has been suggested that side payments may be used to encourage countries to choose cooperative actions. However, even if we include a side payment in our game from the downstream country to the upstream country, the side payment action for the downstream country is probably dominated by a "no side payment" action.[2] To illustrate this situation, consider the simple two-by-two international river game in

[2] Side payments are not usually modeled in the existing literature. They should be modeled because the question of whether countries will make side payments must be addressed in the same noncooperative modeling framework as the international externality. In our example, we suppose that the side payment is a direct utility transfer. However, we could also have assumed that the side payment amounts differed across countries. The direct utility transfer assumption does not significantly affect our results.

Figure 1. The upstream country chooses between its actions—*Share* or *Not Share* water with the downstream country. The downstream country chooses whether it makes a side payment to the upstream country.

		Downstream Country	
		Side Payment	**No Side Payment**
Upstream Country	**Share**	1 (Upstream), 1 (Downstream)	-1 (Upstream), 3 (Downstream)
	Not Share	2 (Upstream), -2 (Downstream)	0 (Upstream), 0 (Downstream)

Figure 1. An International River Game with a Side Payment

The values in the matrix are simple representative payoffs[3], suggesting an order of preference. For example, the outcome most preferred by the upstream country is *Not Share* and *Side Payment*, and its least preferred outcome is *Share* and *No Side Payment*. These payoffs also suggest that a nation's payoffs across the outcomes are of the same order of magnitude in its underlying utility function. This is the case because the utility functions are based on expected utility theory and are only unique to an affine transformation.

The payoffs show that the upstream country's dominant strategy is to choose *Not Share*, because sharing the water always costs it some welfare. The downstream country's dominant strategy is *No Side Payment* because making the side payment always reduces its welfare. These two dominant strategies jointly determine the dominant equilibrium (and Nash Equilibrium) of this game, in which both countries receive a payoff of 0. This is the familiar "prisoner's dilemma" outcome, wherein the equilibrium outcome of the game is not optimal. Both countries could receive higher payoffs if they could agree to choose cooperative actions (i.e., *Share* and *Side Payment*). However, each country has an incentive to cheat if it believes that the other country will choose a cooperative action. This results in mutual defection and lower payoffs.

One solution to this dilemma is to model the game as a repeated game. This extension is a useful modeling abstraction for international relations because nations rarely interact on a "single shot" basis. They interact over time, which gives nations an opportunity to condition actions in one time period on observed, past actions. Interaction over time introduces the possibility of rewarding cooperative actions such as *Share* and punishing defection moves such as *Not Share*. In a repeated game

[3]The top left payoffs in each cell are the Upstream Country's payoffs; the bottom right payoffs belong to the Downstream Country.

context, cooperative outcomes such as *Share* and *Side Payment* can belong to the set of subgame perfect equilibria, which characterizes the equilibria of repeated games. This is because any strategy profile (i.e., combination of strategies across the players such as *Share* and *Side Payment*) that guarantees players average payoffs above their minmax payoffs can be a subgame perfect equilibrium.[4]

The set of subgame perfect equilibria can be easily shown using a graph.[5] If players can coordinate their actions and we consider average discounted payoffs for the entire game, then the set of feasible payoffs in an infinitely repeated game is the convex hull of the set of pure strategy outcome payoffs. In Figure 2, this is the large set with the vertices that are the payoff vectors in Figure 1. The vector of players' minmax payoffs is [0, 0], and the set of individually rational payoffs is the set of all vectors [p_1, p_2] such that $p_1 > 0$ and $p_2 > 0$. The individually rational payoffs, which comprise the set of subgame perfect equilibrium payoffs for sufficiently patient players, are shown by the shaded subset of the feasible payoffs. This approach shows that there are a number of payoff vectors that are consistent with subgame perfection, including the vector [1, 1], which is consistent with repeated play of *Share* and *Side Payment* actions

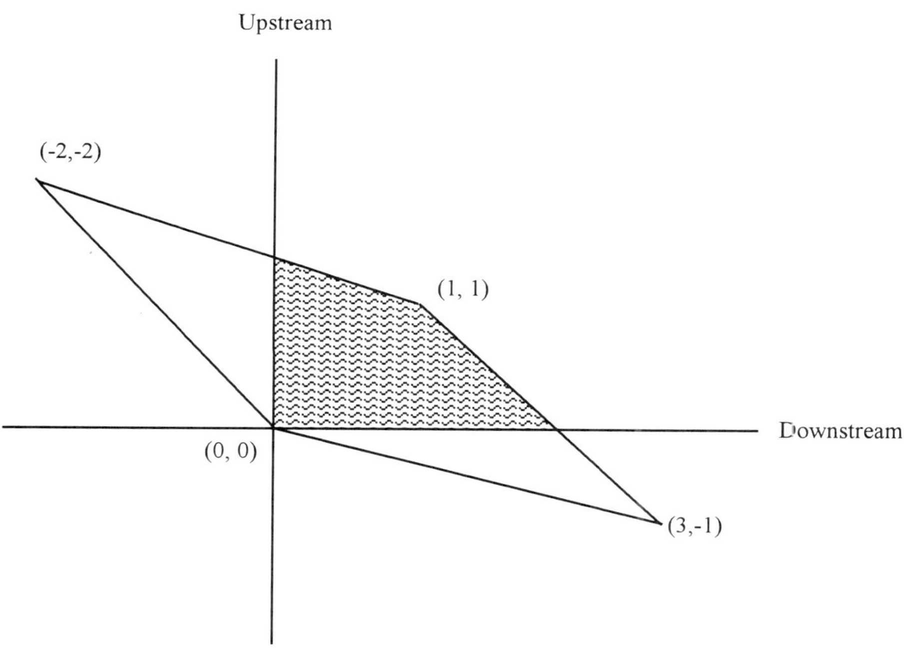

Figure 2. The Set of Feasible Payoffs in an Infinitely Repeated Game

[4] See for example, Fudenberg and Tirole (1991), for criteria that must be satisfied for subgame perfect equilibrium to be an applicable solution concept. We assume throughout the chapter that discount rates for future payoffs are low enough that perfect equilibrium payoffs include all individually rational payoffs.

[5] For a more detailed description of the graphical analysis, see Fudenberg and Tirole (1991, p. 152).

Although the repeated game approach has apparently resolved the unidirectional externality, the *Share* and *Side Payment* outcome represents an application of the "victim pay" principle because the downstream country is essentially "bribing" the upstream country to share water. This cooperative outcome is unsatisfactory for two reasons. First, the international community prefers a "polluter pay" regime, represented by the outcome *Share* and *No Side Payment* in our game (United Nations 1972). As we have shown, such outcomes are difficult to justify as rational outcomes in common property resource disputes that have a unidirectional character. Second, nations are often reluctant to implement "victim pay" outcomes when they are the "victim," because they risk earning a "weak negotiator" reputation (Mäler 1990).

4. ISSUE LINKAGE AND INTERCONNECTED GAMES

Fortunately, "victim" nations recognize that they can improve their negotiation position by linking issues.

> *Linkage between unrelated or only loosely-related issues in order to gain increased leverage in negotiation is an ancient and accepted aspect of diplomacy—a means of widening the threats of rewards at stake in international bargaining, with the hoped-for payoff of increasing the incentive to accede to a government's demands.* (Wallace 1976)

Mäler (1990) and Barrett (1991) noted that issue linkages play an important role in achieving solutions to international externalities, particularly cases that involve victim pay outcomes.

To model issue linkages in a game theoretic context, we turn to interconnected games, which allow countries to link independent issues within a repeated game. The independent issues are represented by games that could be played in isolation. The isolated games are interconnected by allowing countries to condition actions in one game on past outcomes in another. This modeling approach sustains outcomes (e.g., the *Share* and *No Side Payment* outcome) that are not equilibrium outcomes of the isolated games (Folmer et al. 1993; Folmer and Van Mouche 1994). Furthermore, under certain conditions, the interconnected game can give the countries higher payoffs than they would receive in the isolated games (Ragland 1995; Cronshaw and Ragland 1996).

The interconnected game links water and nonwater issues to expand the range and benefits provided by negotiations across multiple issues. A United States–Mexico case study (Ragland 1995) illustrates the application of an interconnected game to an existing international agreement that has an issue linkage. The two countries negotiated the International Boundary Waters Treaty of 1944 for water rights in two river basins, the Colorado and the Lower Rio Grande, when agreements could not be reached in independent negotiations for each basin. Mexico recommended linking the negotiations to use its position as the upstream country in the Lower Rio Grande basin as leverage in the Colorado basin, where it was the downstream country. Isolated game theoretic modeling of the Colorado River would not have predicted the fairly large water share allocated to the downstream country, Mexico, in the International Boundary Waters Treaty of 1944. Instead, the isolated game would have predicted either that Mexico should receive little or no allocation of water, or that Mexico would be forced to bribe the United States for a share of the water. Neither of these events occurred because Mexico was able to link negotiations for the Colorado River with simultaneous negotiations for Lower Rio Grande water. This linkage improved Mexico's negotiation leverage because it controlled significant headwaters for the Lower Rio Grande River, which was used for irrigation in Texas. The interconnected game can model this issue linkage, and predict the correct outcome.

5. CENTRAL ASIA INTERNATIONAL GAME CASE STUDY

In this section, we apply a similar analysis to a Central Asia case study region for which there is a clear upstream-downstream relationship and for which there are currently no enforceable or working agreements. We analyze whether the downstream country has any leverage by looking for issue linkages. By applying the interconnected game analysis to issue linkages, we show that there exist outcomes beneficial to both parties.

The Aral Sea Basin in Central Asia provides an interesting case study on negotiated agreements primarily because of the unique circumstances of the countries involved. The five Central Asian republics (Kazakhstan, Kyrgystan, Tajikistan, Turkmenistan, and Uzbekistan) are all part of a large watershed for which the Amu Darya and Syr Darya Rivers are the primary water sources. Until recently, these countries were part of the former Soviet Union and water disputes were dealt with by Russian authorities. Since the breakup of the former Soviet Union, conflict over water resources in the region has grown and the unsustainable use of the large rivers in the areas has worsened. Water resource management is of primary concern as regional conflicts have become international conflicts.

By 1993 the Aral Sea and its surrounding region had suffered the following ill effects: salinity had increased three fold; sea level had fallen by 50 feet; surface area had shrunk by 75 percent; crop yields had declined; water quality had degraded from agricultural chemicals, industrial pollution, and upstream sewage; and pollution of drinking water had led to high rates of anemia, respiratory and stomach problems, typhoid, dysentery, birth defects, and infant mortality (Bedford 1996).

The combination of these environmental and health impacts and the economic hardships faced since the fall of the Soviet Union has placed incredible pressure on the fresh water resources of the basin. The Aral Basin crisis has focused specific international attention on the overuse of the two tributaries that feed the Aral Sea: the Syr Darya and Amu Darya. A third river in the Aral Sea Basin, the Zeravshan, is an additional source of conflict and is a potential source of issue linkage. Overuse is expected to continue and/or worsen as upper basin states develop new areas and as populations grow. The countries riparian to these rivers are currently looking for ways to resolve conflicts over their limited water resources. Solutions aimed at reducing withdrawals from these rivers require an international agreement on how much water each republic is entitled to use. Prior to the breakup of the Soviet Union, Moscow allocated water and enforced the allocations (Pomfret 1995).[6]

Central Asia's water resource problems are complex and affect a large area. We briefly summarize some recent attempts to resolve water basin issues in the next section. Then we review the set of issues that potentially could be linked in negotiation. Finally, we abstract from the complexity by focusing on two specific issues in Uzbekistan and Tajikistan for the interconnected game analysis.

5.1 The Aral Sea Water Basin

The Amu Darya and Syr Darya originate in the Pamir mountains in Tajikistan and the Tyan Shan mountains in Kyrgystan, respectively, and flow north and westward across numerous borders until their outflow into the Aral Sea (Smith 1995). The Amu Darya all but disappears before reaching the Aral Sea and the Syr Darya is basically a trickle (Brown 1991). The Zeravshan River headwaters are in the Pamir Mountains in Tajikistan also. The Zeravshan dries up before reaching the Amu Darya. A map of the region and the primary rivers is shown in Figure 3. Smith (1995) and Bedford (1996) detail the hydrology of these rivers and outline water usage. Most of the diversions from these rivers go to irrigated agriculture, with approximately 8 million hectares under irrigation. Cotton, a large water-consuming crop, occupies many of those hectares. Uzbekistan is the largest user of water for irrigated agriculture. Much of the population in the region also relies on the flows of the Amu Darya and Syr Darya rivers for drinking water, and there are some industrial users (Pomfret 1995).

The riparian states to these tributaries, in recognition of their domestic dilemmas, international attention, and implications to regional stability, have formed several multilateral organizations to address the issue of fresh water scarcity and the growing potential for conflict over its rights and use. To this end, several commissions have been created. These include the Interstate Coordinating Commission for Water Resources (ICCW) and two river basin authorities (BVOs), one for each the Syr Darya and the Amu Darya (Bedford 1996; Smith 1995). The ICCW was charged with peacefully negotiating water allocations among the basin states of the Syr Darya and Amu Darya. Agreements have not only been slow, but low-level conflict among local groups over drainage canals and water rights continues. To further complicate matters, the BVOs have not been recognized by the individual state legislators and thus have little authority. Although the BVOs have theoretical control over water resources across state borders, a lack of adequate funding and legal power renders these organizations incapable of basin-wide management (Bedford 1996).[7]

In addition, leaders meeting twice in recent years, in Kzyl-Orda in March 1993 and in Nukus in January 1994, have created:

---

[6] Pomfret (1995) also notes that the ministry in Moscow has detailed hydrologic information, which it does not want to share with the Central Asian governments. Whether this data would help reach an agreement or increase conflict is uncertain.

[7] See Bedford (1996) for a detailed discussion of the present institutional framework for international water management.

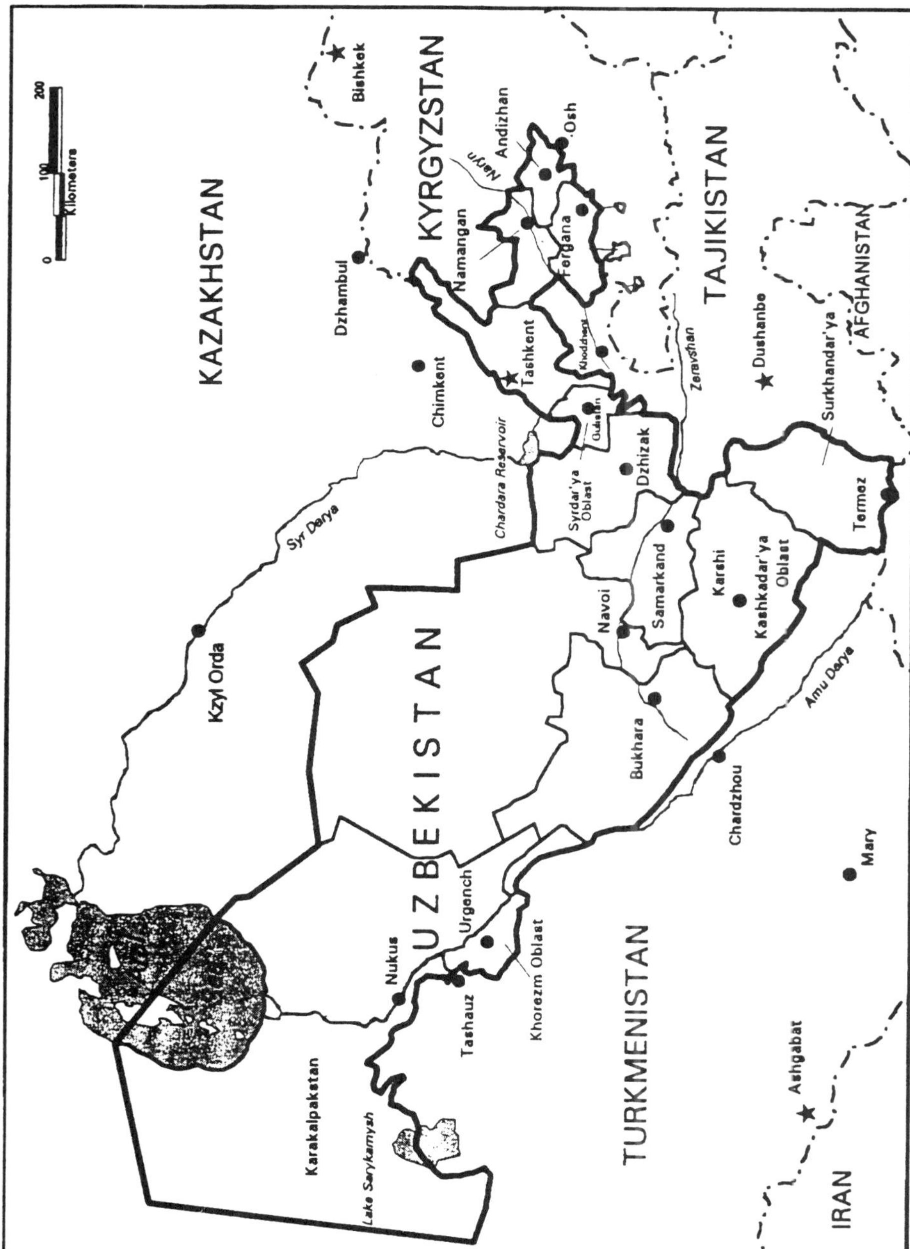

Figure 3. The Aral Sea Water Basin
Source: World Bank (1993), redrawn with permission from the World Bank

- An interstate council with its executive based in Tashkent, Uzbekistan
- An international fund for the Aral Sea based in Almaty, Kazakhstan, to address issues concerning the Sea
- An interstate commission for social-economic development and scientific, technical, and economic cooperation.

Fearful that Central Asia will become a forgotten land since the breakup of the Soviet Union, the five nations have requested and received help from international bodies. The World Bank has offered to loan $200 million to the Central Asian governments for Aral projects over the next three years, and has proposed an action plan detailing 19 projects.

The creation of the international organizations and commitments of funds have not yet generated stable, effective, enforceable agreements over sharing the waters of the Syr Darya and Amu Darya. Coordination, enforcement and financial considerations continue to be problematic.[8] With a limited and variable supply of water, the average per capita quantity of which is continually dwindling from population and growth pressures, the Republics in conjunction with the ICCW and the two river basin authorities (BVOs) may want to consider linking water-sharing with other disputed claims. These other issues which may be linked include air pollution, other rivers and water resource development projects (in particular, a proposed dam in Tajikistan), trade, and trans-boundary support of irredentist factions.

5.2 Issues for Potential Linkages

To date, the multilateral efforts to prevent conflict and to share the Aral region's scarce water resources have not been effective. Several major factors precluding peaceful sharing include extreme dependence on water for food and economic growth, lack of indigenous water management institutions (Smith 1995), and water agreements that are not enforced.

These factors suggest that each nation remains an unrestricted individual actor unwilling to be controlled by centralized coordination and agreement. Better results may be forthcoming if the countries are able to link water-sharing agreements to other valued actions and resources. As observed in many international agreements, it seems reasonable to select countries for which an agreement might be negotiated.

We have chosen to begin an analysis with the countries of Uzbekistan and Tajikistan because these two countries have interests in all three river basins and there seem to be several obvious issue linkages. We are initially evaluating the situation from a two-player perspective using the two largest players. The exclusion of Turkmenistan in this initial analysis is intentional due to our belief that Turkmenistan and Uzbekistan have similar interests in relation to Tajikistan's actions. Thus we are able to abstract from an n-player game to a two player game. Uzbekistan and Tajikistan also represent large

---

[8] D.R. Smith, personal communication, 1996.

downstream and large upstream users, respectively, and have rapidly growing populations. In addition, the two countries have recently signed a cooperation agreement (Economics Intelligence Unit Limited 1994). This suggests there is interest in coordinating the goals of the two countries. In this case, Uzbekistan and Tajikistan share three rivers plus nonwater interests including access to fossil fuels, minerals such as gold, a common airshed, as well as populations with common ethnicities. The Republics should look beyond water to include all their shared issues as bargaining chips.

We consider four categories for potential issue linkages. First is the total allocation of fresh water on the three rivers shared between Uzbekistan and Tajikistan. Implicit in this category is the potential for cooperation on water resource development projects. The second category is non-water environmental degradation such as the pollution of the common airshed. A third category is the civil war in Tajikistan. A final category is economic trade. We select two of these categories for the preliminary interconnected game analysis.

5.3 Water Issues

Uzbekistan (downstream) and Tajikistan (upstream) share three rivers: Syr Darya, Amu Darya, and Zeravshan. Though the rivers flow through both nations, the amount diverted by each is unequal and arguably inequitable. Uzbekistan diverts 59 percent of the water from these basins, compared to Tajikistan's 12 percent, yet is only twice as populous as Tajikistan. Perhaps the most illustrative statistic of the imbalance is the disparity between average per capita consumption levels. In Uzbekistan, exogenous (external) water resources amount to 5,315 cubic meters per person per year (m^3/p/a), and endogenous (internal) water resources to just 459 m^3/p/a. About 91 percent of Uzbekistan's water originates outside its borders (Smith 1995; Population Action International 1994).

Tajikistan is less reliant on an external source of water, with only 50 percent flowing into the country. On a per capita basis, Tajikistan seems to be more water rich, supplying an endogenous amount of 17,731 m^3/p/a and 8,846 m^3/p/a exogenously.

These figures indicate that Uzbekistan is more reliant on water from outside its borders and therefore carries more risk. These per capita figures can help determine utility values for each nation. Tajikistan's marginal value of water may be less than Uzbekistan's because of its greater per capita supply, but its marginal utility for other goods may be higher, such as in units of commodities traded or food. Linkages with water, combined with varying marginal utilities, lay the groundwork for the interconnected game below.

In structuring the games, water allocations theoretically may be shifted among any or all of the three rivers (given real world constraints such as variability, storage, and third-party diversions). However, the hydrological sequence of the Amu Darya poses significant transfer problems because of the large distance and third parties separating Uzbekistan and Tajikistan. Also, given the small amounts consumed by industry and domestic purposes, we assume only irrigation water will be assumed to be negotiable.

Tajikistan also has the capability of diverting more water from the Amu Darya and is entertaining the notion of building a dam on the upper Zeravshan which would primarily be used for hydropower production. Nurov (1992) suggests there is considerable potential for hydropower production in Tajikistan. He estimates that the rivers flowing through Tajikistan have the potential to produce 60 million kilowatts per hour and 7.1 million kilowatts by the year 2000.

5.4 Air Pollution Issue

Tajikistan's economy is highly dependent on cotton production, but Tajikistan's economy is also dependent on its metal mining and processing industry. The aluminum industry produces air pollution hazardous to the population in the shared airshed between Uzbekistan and Tajikistan. One component of this industry is Turusunzade, Tajikistan's aluminum smelter, which has a capacity of 500,000 tons per year, making it one of the largest smelters in the world. The plant is the most modern aluminum smelter in the former USSR, accounting for 15 percent of the former USSR's aluminum production (Pomfret 1995). Turusunzade also employs more than 12,000 people and supports a community of about 100,000 (World Bank Country Study 1994).

There are large environmental concerns associated with the aluminum industry. Turusunzade produced only 250,000 tons in 1993 and 1994 (Mining Annual Review, 1995), but intended to increase production to 272,000 tons in 1995, increasing the air pollution. Uzbekistan is most affected by the pollution as it drifts across the border within the common airshed. The main threats to health of the Uzbeks from the Tajik aluminum plant are from exposure to hydrogen fluoride and a variety of organic compounds (World Bank Country Study 1993).[9]

Improvements to the anode lines at Turusunzade would reduce contamination problems and would also reduce the amount of cryolite used in production, which would in turn reduce fluoride emissions. This is an expensive process which the Tajiks were unable to finance (World Bank Country Study 1994).[10] To control the pollution, Uzbekistan and Tajikistan signed an agreement in February 1995 mandating that Tajikistan install automated systems for pollution control and be held responsible for economic damage caused by the pollution. Uzbekistan would pay for the control systems and their installation. The agreement prescribes measures to "work out a single agreed scientific-technological program for reducing the harmful effects of the Tajik aluminum plant."[11] The agreement was signed for a period of three years with an extension to five years, with unanimous agreement.

---

[9] It should be noted that Tursunzade aluminum plant consumes about half of Tajikistan's electricity output. The Nurek Dam on the Vaksh River is the largest hydropower facility in Central Asia, and most of its output goes to the aluminum plant (Pomfret 1995).

[10] World Bank (1994) estimates of retrofitting costs are approximately $500 per ton of capacity, which implies a $200 million dollar investment.

[11] FBIS-SOV-95-038.

5.5 Civil War in Tajikistan

Tajikistan became an independent state in 1991 and since that time has been impaired by a Civil War. Some of the conflict stems from tensions between Islamic fundamentalists and communists. Pomfret (1995) suggests, however, that much of the tensions arise from old clan loyalties and a fear of Turkic domination in Central Asia, "especially reflecting Uzbek-Tajik conflicts."[12] The southern regions of Tajikistan have large Uzbek populations. The rebel groups in southeastern Tajikistan have been supported by helicopters and aircraft supplied by Uzbekistan (Pomfret 1995; Smith 1995). Smith also suggested that Uzbekistan may hope that prolonged political instability in Tajikistan will slow the construction of new water diversion projects.[13]

5.6 Trade Issues

Free trade across the border between Tajikistan and Uzbekistan since 1991 has been difficult.[14] Natural gas, oil, and gold are plentiful in the region. Trade of food, the product of irrigation water, could benefit both countries. Tajikistan has a strong nonferrous metal mining sector. It also produces oil, gas, coal, and industrial minerals (Mining Annual Review 1995). An interconnected game including water in one game and the trade for food in another may produce results preferred to bargaining over the water alone.

5.7 Other Issues

One other feature that Uzbekistan and Tajikistan share is their landlocked nature. Unlike the other issues, this feature is not unique to these two Central Asian countries. Currently, most transport and communication flows northward, including the flows of the rivers (Pomfret 1995). Establishing rail routes to the sea and building oil pipelines are likely to be important issues requiring regional coordination. Finally, establishing free trade or customs unions between the republics is another area of potential linkage. This final link will not be possible until each country has a stable independent currency and inflation is curbed.[15]

5.8 Interconnected Game Analysis

For our interconnected game analysis, we chose to focus on two potentially important issues from those outlined above. One linkage is suggested by Tajikistan's desire to develop the upper Amu Darya Basin. In this first isolated game, Tajikistan has the

---

[12] Pomfret (1995) details the political instabilities in Tajikistan.

[13] D.R. Smith, personal communication, 1996.

[14] J. Zindler, personal communication, 1996.

[15] Inflation has been in the triple digits in all five Central Asian republics (Pomfret 1995).

choice of developing the river (diverting additional flows) or not. Tajikistan has the potential to divert an additional 7.4 m^3 (6 percent of the Amu Darya's mean flow).[16] Uzbekistan's choices are whether or not to support the rebel groups in Tajikistan. Clearly, Uzbekistan would prefer the river to remain undeveloped, and we assume Uzbekistan thinks some instability in the region will keep Tajikistan from diverting additional flows. Figure 4 shows an example of what the isolated game payoffs might look like. The payoffs represent preference orderings and are not based on empirical analysis.[17] This game has an iterated dominance solution of *Divert* and *Support*. The minmax payoff vector is [0, 0].

		Tajikistan	
		Not Divert	**Divert**
Uzbekistan	**Not Support**	2 1	-2 1.5
	Support	1 -1	0 0

Figure 4. The Amu Darya River Game

Figure 5 shows how the situation surrounding the aluminum plant might be modeled as an isolated game in which Tajikistan chooses whether to abate air emissions by installing scrubbers and Uzbekistan chooses whether to provide aid (financial and technical) for the cleanup. Although a temporary agreement has been concluded on the cleanup of the plant, we show that if this issue had been linked with flows on the Amu Darya, benefits to both sides could be higher.

In this game, Uzbekistan prefers Tajikistan to clean up and will receive benefits whether or not financial aid is given to Tajikistan. If Tajikistan does not install scrubbers, however, Uzbekistan prefers not to provide financial assistance. This game has a prisoner's dilemma payoff structure and the equilibrium is *Not Abate* and *No Side Payment*. Again, the minmax payoff is [0, 0]. It is important to reiterate that the payoffs we show are representative in that they show *order of preference* and they suggest that the payoffs or utilities are of the same order of magnitude across the two isolated games.

---

[16]We acknowledge that Turkmenistan diverts substantial amounts of the Amu Darya between Tajikistan and Uzbekistan. We emphasize, however, that the Tajik's diversions will hurt both Uzbekistan and Turkmenistan. Tajik diversions may hurt Uzbekistan by decreasing flows available for irrigated agriculture *and* reducing flows available for replenishing the Aral Sea.

[17] It has been proven (Ragland 1995) that as long as the payoffs are of the same order of magnitude and the games have asymmetric prisoner's dilemma payoff structures, then the theoretical payoffs will suffice to illustrate the benefits of issue linkage.

		Tajikistan	
		Abate	Not Abate
Uzbekistan	Side Payment	3, 1	-3, 2
	No Side Payment	4, -1	0, 0

Figure 5. The Air Pollution Game

Figures 6 and 7 show the feasible and equilibrium payoffs for the Amu Darya River game and the Air Pollution Game, respectively. The subsets of individually rational payoffs will be those within the positive quadrant. Figure 8 superimposes the set of all possible aggregate payoffs associated with isolated plays of the games (dashed lines).

As long as the payoffs for the two games represent preference orderings and are the same order of magnitude, the games can be interconnected. We argue, for example, that the economic benefits of diverting the Amu Darya to Tajikistan is of the same order of magnitude as the cleanup. Figure 9 shows the stage game payoff matrix for the interconnected game. The pure strategy equilibrium of the stage game has Tajikistan choosing *Divert* and *Not Abate* and Uzbekistan choosing *Support* and *No Side Payment*. The minmax payoff vector is [0, 0], so any strategy profile that guarantees the countries positive average payoffs can be supported as a subgame perfect equilibrium strategy of the interconnected game. Note in Figure 9 the strategy profile in which Tajikistan chooses *Divert* and *Abate*, and Uzbekistan chooses *Not Support* and *No Side Payment*. If this strategy profile is played repeatedly, each country receives a positive average payoff, so this outcome could be a subgame perfect equilibrium outcome of the interconnected game.

We obtain the interesting result that this outcome does not constitute equilibrium outcomes in the isolated games, even under repeated play. Consequently, even though there are side payments in progress, Uzbekistan would clearly be better off by linking air quality to river flows. By linking issues within the interconnected game structure, we have shown an outcome that resolves the water and air externality problems and avoids side payments.

Figure 10 shows the feasible and equilibrium payoffs for the interconnected game (solid line) superimposed on the aggregate payoffs for isolated plays of the games. By allowing the countries the option of conditioning their actions in the Amu Darya River Basin to outcomes within the common airshed, the set of equilibrium outcomes has been expanded. For the scenarios and payoffs provided above, the interconnected game generates strict dominance over the aggregate payoffs.

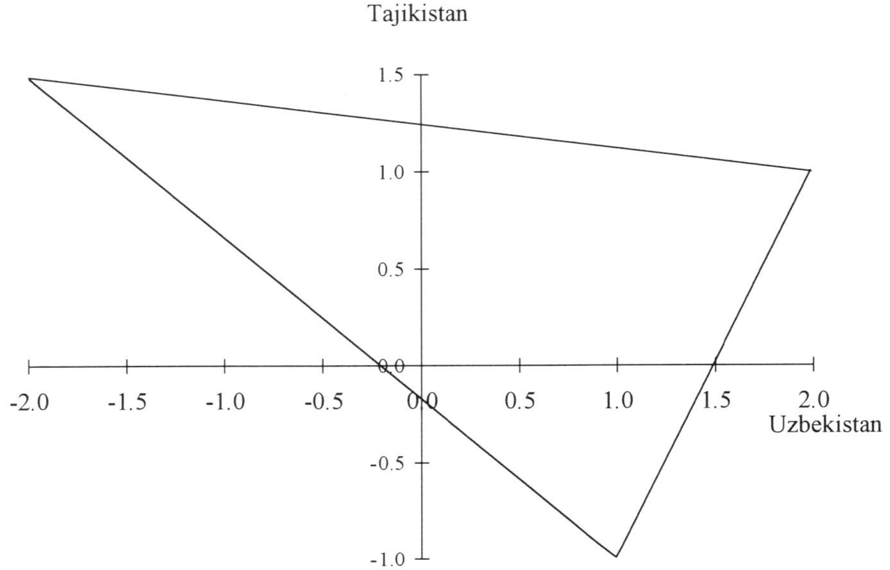

Figure 6. Feasible and Equilibrium Payoffs for the Amu Darya River Game

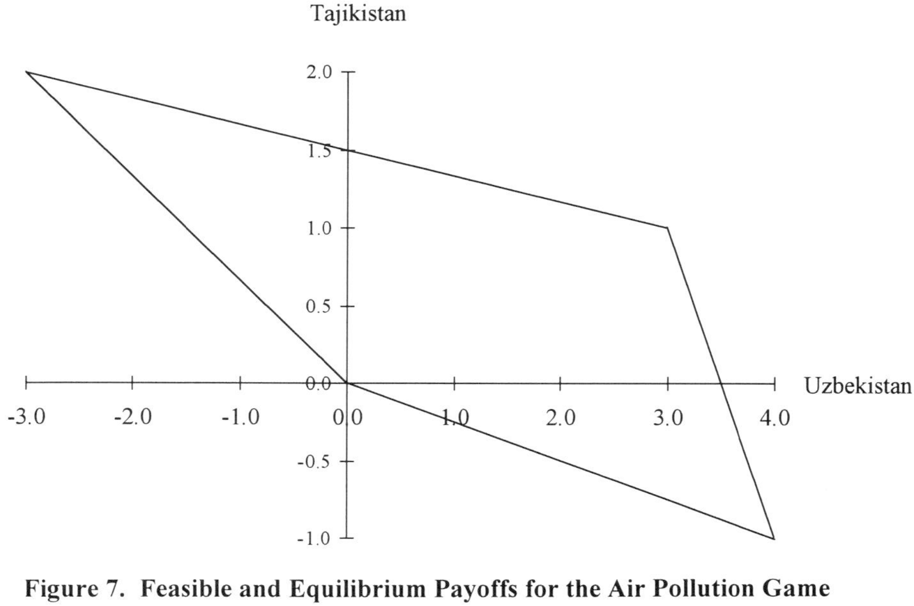

Figure 7. Feasible and Equilibrium Payoffs for the Air Pollution Game

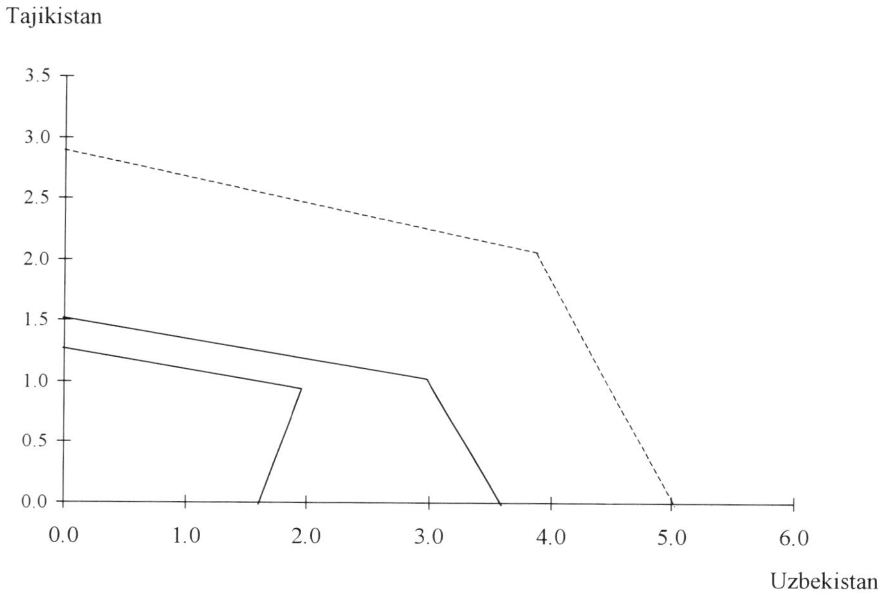

Figure 8. Aggregate Payoffs

		Tajikistan			
		Not Divert Abate	Not Divert Not Abate	Divert Abate	Divert Not Abate
Uzbekistan	Not Support Side Payment	5, 2	-1, 3	1, 2.5	-5, 3.5
	Not Support No Side Payment	6, 0	2, 1	2, 0.5	-2, 1.5
	Support Side Payment	4, 0	-2, 1	3, 1	-3, 2
	Support No Side Payment	5, -2	1, -1	4, -1	0, 0

Figure 9. The Tajikistan-Uzbekistan Interconnected Game

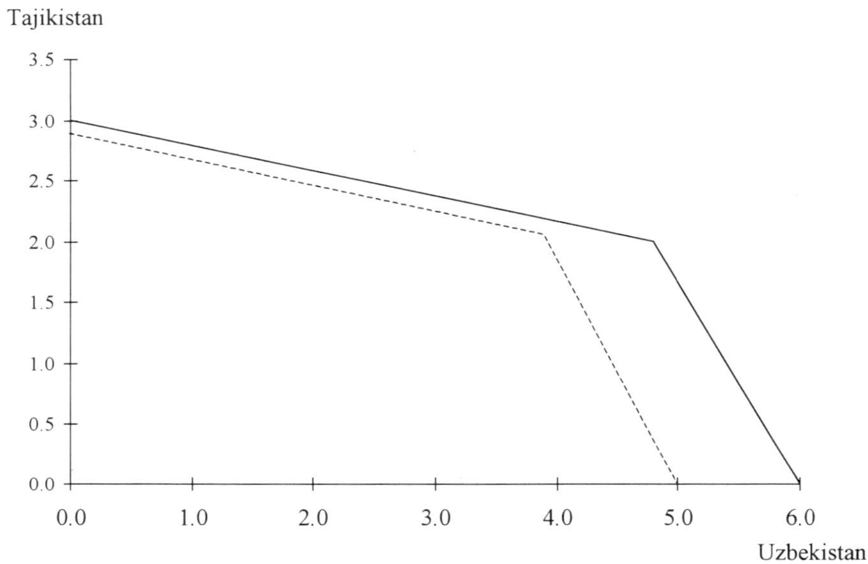

Figure 10. Feasible and Equilibrium Payoffs for the Interconnected Game versus Isolated Plays of the Games

6. MIDDLE EAST INTERCONNECTED GAME CASE STUDY

This two-country case study examines the potential for a stable water-sharing agreement between Turkey and Syria, two of the three riparian states of the Euphrates River. While Iraq should be included in comprehensive long-term management of the river, this chapter is concerned with bilateral negotiation strategies between Turkey and Syria. As in the Central Asia case, we are abstracting to a two-player game using the two largest players. In this analysis, linkages between Euphrates River allocations and other issues are framed within an interconnected game wherein these nations can forge a stable water-sharing agreement and reduce tensions over other issues in the basin. A map of the region is shown in Figure 11.

6.1 The Euphrates River Basin

Precipitation is extremely maldistributed among the riparians of the Euphrates River. The Euphrates River originates in Turkey, which contributes 90 percent percent of the river's 33 cubic kilometer (km^3) average annual flow. The river flows southeast into Syria, which contributes the remaining 10 percent (Gleick 1994). Iraq, the furthest downstream riparian, contributes no water except for the runoff from infrequent rains and irregular transfers of floodwater (Bakour and Kolars 1995). The Euphrates converges with the Tigris in southern Iraq, forming the Shatt-al-Arab waterway and a large delta before draining into the Persian Gulf.

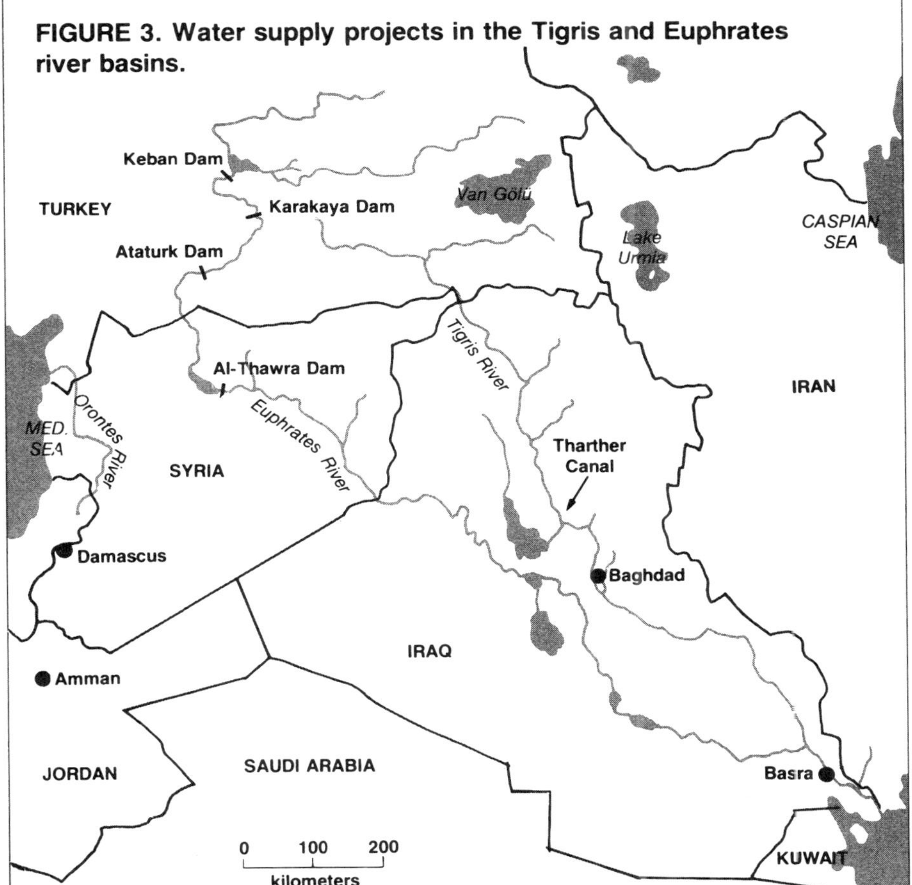

SOURCES: Redrawn from maps of J. Waterbury, "Dynamics of Basin-Wide Cooperation in the Utilization of the Euphrates" (paper prepared for the conference "The Economic Development of Syria: Problems, Progress, and Prospects," Damascus, 6–7 January 1990); and General Staff Map Section, Director General of Military Survey, Ministry of Defence, United Kingdom, 1991.

Figure 11. The Euphrates and Orontes River Basins
Source: Waterbury, 1990; Ministry of Defense, 1991
Reprinted with permission from Pacific Institute for Studies in Development, Environment, and Security, Oakland, California

All three nations are highly dependent upon the Euphrates, and each has attempted at various times to control the river through the construction of on-stream dams and reservoirs. Each nation has plans to divert ever more water to supply rapidly growing populations and to fuel their agricultural and economic hopes. Of greatest concern to Syria is Turkey's massive water development complex to harness the water and hydroelectric potential of the Euphrates and Tigris rivers called the Southeast Anatolia Project, also known by its Turkish acronym GAP.

GAP plans include the construction of 21 dams and 17 hydroelectric projects, including the rock-fill Ataturk Dam, the fifth largest of its kind in the world (McCaffrey 1993). Estimates indicate that GAP's reservoirs will supply irrigation water to between 1.5 and 2.0 million hectares of farmland in the region, effectively transforming parched soil into the Middle East's bread basket. Slim (1993) notes, "Once complete, GAP will more than double the amount of irrigated land in the country, increase agricultural production in the southeast twentyfold, and generate four times the amount of hydropower now produced in Turkey."

GAP provides Turkey with the ability to control the Euphrates' quantity and quality. Turkish diversion for irrigation may be almost half the total flow of the Euphrates, or 15.5 km^3, leaving just 50 percent for both Syria and Iraq. In addition, 3.3 km^3 of irrigation return flow is expected to re-enter the river upstream from Syria. The prospect of reduced and degraded flow has concerned Syria and Iraq, who have limited leverage over Turkey, their more economically and militarily powerful neighbor. Syria's position is aggravated by a high 3.8 percent population growth rate (Bakour and Kolars 1995). Although per capita water availability, a measure of water stress, was about 2,089 m^3 in 1990, the UN Low Projection for 2025 sees future per capita availability declining below the generally accepted water stress level of 1000 m^3 per capita, assuming constant inflow from the Euphrates (Population Action International 1994). Any increased use by Turkey could further reduce per capita water availability.

Allocation of water rights has long been contentious. Since the 1960s when plans emerged to develop the Euphrates, Turkey, Syria, and Iraq have negotiated unsuccessfully to forge a lasting agreement. Tripartite talks in 1965 produced a flawed agreement which over-subscribed the Euphrates by approximately 5 km^3. Since the 1970s when Turkish development began to reduce deliveries to Syria and Iraq, ill will has increased among the riparians. In 1990 Turkey actually restricted water flow to Syria for a month in order to fill the newly finished Ataturk Dam, setting off warning lights in Damascus.

Since 1990, Turkey has agreed to release to Syria an average of 500 cubic meters per second (m^3/s), which is equivalent to 15.8 billion cubic meters (bcm) per year, or half the average annual flow. While this quantity would be sufficient for Syria's current demand, Syria has agreed to share this water with Iraq, which has also demanded 500 m^3/s from Turkey. Additionally, Syria has implemented its own plans for agricultural development called the Syrian Euphrates Development Project, which includes additional irrigation of up to 620,000 hectares (Bakour and Kolars 1995). The numerous pressures on Euphrates consumption has compelled Syria into demanding additional cross-border releases from Turkey.

In the meantime, Turkey continues to develop GAP in the hope that economic growth from agricultural development in southern Turkey will mollify the rebellious Kurds.

Estimates indicate that upon full development of GAP in 2040, cross-border flows to Syria will diminish to 9.07 bcm per year, which is less than the 50 percent allocation (Bakour and Kolars 1995).

6.2 Interconnected Game Analysis

The interconnected game links water and nonwater issues to expand the range and benefits provided by negotiations across multiple issues. One linkage is suggested by past negotiation efforts. Syria has been supplying refuge to the irredentist Kurdish rebels fighting a secessionist battle in Turkey under the Marxist Kurdistan Workers Party (PKK) flag (Gruen 1993). Turkey has insinuated that Syria supports this rebel group as a bargaining chip for water.

We represent the implicit linkage using the stage game in Figure 12.[18] As in the Central Asia case, we assume that Syria is better off when Turkey shares water in the Euphrates, and Turkey is better off when Syria ceases to support the Kurdish rebels. However, Turkey has a dominant strategy: to not share the water. Syria's best response to this strategy is to support the rebels. Consequently, the stage game iterated dominance equilibrium is *Support* and *Not Share*. Although the *Not Support* and *Share* outcome could be an equilibrium outcome under repeated play, this particular issue linkage has failed to produce that outcome. Perhaps Syria's leverage is too weak in this linkage; additional linkages may improve its leverage. We add a linkage with the Orontes River dispute in our interconnected game example.

Syria \ Turkey	Share	Not Share
Not Support	2, 1	-2, 3
Support	1, -1	0, 0

Figure 12. The Euphrates River Game

The Orontes River is a small trans-boundary river which flows north from Lebanon, into Syria and finally into Turkey, where it empties into the Mediterranean. Although the quantity difference between the Orontes and Euphrates is quite dramatic (1.1 bcm versus 32.8 bcm) (Bakour and Kolars 1995), two qualities make this issue compelling:

[18]Again, the payoffs are meant to be representative of preference orderings and are not based on empirical analysis due to lack of data.

Turkish interest in utilizing the resource, and a notable Syrian opportunity to gain leverage as an upstream riparian.

Turkey depends on the Orontes for water supply to its coastal city of Antioch in the Hatay province, where the river empties into the Mediterranean. Several factors complicate the Orontes. First, Antioch is the base for the Turkish navy. Second, the Turks have long irrigated this area using Orontes water. Third, ownership of the Hatay province itself is claimed by both Turkey and Syria. Although both quantity and quality concerns are important issues for the Orontes Basin, we will focus on quality for this analysis.

Gleick (1994) and others report that Orontes inflow into Turkey is severely contaminated by agricultural return flow from Syrian irrigation in the Ghab valley and by urban industrial pollution. Syrian utilization has thus limited the use of the Turkish portion of the Orontes River. Figure 13 shows that this circumstance might be modeled as an isolated game in which Syria chooses whether to abate pollution, and Turkey decides whether to compensate Syria with a side payment. In the one-period game, Syria's dominant strategy is to *Not Abate* Orontes River pollution, and Turkey's dominant strategy is *No Side Payment.*

		Turkey	
		Side Payment	**No Side Payment**
Syria	**Abate**	1, 2	-1, 4
	Not Abate	2, -2	0, 0

Figure 13. The Orontes River Game

Figure 14 shows the interconnected game. The minmax payoffs for the isolated games and the interconnected game are (0, 0), so any strategy profile that gives Turkey and Syria positive average payoffs can be a subgame perfect equilibrium. Consider a trigger strategy profile in which Syria chooses *Abate* and *Not Support* and Turkey chooses *No Side Payment* and *Share* in every period of a repeated game, but both revert to playing the isolated game Nash equilibria if someone defects (i.e., Syria plays *Not Abate* and *Support*, and Turkey plays *No Side Payment* and *Not Share.* The average expected payoffs are (1, 5), so this strategy profile can be a subgame perfect equilibrium.

Figure 15 shows the feasible and equilibrium payoffs for the interconnected game (solid line) superimposed on the aggregate payoffs for isolated plays of the games (dashed lines). Again, by allowing the countries the option of conditioning their actions in the Euphrates River Basin to outcomes in the Orontes River Basin, the set of equilibrium outcomes has been expanded.

Syria \ Turkey	Share Side Payment	Not Share Side Payment	Share No Side Payment	Not Share No Side Payment
Not Support Abate	3, 3	-1, 5	1, 5	-3, 7
Support Abate	2, 1	1, 2	0, 3	-1, 4
Not Support Not Abate	4, -1	0, 1	2, 1	-2, 3
Support Not Abate	3, -3	2, -2	1, -1	0, 0

Figure 14. The Syria-Turkey Interconnected Game

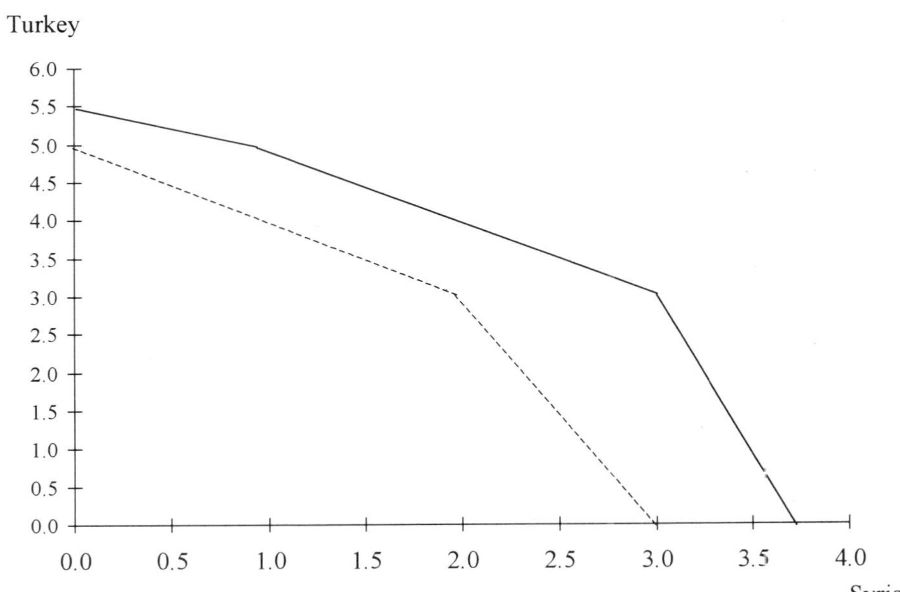

Figure 15. Feasible and Equilibrium Payoffs for the Interconnected Games versus Isolated Plays of the Games

7. CONCLUSIONS

The unidirectional characteristic of international river basins makes game theoretical analysis in isolation from other international issues an unsatisfactory method of analysis. We find that the interconnected game approach generates better modeling results. Previous modeling (Ragland 1995) shows how the interconnected game successfully modeled linkages and predicted the correct outcome for the International Boundary Waters Treaty of 1944. Our Central Asia and Middle East case studies illustrate current water disputes that may be more easily resolved if countries can link the disputes with other issues. For the Central Asia case, we identified several issues pertaining to water resource problems and other resource or trade problems that might be linked. Two were chosen for interconnected game modeling, but there are several alternatives (e.g., linking the Amu Darya with Syr Darya flows, development on the Zeravshan, or grain trade). The Middle East case explored linkages between water allocation in one basin and pollution control in another. There is evidence that linkages are important. The rebel support linkage has been tried in the past (by "gentlemen's agreement"); however, the linkage may not have been strong enough. Giving Syria additional leverage in the form of pollution control may prove successful. The inclusion of Iraq is an additional alternative. Some of these other linkage possibilities will be explored in a future paper.

REFERENCES

Bakour, Y., and J. Kolars. 1995. The Arab Mashrek: Hydrologic History, Problems and Perspectives. In P. Rogers and P. Lydon, eds., *Water in the Arab World: Perspectives and Prognoses*. Division of Applied Sciences, Harvard University.

Barrett, S. 1995. Free Rider Deterrence in a Global Warming Model. Mimeo, Environment Directorate, OECD.

Barrett, S. 1994. Conflict and Cooperation in Managing International Water Resources. Mimeo, Country Economics Department. Washington, D.C.: The World Bank.

Bedford, D.P. 1996. International Water Management in the Aral Sea Basin, *Water International* 21: 63-69.

Biswas, A. 1991. Water for Sustainable Development, A Global Perspective, *Development and Cooperation* 5: 17-20.

Bohm, P. 1990. Efficiency Issues and the Montreal Protocol on CFCs. The World Bank Environment Working Paper No. 40. Washington, D.C.: The World Bank.

Brown, L.R. 1991. The Aral Sea: Disaster Area and Interdisciplinary Solution, *Interdisciplinary Science Review* 16: 345-350.

Carraro, C., and C. Siniscalco. 1991. The International Protection of the Environment: Voluntary Agreements among Sovereign Countries. Mimeo, Fondazione Eni Enrico Mattei, Italy.

Cronshaw, M.B. and S.E. Ragland. 1996. Interconnected Games. Mimeo, University of Colorado, Boulder.

Economics Intelligence Unit Limited. 1994. EIU Country Reports. Fourth Quarter.

Folmer, H., P. van Mouche, and S. Ragland. 1993. Interconnected Games and International Environmental Problems, *Resource and Environmental. Economics* 3: 313-335.

Folmer, H.P., and P. van Mouche. 1994. Interconnected Games and International Environmental Problems: II. (Forthcoming).

Fudenberg, D., and J. Tirole. 1991. *Game Theory*. Cambridge: The MIT Press.

Gleick, P.H. (ed.). 1993. *Water in Crisis: A Guide to the World's Fresh Water Resource*. New York: Oxford University Press.

Gleick, P.H. 1994. Water, War and Peace in the Middle East, *Environment* 36: 3.

Gruen, G.E. 1993. Recent Negotiations over the Water of the Euphrates and Tigris. In Proceedings of the International Symposium on Water Resources in the Middle East: Policy and Institutional Aspects, Urbana, Illinois, October 24-27, 1993. International Water Resources Association, Urbana, Illinois.

Hoel, M. 1990. Emission Taxes in a Dynamic Game of CO_2 Emissions. Mimeo, University of Oslo, Norway.

Hoel, M. 1991. Global Environmental Problems: The Effects of Unilateral Actions Taken by One Country, *Journal of Environmental Economics and Management* 20: 55-70.

Mäler, K.G. 1989. The Acid Rain Game. In H. Folmer and E. van Ierland, eds, *Valuation Methods and Policy Making in Environmental Economics*. New York: Elsevier.

Mäler, K.G. 1990. International Environmental Problems, *Oxford Review of Economics Policy* 6: 80-108.

Markusen, J.R. 1975. International Externalities and Optimal Tax Structures, *Journal of International Economics* 5: 15-29.

Martin, E., R.H. Patrick, and B. Tolwinski. 1993. A Dynamic Game of a Transboundary Pollutant with Asymmetric Players, *Journal of Environmental Economics and Management* 25: 1-12.

McCaffrey, S.C. 1993. Water, Politics and International Law. In P.H. Gleick, ed., *Water in Crisis: A Guide to the World's Fresh Water Resource*. New York: Oxford University Press.

Mining Annual Review. 1995. July: 190.

Ministry of Defence, United Kingdom. 1991. General Staff Map Section, Director General of Military Survey.

Nurov, A. 1992. The Surrounding Regions of Aral: Matters for Concern and Consideration. In Environmental Management of the Aral Sea Region, Report of the International Symposium held at Unu Headquarters Building, Tokyo, Japan.

Pomfret, R. 1995. *The Economies of Central Asia*. Princeton: Princeton University Press.

Population Action International. 1994. *Sustaining Water: Population and the Future of Renewable Water Supplies*. Washington, D.C.: Population Action International.

Ragland, S.E. 1995. International Environmental Externalities and Interconnected Games. Ph.D. dissertation, University of Colorado, Boulder.

Slim, R.M. 1993. Turkey, Syria, Iraq: The Euphrates. In *Culture and Negotiation: The Resolution of Water Disputes*, G.O. Faure and J.A. Rubin, eds. Culture Newbury Park: Sage Publications.

Smith, D.R. 1995. Environmental Security and Shared Water Resources in Post-Soviet Central Asia, *Post-Soviet Geography* 36: 351-370.

Tahvonen, O., V. Kaitala, and M. Pohjola. 1993. A Finnish-Soviet Acid Rain Game: Noncooperative Equilibria, Cost Efficiency, and Sulfur Agreements, *Journal of Environmental Economics Management* 24: 87-100.

Trofimova, I., S. Koublanov, E. Shmurak, I. Buksha, V. Meshkova, A. Sidorov, A. Shereshevsky, L. Dmitrenko, L. Sinitskaya, P. Gozhik, Ya Fedorin, M. Yakimchuk, V. Vernigorov, and I. Makridicheva. 1996. Vulnerability and Adaptation Assessments for Ukraine. In J.B. Smith, S. Huq, S. Henart, L.J. Mata, I. Memesova, and S. Toure, eds., *Vulnerability and Adaptation to Climate Change: A Synthesis of Results from the U.S. Country Studies Program*. Dordrecht: Kluwer Academic Publishers.

Tulkens, H. 1979. An Economic Model of International Negotiations Relating to Transfrontier Pollution. In *Communication and Control in Society*, K. Krippendorff, ed. Gordon and Breach Science Publishers.

United Nations. 1972. Report of the United Nations Conference on the Human Environment. United Nations Document A/Conf/48/14/REV 1.

van der Ploeg, F., and A.J. de Zeeuw. 1990. International Aspects of Pollution Control. Mimeo, Tilburg University, The Netherlands.

Wallace, W. 1976. Atlantic Relations: Policy Co-Ordination and Conflict, *International Affairs* 52: 163-179.

Waterbury, J. 1990. Dynamics of Basin-Wide Cooperation in the Utilization of the Euphrates. Paper prepared for the conference The Economic Development of Syria: Problems, Progress, and Prospects, Damascus, January 6-7, 1990.

World Bank Country Study. 1994. Tajikistan. Washington, D.C: The World Bank.

World Bank Country Study. 1993. Uzbekistan: An Agenda for Economic Reform. Washington, D.C: The World Bank.

PART II

DOMESTIC TRANS-BOUNDARY WATER CONFLICT AND COOPERATION

5 INTERNAL WATER DISPUTES: CAUSES AND SOLUTIONS

David Zilberman
University of California, Berkeley, USA
Douglas Parker*
University of Maryland, College Park, USA

Water resource management issues have been major topics of political debate and dispute throughout the world. Most of the important concepts and techniques in environmental and resource economics (benefit-cost analysis, option value, etc.) have been developed to address water resource management problems. The discussion of water disputes cannot be removed from the context of location and time. Some disputes become settled just as new ones arise. Therefore, the first part of this chapter is an overview of the evolution of the relationships among three major groups—environmental, municipal, and agricultural users of water who have competed for the use and control of these resources over the last 150 years. The rest of the chapter addresses some of the current, internal water policy issues and suggests methodological perspectives and solutions. In particular, we discuss obstacles to the introduction of water markets and address issues of groundwater and water quality. While our discussion is general and relies on the results of theory, we provide examples that are mainly taken from situations in California and the western United States.

1. THE MULTIPLE DIMENSIONS OF WATER RESOURCE MANAGEMENT PROBLEMS AND THE EVOLUTION OF WATER RESOURCE INSTITUTIONS AND POLICIES

Water resources have multiple uses. Some uses are consumptive (washing, drinking, industrial and chemical processes) while others are nonconsumptive (primarily instream uses for recreation, hydroelectric power generation, and navigation). Each water use serves a general policy purpose that benefits certain constituencies. Irrigation, for example, serves the objective of producing cheap and abundant food. It benefits consumers and producers who use the irrigation water, and agribusinesses that provide them inputs needed to produce their product.

---

* At the time of the research reported here, Douglas Parker was a faculty member at the University of California, Berkeley.

Similarly, hydroelectric power serves the objective of providing cheap and clean energy and significantly benefits producers and consumers in regions receiving the energy. Navigation reduces transportation costs and significantly increases commerce; thus, all parties benefiting from commerce are direct or indirect beneficiaries. Environmentalists, recreationists, and fishermen benefit when significant amounts of water are retained in streams instead of diverted.

Development of projects that provide more opportunities for irrigation, navigation, and hydroelectric power generation have been perceived to be an important component of "regional development." They provide expanded opportunities for economic development in primarily rural regions. In addition, the development of frontiers in sparsely populated areas has been a major component of national security policies in many countries. Settlements in the western United States, various regions in Israel, and Australia have been major components of the expansion and solidification of the state. Therefore, water projects that develop frontier areas to better connect them with the rest of the country have had the advantage of meeting much national security objectives.

Rausser and Zusman (1991) introduced a reasonable approach to model political choices. They argued that political decisions are the result of cooperative games among various interest groups, and the outcomes reflect the relative power of those groups. This model can be used in providing a rationale to the econometric framework developed by McFadden (1972), which quantitatively relates policy decisions to the policy objectives that they serve. Thus, policy objectives that have stronger political support will have more weight in determining policy decisions. National security, cheap food, and economic and regional development have been dominant policy objectives in the past (and, at present, are still very important). They have been given high priority by many diverse groups with significant political weights (farmers and agribusinessmen, businessmen, developers, and consumers). Water resource policies have served mostly these objectives. To meet these objectives, the government enacts policies and regulations that facilitate utilization of water resources for irrigation, municipal and industrial expansion, navigation, and hydroelectric power generation. Many of these policies have a significantly negative effect on the environment. However, since the role of environmental considerations was not as important 50 years ago as they are now, environmental quality objectives were compromised in the development of water resources.

A state can assure that public goods are provided in two ways: It can finance and provide them itself, or it can provide incentives for individuals to produce public goods. A developed state that has the capacity to raise funds through taxation can use its resources to finance public goods. States in early stages of development or without significant financial resources may use their power to provide property rights as mechanisms to finance activities that they deem beneficial. In the nineteenth century, the U.S. government was not able to finance much of the water development activities directly, so it provided legal frameworks (e.g., prior appropriation arrangements) that led individuals to settle lands and establish water projects.

Prior appropriation was established as the dominant water rights system for the western United States. It is a queuing system based on the principles of "first-come, first-served" and "use it or lose it." It facilitated conveyance of water from regions of origin to regions where water resources had higher value added in agricultural production, gold mining, and other activities (Gardner 1983). Similar water rights regimes have been established throughout the world. There were other arrangements and institutions that complemented the prior appropriation system in providing incentives to encourage settlement of the western United States in the nineteenth century, including homesteading, provision of land grants to the railroad, and enactment of the Mining Act which provided prospectors the right to all the minerals they discovered. These institutional and legislative arrangements have one common element—to provide individuals who conduct activities that are advantageous to the state with properties rights or valuable assets that are needed for these activities.

In the late nineteenth and twentieth centuries, the state and federal governments, when they had financial resources to develop water resources themselves, created several government agencies to engage in the construction of water projects (e.g., the U.S. Bureau of Reclamation and the U.S. Army Corps of Engineers). These agencies were responsible for the construction and management of most of the gigantic water projects that dominated water resource development in the first 50 years of the twentieth century. Their bureaucracies have strived continuously to promote development of water resources. The evolution of water resource institutions and policies and many of the debates and conflicts in water resource management today are the results of institutions and legislation that were established in the past.

2. THE ENVIRONMENTAL MOVEMENT AND WATER POLICY DEBATE

The increased concern over environmental issues resulted in the establishment of the U.S. Environmental Protection Agency and, in the 1970s, enactment of strong environmental legislation (e.g., the Clean Water Act and Clean Air Act). Prior to that, environmental considerations were not given high priority, and desirability of water resource development was not challenged. The major issues of debate centered on allocation of water between urban and agricultural sectors, the size of irrigation and hydroelectric power projects, and the manner in which projects would be financed. Environmental water projects that manipulated natural water resource systems were inherently perceived as an undesirable activity. Beginning in the 1970s, the main line of reasoning of environmentalists was that development may cause more harm than good. They challenged the pro-development paradigm of the water establishment. Thus, the introduction of the environmental perspective made water policy debates more aggressive and confrontational.

The approach of environmental groups to water policy reform emphasized three major elements: (1) the use of economics to support strong opposition to new traditional water development projects, (2) pursuit and advocacy of restoration projects, and (3) argument for legislation and arrangements that divert water resources for environmental activities.

2.1 The Role of Economic Analysis in Slowing Traditional Water Development

Large investments in water resource development by government and international organizations led to the establishment of industries and bureaucracies that specialized in executing, designing, and managing water projects. While there is no doubt that many water projects have been extremely worthwhile, many could not be justified on pure economic grounds. [1] Reisner (1986) argues that, over the years, many water projects of dubious value were constructed, to a large extent, in order to perpetuate the survival of water resource agencies that contributed to the well-being of water resource developers (these developers contributed to the well-being of political decision makers). Furthermore, the capacity to obtain external funds to support construction of water projects in one's district became a badge of honor for politicians in the United States and other countries, making water projects a very popular form of "political pork."

Economic research in the 1960s and 1970s used economic efficiency to argue against proposed projects that were later executed. Thus, the economic and environmental considerations were quite frequently complimentary in their negative view of many proposed water development projects.

Another factor that weighed against publicly funded water development projects was the gradual worsening of the financial situation of many governments. From the late 1970s on, elimination of proposed water projects became a major component in efforts to reduce government deficits. However, at least in the United States, President Carter's attempts to eliminate many proposed water projects were met with strong objections, and many projects were approved through the efforts of politicians whose regions were scheduled to benefit from the proposed spending.

To improve the economic quality of government agencies' selection of projects, an analytical procedure for evaluating water projects, based on the "benefit-cost approach," was introduced in the 1970s and revised in the 1980s. This procedure, described in "Principles and Guidelines," was developed by economists employed by water agencies. Zilberman et al. (1994) reviewed the "Principles and Guideline" and found that it embodied many sound economic principles. Yet, because it was written and executed by water agencies, the procedures it prescribed tended to favor construction of new structural solutions. For example, the use of "contingent valuation" is prescribed to assess the nonmarket environmental benefits of proposed projects but not the cost of the projects. Similarly, "Principles and Guidelines" does not underscore the need to find nonstructured (institutional, economical) solutions for water resource problems. In spite of the "pro-development" bias of "Principles and Guidelines," its introduction preceded a period where there was a significant

---

[1] See discussions in *Cadillac Desert* (1986) where the author, Marc Reisner, is not very enthusiastic about the water industry. Even he sees the value of some of the early projects. For an economic critique of some water projects, see Gardner (1983).

reduction in new federal water development projects. Some environmental activists tend to downplay the value of economic analysis for environmental policy making. However, when it comes to new water project proposals, more rigorous economic scrutiny appears to support the environmentalist's perspective.

Indeed, environmentalists have relied on economic argumentation in their efforts to curtail traditional water development projects. Economics, however, has not been very useful in its environmental efforts to initiate water resource restoration projects. The efforts championed by environmentalists (and supported by economists) to slow the construction of water projects led to severe conflicts among environmentalists, agriculturists, and developers in the urban sectors. The confrontation between the various groups occurred at the legislative bodies, at the federal and state levels, and through the ballot (in western states). In most cases, it resulted in decisions against new water development activities.

Thus, the slowing expansion of water resource development led to tension between the two parties that had previously collaborated to promote water development, namely, agricultural and urban water uses. Water allocation between the parties became a zero sum game and, at least in California, each of the parties collaborated with environmentalists in inducing water conservation by the other party. Agricultural and environmental pressure led to the introduction of water use metering, stricter water use standards, and mechanisms such as block pricing for water use in the urban sector. Demands by environmentalists and urban water interests led to a reduction in agricultural water subsidies, pressure to introduce efficient water pricing schemes, and incentives for water conservation in agriculture. The pronounced increase in environmental considerations of water resource policy making also led to more emphasis in alternative types of water and water restoration projects.

2.2 The "Elusive" Restoration Projects

Environmental economists have assumed that negative environmental side effects of water resource development will lead to "irreversible" outcomes. Incorporation of these considerations in water project analysis suggests that the execution of water projects may be delayed (or size of projects may be reduced) to accommodate uncertainties about environmental conditions, technology, and preferences (Dixit and Pindyck 1984; Zhao and Zilberman 1997).

While some development activities resulted in devastating irreversible outcomes, many ecosystems proved to be resilient and responded very well to restoration efforts (e.g., Lake Erie, the Rhone River in France, and the Hula Lake in Israel). Indeed, Zhao and Zilberman (1997) suggest that the results of project evaluations under uncertainty should be modified to account for costly restoration possibilities. In practice, however, designs and decisions about restoration efforts are quite tricky. One reason is that parties can interpret restoration projects differently. Some view them as rural revitalization projects that release water resources from agricultural and industrial uses to construct an environment that is conducive to recreational activities. In essence, individuals with this perspective view recreation and tourism as having the higher value of water resource use and support construction of new types of water development projects under the guise of "restoration." Environmentalists, on the other hand, take the restoration term more literally. However, there are two major questions about restoration projects that need to be addressed: (1) What does society want to restore? and (2) Is a return to the past feasible?

In the ongoing debate over the restoration of the San Francisco Bay and Delta, some advocate restoring much of the habitat to its pre-water-project state. But why stop there? There may be people who desire restoration to the presettlement state. Viewing past states of nature as reference points may provide new perspectives to facilitate a debate. However, it seems quite futile because nature evolves, and attempts to cling to a historical moment are romantic but impractical.

Another approach of restoration efforts is to establish a criteria for a "healthy" ecosystem and to design strategies to reach the desired level of health at least cost. This type of approach is consistent with a standards-and-taxes approach of Baumol and Oates (1974). Namely, professionals set a target level and least-cost policies are designed to attain it. However, a "healthy" ecosystem is a rather elusive concept. A more effective policy process would be for decision makers to consider tradeoffs between various levels of ecosystem health (measured in a manner that is comprehensible to nontechnical individuals) and the costs that they entail. One way to produce such a tradeoff relationship is to establish quantitative indicators of an ecosystem's health that will convey the relevant information for decision making processes. But establishment of useful indicators of an ecosystem's health is quite an intellectual challenge that reveals the scientific community's limited capacity for interdisciplinary communication.

Economists and quantitative policy analysts, in an ideal world, would like to be able to use information about the value of environmental health indicators as a basis for input of benefit-cost analysis. They would like to deduce from environmental indicators information about goods and services that the ecosystem can provide. Biological and physical sciences may use variables that measure the state of important biological and physical processes as ecosystem health indicators, even though it is frequently very difficult to infer the benefits associated with various measures of a "healthy" ecosystem. Nevertheless, there has been a gradual transition from the use of physical measures of an ecosystem's health (concentration of toxic chemicals in water) to biological indicators (population of species, e.g., the Chinook salmon or the Spotted Owl in the Pacific Northwest and California). However, even when biological indicators are being used to monitor an ecosystem's health, it is very difficult to infer the benefits that they provide.

Water development agencies (e.g., the Army Corps of Engineers and Bureau of Reclamation), which implemented some of the past projects that have been objectionable to environmentalists, have embraced recent environmental objectives and become actively involved in designing and promoting environmental restoration and other projects that aim to improve environmental quality. Unlike the popular perception, these agencies do not have a pro- or anti-environmental outlook. Their first interest is to pursue policies that will assure their survival and contribute to their growth. As water resource restoration gains in prominence, we will witness collaborative efforts by government agencies to address environmental and water quality issues. For example, in California a task force called Cal-Fed was formed through a joint action of the U.S. Environmental Protection Agency, U.S. Fish and Wildlife Service, the California Department of Water Resources, and other agencies to work on a comprehensive plan for the restoration of the San Francisco Bay and Delta. It seems to us (from our experience in California) that such a task force should attempt to work closely, but not exclusively, with environmental groups and form as broad a consensus as possible for their plans.

Thus far, voters in California and the political forces in many states have been quite supportive of water resource restoration plans. Unlike traditional water resource development projects, they do not have to pass a rigorous economic efficiency test. They are not subjected to the procedures that are detailed in "Principles and Guidelines." Obviously, some proposed water resource restoration projects can pass the economic efficiency test with flying colors. However, many more are likely to be unjustified from an economic efficiency perspective, and there is a risk that restoration projects will become "political pork" like the traditional water resource projects.

Currently, there is broad support and relatively little objection to proposed water resource restoration projects because they do not require much transfer of water from other uses and, if they do, they generally compensate the party that loses its water. In many cases financial considerations remain the main obstruction to these proposed projects. The performance of existing and proposed restoration projects, as well as the capacity to maintain long-term political support for such projects, will require that their design and approval be subject to rigorous economic analysis. Government agencies, and particularly economists in those agencies, are challenged to develop effective procedures for economic analysis of restoration projects that will either be incorporated into "Principles and Guidelines" or introduced as a parallel procedure. Economists and other social scientists working in resource management should provide the intellectual underpinnings for such project assessment processes.

2.3 The Transfer of Water for Environmental Activities

Traditional water projects diverted water resources from their natural state of providing mostly environmental benefits to offering other sorts of benefits—agricultural, industrial, municipal, etc. In a way, the ultimate form of water resource restoration is the reallocation of water from nonenvironmental activities to environmental quality-improvement purposes. Environmentalists have worked hard to establish such arrangements.

Indeed, over the last few years, we have seen a growing number of arrangements where water resources were redirected for environmental quality-improving activities. Perhaps the most outstanding example of this phenomenon is the Central Valley Project Improvement Act which recognizes environmental quality improvement as a "beneficial" use of water resources and requires that one of the objectives of the Central Valley Project is to provide water for such beneficial environmental use. Furthermore, it allocates a large volume of water for environmental quality improvement purposes. This Act provided legitimization for the pursuit of environmental objectives by a major water development agency (the Bureau of Reclamation) and reaffirmed the elevated status of environmental projects.

There has been an emergence of purchasing fund schemes where government agencies purchase water rights from agricultural and municipal users and reallocate them for environmental purposes. This is another embodiment of the growing tendency to divert water away from other uses toward environmental ones. In addition to the transfer of water for environmental purposes, the Central Valley Project Improvement Act also sanctions water-trading activities and the introduction of mechanisms to encourage water conservation. The high inefficiency of the water allocation system in the West, which was based on the prior appropriation rights system, provided low-cost opportunities to divert water to agricultural uses. Sunding et al. (1997) demonstrate the high heterogeneity of water use efficiency in agriculture and argue that the introduction of trading could significantly lower the costs of a reduction in agricultural water supplies. If the water that was moved away from agriculture was previously used on low-valued crops such as pasture, the overall losses would be quite limited. Sunding et al. (1997) estimate that about 20 percent of the water used in California generates 60 percent of the benefits and that the bottom 10 percent of the water used in agriculture generates less than 2 percent of the aggregate benefits. The cost of reducing the overall agricultural water supply is likely to be low if the cuts are distributed so that the most inefficient activities are eliminated. The ability to trade water and move it freely across locations is essential for maintaining efficient use of water in agriculture and for meeting aggregate agricultural water supply cuts at least cost. Therefore, a reform that reduces agricultural water supply while introducing trading and market-like mechanisms will be accepted with relatively less resistance.

3. THE GRADUAL TRANSITION TO WATER MARKETS

Water allocation mechanisms in the United States are very diverse and range from relatively unrestricted water markets (in New Mexico) to almost a complete ban on transfers in several states. In most regions, owners of water rights are not allowed to sell their rights permanently, especially to buyers outside the water basin. In some locations, water rights are attached to land rights, and the only way to acquire water is to buy the land to which it is attached. Seasonal water transfers and trades are much more widespread than permanent water trades, and seasonal transfers within water districts are the least restricted. Economists have pointed out for many years the basic inefficiency of banning transfer through various water rights regimes (Gardner 1983; Burness and Quirk 1979; and Anderson 1984) and have argued for the expansion of water trading. Indeed, water laws throughout the United States are evolving towards more efficient allocation schemes that expand trading and transfer possibilities.

Frequently water reforms are triggered by crises situations. As Rausser and Zusman (1991) argue, a crisis that threatens the welfare of several interest groups is needed to induce them to change the status quo. For example, during the last two years of the recent (1987-1992) drought in California, a new institution, the California Water Bank, was introduced to enable water trading between farmers with ample water supplies and others who suffered water shortages. The Water Bank enabled growers of some of the highest-value crops in California to survive the drought and significantly limited its economic impact. California agriculture came out of the drought in good shape and with more good will towards the introduction of water markets and trading. The drought experience increased the political acceptability of the Central Valley Improvement Act which reduced agricultural supply and expanded water trading possibilities. The experience in California is not unique, and California is ahead of many states in terms of the extent to which water trading is utilized.

3.1 Alternative Water Trading and Pricing Systems

Water reforms in some cases involve reducing aggregate supply for a water district or a region while simultaneously allowing water trading. A new water pricing and trading system has to meet several criteria to be efficient and effective (Brill et al. 1997). First, it has to lead to marginal cost pricing where the marginal cost that users pay is equal to the district's marginal cost of water. Second, the districts are public utilities that are barred from making a profit. Therefore, the pricing scheme has to meet a balanced budget constraint. The total revenues of the district has to be equal to the total cost. Third, the allocation system has to equitable which in most cases imply that individuals will have initial rights that are proportional to their historical water rights. Three types of water pricing are being considered:

- *Passive Trading Systems*. According to this system, each user is assigned a right to a proportion of the district's water and the responsibility for the same proportion of the district expenditure. The user's consumption may deviate from the initial allocation and be charged according to the district's marginal cost of water. Thus, if a farmer is assigned 10 percent of the district's water and the district's annual expenditure is $200,000, then the farmer is initially assigned the responsibility of $20,000. But if the farmer consumes 50 acre feet above his initial share, and the marginal cost for the district is $100 per acre foot, then the farmer will have to pay the district $25,000. On the other hand, if another farmer consumes 100 acre feet less than his initial allotment, he will have to pay the district only $10,000. Under this scheme, all the transactions and accounting will be between the users and the district. This type of approach is also called the buy-back approach (because the district buys back from the farmers any extra water using marginal cost pricing, and the farmer may buy from the district extra water paying marginal cost).

- *Active Trading Systems.* Under this approach, each user is assigned a right to a certain amount of water based on his proportional use in the past. Thus, if a user consumes 10 percent of the district's water in the past and the water supply of the district is 100,000 acre feet, the user will get the right to 10,000 acre feet. The users pay the district for their initial allotment at the average cost of water to the district and can sell the right to consume the water for a year or longer in a water market, directing the district to deliver the water to the buyers. The transaction between the owner of the rights to the water and the buyer of these rights is done without district involvement. The district is only told where to deliver the water but is not necessarily informed of the water price.

An active market may entail more transaction costs because it requires that sellers and buyer find one another. Over time, some traders may specialize in playing the role of speculators who buy and keep water rights for a while and then sell them, thus making the transactions smoother and reducing the cost of trading. Active markets may be especially appropriate in situations where farmers in several districts trade water simultaneously or when farmers do not want the district to control trading. Olmstead et al. (1992) report on the experience of WaterLink, a computerized water market introduced in the Westlands water district in California. One interesting observation from this market is the emergence of repeated trading between certain partners. In this market, farmers are not required to report the price paid for each water transaction; they only have to report to the district information about the physical transaction. The lack of requirement to disclose the prices associated with each transaction may make the market less efficient. Hopefully, water price disclosure mechanisms will be introduced to active markets in the future.

As trading becomes more common, districts may wish to continue to trade on behalf of the farmers and remain their only water supplier, while farmers may wish to conduct their own transactions. One may foresee situations where districts do their political part to prevent trading by individual farmers. This will likely be the case if trading activities by individual farmers result in negative side effects to the district and its other members.

- *Block Pricing.* An alternative approach to achieve greater efficiency is block pricing. Users are charged at a relatively low rate for per-acre consumption of water below a certain threshold level and a certain volume of water per acre, and consumption above this level will be charged the marginal cost of water to the district. Under certain conditions, block pricing leads to efficiency, especially, for example, in the case where users are rather homogeneous and all of them consume water above their initial allocation. However, farmers may be extremely heterogeneous in their productivity. Some of the inefficient ones may use all their initial allocation at a low price instead of selling this water to some other users who have a much higher water use efficiency. Using an example from Israel, Brill et al. (1997) show that this inefficiency of block pricing may be quite significant.

Block pricing, though, may be a very effective tool whose introduction, relative to the existing system, will lead to significant efficiency gains. This is especially the case in the urban sector where water trading between municipal customers, especially small ones, is costly. In this case, introduction of block pricing may represent a large improvement. Indeed, block pricing has been used successfully by the electric utility industry, and the introduction of this system to municipal water users should be encouraged. When it comes to large water users, such as agricultural users and some of the big industrial users, introduction of block pricing for purchase of water from the district, while allowing trading of water after purchase, will lead to efficiency. In this case, trading among farmers will lead to a water price that is equal to the marginal cost of water, and block pricing will be another form of passive trading.

3.2 Obstacles to Water Markets

Several obstacles have to be overcome before efficient water markets are established. They include:

- *Transaction Costs.* Little attention was paid to monitoring of water use and establishment of accurate properties rights under the prior appropriation system. The establishment of property rights may be a costly process that may hamper the speed and the extent of water reforms. A transition to water markets also requires monitoring of transactions and protection against stealing. These activities may be very costly. As long as the direct efficiency gains associated with the introduction of markets, due to better resource allocation, do not outweigh the extra transaction and other costs, introduction of markets may not be justified on economic grounds.

- *Political Economic Consideration.* Under the prior appropriation rights system, historical users of water maintain their rights through continued use. However, it is not clear who will own the right to sell water once trading is allowed. Senior rights owners will strongly oppose introduction of trading if the state, through water resource agencies, becomes the owner of the water and thus receives the proceeds from water sales. However, under most circumstances, senior rights owners will favor water reforms that make their water rights transferable. However, such reform may not be looked upon favorably by other parties, environmentalists, some city dwellers, and others who consider farmers with senior rights to be overly privileged and thus feel that the introduction of transferable rights and other mechanisms will provide windfall profits to a privileged group.

- *Third-Party Effects: Impacts on Employment and Input Suppliers' Income.* "Third-party effects" is a term used to refer to the presumed negative effects of water transactions on other members of the community and is used in attempts to block trading. Important categories of third-party effects are (1) impacts on employment and input supplies, and (2) impacts on secondary users of water.

While owners of water rights pocket the proceeds from water sales, the transfer of water is alleged to reduce the volume of the local agricultural industry and, thus, also causes problems of unemployment, reduced incomes of agricultural input supply providers and agricultural produce processors, reduced revenues of local governments, etc. In many cases, water transfers indeed damage the local economy, but in some cases the impact of the transaction may not be negative. In particular, if only part of the water is transferred and farmers adopt water-conservation technologies, farmers may actually increase purchases of agricultural inputs and increase outputs. Even in cases when the sale of large volumes of water occurs outside of the region and affects agricultural production, there may be a positive externality to the region through taxation and spending if the resources gained by the seller of the water are used for purchases in the region of origin.

Nevertheless, in many cases, water transfers will significantly reduce welfare at the region of origin. It is not unusual that the well-being of some individuals in a region may worsen when one of the resources within the region diminishes. What is unusual is that, in the case of water allocation, decisions are made in both the economic and political arenas, and individuals who may expect to lose from water trading may attempt to block it. Furthermore, potential losers from water transfers may use their political muscle to obtain compensation for their losses if trade is allowed.

• *Third-Party Effects: Impacts on Secondary Users of Water.* Since irrigation efficiency of many agricultural users has been quite low (n average, 60 percent of applied water in California ends up as evapo-transpiration, and the rest, 40 percent, ends up as deep percolating water or runoff), the unutilized water of one user may become the source of water for another. In other words, one farmer may consider his neighbor's runoff as part of his annual water supply or, in other cases, a municipality may consider groundwater generated by deep percolation from agricultural activities as its permanent source of water. Thus, increasing the water use efficiency of a farmer or a group of farmers, which may enable them to increase their volume of production or sell their water outside the region, may negatively affect the well-being of operators who rely on their residual water. Third-party effects may actually be a source of inefficiency because in many cases the value of the product produced with the residual water may be higher (especially if transportation costs are taken into account).

This type of third-party effect is a positive externality that has not been internalized by marketing arrangements in the past, and its elimination may be a source of inefficiency. There are two ways that this problem can be addressed. First, Burness and Quirk (1979) suggest that only the effective water (applied water that is actually utilized by the crop) be considered as the property of the original water user. That will not only significantly reduce the amount of water that they can sell outside the region, but it will also provide an incentive against water conservation technologies.[2] An alternative solution to this type of third-

[2] Third-party effects have to be recognized when considering the efficiency of conservation

party effect is to internalize the externality by considering all of the applied water as the property right of the original user and to allow them to charge the third party for the use of the residual water. Monitoring problems may make this arrangement difficult to implement.

• *Lack of Infrastructure.* Canal systems were constructed to accommodate historical water rights systems. The traditional canal systems suffer from three limitations. First, there is a severe problem of conveyance losses. Many canals are not lined, and the conveyance losses are actually the main source of groundwater accumulation that may benefit third parties. Second, canals have not been very well protected and, once markets are established, stealing of water may be relatively easy. Third, the flexibility allowed by existing systems is very limited. In many cases, inability to deliver water to new places may make the introduction of markets very costly and difficult. Allowing farmers to buy and sell water will require improved infrastructure in terms of well-designed canal systems. The construction of canals will require public financing or public approval because of zoning laws or environmental regulations. Parties that may lose from trading may use the political process (where public funding of a canal project is needed) and/or the political and legal processes (where public approval of canal construction is needed to facilitate trading) to obstruct the building of infrastructure.

In California, one of the major obstacles to trade is the lack of canal facilities to transfer water from the east side of the San Joaquin Valley, where water is abundant, to the west side of the Valley. In the past, environmentalists objected to such activities because they believed that a lack of conveyance facilities was a major barrier to increased importation of water from regions where it provided environmental benefits. With trading, the canal infrastructure may eventually increase water available for environmental purposes. Furthermore, the benefits of water used for environmental purposes may grow with increased flexibility.

• *Groundwater Management Issues.* The introduction of water markets has been addressed mostly in surface water context. However, in many regions, some water users rely on both ground and surface water, while other regions rely solely on either. Access to groundwater by some of the members may add another complexity to transition to markets of water districts that allocate surface water. Water use patterns were determined according to surface water allocation schemes (prior appropriation) and availability of groundwater. Some users of groundwater benefited from the "inefficiency" of those with senior rights and significant surface water supplies (third-party effects). In many other cases, surface water contributed to the buildup of groundwater aquifers that had not been used except in periods of drought. In this case, conjunctive use mechanisms provide security to the water districts. With water trading, some of

---

efforts. If the productivity of the water used by the third party is greater than the productivity obtained by the conservation effort minus the cost of the conservation, then conservation is not justified on efficiency grounds.

the individuals with access to groundwater may overpump their groundwater inventories to sell them in the market. This may be the case especially if individuals are pressed financially, or are not secure about the future of groundwater rights. Thus, reform of surface water allocation cannot be done in this situation without effectively regulating groundwater use.

Thus, one of the most important challenges is to develop mechanisms to (1) establish the water rights for groundwater and (2) monitor its use. In cases of shared aquifers, a groundwater authority may need to determine, for example, overall annually pumping levels and establish and enforce individual water user's rights to pump. In this situation, the right for pumping may be transferable and the pumpers may either sell the right to pump or in other cases may sell the water once it has been pumped. The technical difficulties and uncertainties associated with management of groundwater may make reform of groundwater resource management especially challenging.

4. WATER QUALITY DISPUTES

Growing concerns about water quality lead to conflicts and require policy interventions. Water contamination is the result of current and past agricultural, industrial, and municipal activities. There is a clear distinction between policies that reduce and prevent current and future contaminations, and policies that facilitate cleanup of past contamination. The nonpoint source nature of water contamination problems is a major obstacle to effective policy formation. When chemical runoff of many farms is contaminating a river, it is almost impossible to identify the exact source of each incidence of damage. Monitoring schemes that enable identifying the externalities of each producer may be a realistic future option. At the present, pollution control may be achieved by incentive schemes based on observable production activities. The theory of incentive compatibility can contribute to the design of such schemes.

Policymakers have mostly relied on direct controls to address water contamination. For example, dairy waste management, which is one of the major water quality problems in the United States, has been regulated by setting upper bounds on cows per acre, by requiring farmers to ship waste products outside of a region, or by zoning regulations that disallow certain activities close to population centers. Similarly, enforceable best management practices were established to regulate pesticides and nitrate contamination of groundwater. These solutions may backfire. They do not allow flexibility to producers or adjust for heterogeneity, and they do not provide incentives for innovations that may eliminate the problem at a lower cost. Boggess et al. (1993) notice that the cost of direct control in addressing pollution problems may be twice as high as the cost through efficient incentives.

The introduction of efficient financial incentives to address water quality problems has to overcome several obstacles. Polluters prefer direct control to pollution taxes because taxes cost more. Regulators are uncertain about the impact of financial incentives on the environment and pollution taxation is difficult because of monitoring problems. The first problem can be addressed by developing systems of transferable rights for waste generation. When waste products cannot be monitored, it may be worthwhile to measure the amount of waste generated by proxies. For example, in the case of dairy waste management, it may be worthwhile to develop simple formulas relating cows per acre to waste generation, with some adjustment for certain practices that are more efficient. Producers will be given credits or debits according to their activities, and these will be compared to their initial rights. Producers who are very efficient in production of milk may buy more pollution rights while producers who are less efficient may sell them.

This type of approach may not be the easiest to administer, but the accumulated experience with other systems of transferable permits makes us optimistic to that such a system may work. Similarly, when it comes to groundwater contamination by chemical inputs in crop production, one can develop a scoring system based on crop choice, land quality, and irrigation technology that will approximate contribution to groundwater pollution at each field. Farmers will be able to buy the rights that enable them to produce relatively pollution-intensive crops or will switch to less pollution-intensive crops and sell the rights to pollute. Developing a system of scoring that will be used as proxies for actual pollution levels will require much more interdisciplinary knowledge and more interaction between economists, and natural and biological scientists.

There have been several disputes concerning ground quality standards for chemical residue and how to achieve them in contaminated locations. Determining allowable upper bounds of chemical residue concentration in water is not straightforward since there is much uncertainty regarding the environmental and health effects of various chemical residues as well as a diversity of risk assessment procedures. Thus, different parties will come with various estimates based on their assumptions regarding certain coefficients. Increased consistency and the ability to communicate and compare risk information is crucial in establishing a dialogue to provide solutions for water quality regulation problems (see Lichtenberg et al. 1993). Moreover, once water quality standards are established, development of appropriate schemes for sharing the responsibility for cleanup is quite a challenge. In many of these cases, disputes lead to lengthy legal battles. It is not clear what share of the liability should rest with the chemical company manufacturing the products, the distributors who sold it, the farmers who use it, etc.

Obviously, in cases of negligence (in production or use), the negligent party should be liable. But in many cases, the negative side effects of a chemical is discovered years after it has been introduced and neither producers or users of the chemical have caused any damage knowingly. Sunding and Zilberman (1997) argue that assignment of the liability in these cases to the manufacturer may be inefficient and may reduce the likelihood of innovation. It is convenient to use the notion of joint and several liability that results in a "big pocket" approach for assigning liability to pollution. However, "big pockets" are not necessarily big enough, and in many cases the benefits to manufacturers of certain chemicals would not exceed the expected liability risk. This will lead to increased concentration in the chemical industry that will reduce innovation and the new products available to agriculture. Alternatively, companies may fictitiously divide into small units that will be responsible for production and distribution of certain products and, in this case, they will declare bankruptcy once they get involved in a liability dispute. Therefore, it will be useful to develop mechanisms of shared liability by all parties that benefit from an innovation. In many cases, the community that may suffer deteriorating water quality because of chemical use may be the same community that benefited in the first place. So some of the responsibility of the cleanup belongs to the community. This will also prevent the incentive to exaggerate the costs of cleanup and to take an extreme position in litigation associated with cleanup of residues generated in the past.

5. INTERREGIONAL WATER DISPUTES

Traditional water projects transferred water from wet regions with excess supply to drier regions with excess demand. These regions were controversial in many cases and may encounter objections. Similarly, in many situations, wastewater is also shipped from one region to another. There is a growing tendency in certain regions to ship industrial sludge to other regions. A popular solution to the waterlogging problem—when deep percolating irrigation water reaches an impenetrable layer close to the soil and starts to collect causing risk to ongoing agricultural production—is to dig drainage canals, collect the water, and transfer it out of the region, in most cases to the ocean. In these cases, the act of transferring water may cause disputes between the region of origin and the region of destination.

There is a significant difference, however, if the disputes are within the boundaries of one political entity, say, country, or within two different countries. When the two regions are part of the same country or state, they are governed by one well-defined entity which has the authority to make and enforce the final decision. In many cases, a national government, taking into account the benefits of all regions, actually initiates transfer of water between regions or may even facilitate financing the construction of drainage projects. In California, the government was instrumental in building the state water project that moved water from northern California to southern California. It also initiated a drainage project that was designed to move drainage water from the San Joaquin Valley to the San Francisco Bay.

When the two regions are part of the same political entity, the representatives of the two regions utilize the internal political process in influencing the final outcome so that it benefits their region as much as possible. However, there is a political structure that sets clear rules for the game, and all the players have to act accordingly. In most cases, the region that has more political power (i.e., more economic power, population, etc.) comes out ahead.

In California, regions that have lost some of their resources are regions that have less political power. The rural areas of northern California had less power than the developers of Los Angeles. However, as the work of Becker (1983) and others suggests, the minority region may not lose completely, and plans may be modified to take into account their interest so that they can be somewhat compensated.

The environmental organizations, however, generally tend to support the interests of the regions in which resources are harmed as a result of water projects.

Environmental groups support the interests of the wilderness in northern California that was affected by water transfers. They also represent the interests of the San Francisco Bay and Delta where the water quality has been reduced as a result of water projects. Furthermore, the political process in the United States is designed to give geographic regions power. Note that the legislative branch consists of the House, where weight is given to regions according to their population, and the Senate where the states have equal weight. It seems that, during the twentieth century, we went through a gradual process of increased democratization where the interests of regions were well expressed by their representatives. Thus, the capacity to legislate projects that one-sidedly exploited one region has diminished. Environmentalists work with local communities in northern California to use the political process to make it more difficult to transfer water from the north to the south. Similarly, environmentalists worked with the community in the San Francisco Bay and Delta to block a proposed drainage from the Central Valley to the San Francisco Bay.

When it comes to disposal of wastewater, the cost of shipping wastewater elsewhere leads to solutions that aim to reduce and delay waste generation. Dinar and Zilberman's (1991) book provides a summary of the strategies that California may use in order to solve agricultural drainage problems with a minimal amount of drainage export. This strategy includes provision of financial incentives to reduce drainage; incentives for water conservation through the introduction of tiered pricing or an increase in the price of water; and the subsidization of the adoption of water-conserving technologies such as drip and sprinklers; purchase and removal from production of lands in areas with severe waterlogging problems, etc. In addition, a significant amount of effort is placed on development of technological solutions that will reduce the need to export waste products. They include the use of evaporation ponds, plants, mechanisms to recycle the water, procedures to upgrade water, etc.

In some cases where eliminating or reducing wastewater is prohibitively high, states and regions pay other regions to absorb their wastewater. For example, the state of New York is paying farmers in the midwest a significant amount of money for the right to dispose industrial sludge on their land. New arrangements are emerging where farmers are paid by livestock operators to use animal waste as fertilizer. Thus, the inability to cheaply export waste products leads to new market arrangements as well as new technologies that increase overall efficiency of water use.

6. CONCLUSION

This chapter argues that internal policy disputes are the result of an evolutionary process that is path dependent. We are entering an era where environmental considerations play a more important role in water resource allocation. Water is becoming more scarce and markets are being introduced. The introduction of markets is gradual because of transaction cost issues and the need to address distributional considerations and modify existing institutions. Introduction of surface water markets cannot be done efficiently without control and monitoring of groundwater use and without addressing the externalities of existing water use patterns.

Market forces and incentives can and should play an important role in solving water quality problems. Through wide varieties of incentives (taxation of water and input use, subsidization of conservation technologies, and liability rules), water contamination can be reduced and water quality upgraded. Interregional water transfers can and should be used when appropriate, and there is a potential for gains from trade of such activities. Both economic and environmental impacts of such trades have to be assessed to facilitate efficient design. Finally, as we become more concerned about environmental uses of water, we are challenged to develop a way to efficiently manage use for environmental purposes.

REFERENCES

Anderson, T. 1984. *Water Rights, Scarce Resource Allocation, Bureaucracy and the Environment.* Harper & Row Publishers, Inc.

Baumol, W., and W. Oates. 1974. *The Theory of Environmental Policy.* Englewood Cliffs, NJ: Prentice Hall, Inc.

Becker, G. 1983. A Theory of Competition Among Pressure Groups for Political Influence, *Quarterly Journal of Economics* 98: 371-400.

Boggess, W., R. Lacewell, and D. Zilberman. 1993. "The Economics of Water Use in Agriculture." In G. Carlson, D. Zilberman, and J. Miranowski, eds, *Agriculture and Resource Economics.* New York: Oxford University Press.

Brill, E., E. Hochman, and D. Zilberman. Forthcoming. Allocation of Water at the Regional Level, *American Economic Review.*

Burness, H., and J. Quirk. 1979. Appropriative Water Rights and the Efficient Allocation of Resources, *American Economic Review* 69: 25-37.

Dinar, A., and D. Zilberman. 1991. *The Economics and Management of Water and Drainage in Agriculture.* Boston: Kluwer Academic Publishers.

Dixit, A., and R. Pindyck. 1994. *Investment Under Uncertainty.* Princeton: Princeton University Press.

Gardner, B.D. 1983. "Water Pricing and Rent Seeking in California Agriculture." In T. Anderson, ed., *Water Rights: Scarce Allocation, Bureaucracy, and the Environment.* Cambridge, Mass.: Ballinger Publishing Company.

Lichtenberg, E., R. Spear, and D. Zilberman. 1993. The Economics of Re-entry Regulation of Pesticides, *American Journal of Agricultural Economics* 75: 946-958.

McFadden, D. 1972. The Revealed Preferences of Government Bureaucracy: Empirical Evidence, *Bell Journal of Economic* 7: 55-72.

Olmstead, J., D. Sunding, D. Parker, R. Howitt, and D. Zilberman. 1997. Water Marketing in the '90's: Entering the Electronic Age, *Choices*, Third Quarter: 24-28.

Rausser, G. C., and P. Zusman. 1991. "Organizational Failure and the Political Economy of Water Resources Management." In A. Dinar and D. Zilberman, eds., *The Economics and Management of Water and Drainage in Agriculture.* Boston: Kluwer Academic Publishers.

Reisner, M. 1986. *Cadillac Desert: The American West and Its Disappearing Water.* New York: Viking Penguin, Inc.

Sunding, D., D. Zilberman, N. MacDougall, R. Howitt, and A. Dinar. 1997. "Modeling the Impacts of Reducing Agricultural Water Supplies: Lessons from California's Bay/Delta Problem." In D. Parker and Y. Tsur, eds., *Decentralization and Coordination of Water Resource Management.* Boston: Kluwer Academic Publishers.

Sunding, D., and D. Zilberman. Forthcoming. Allocating Strict Liability in a Multi-Market Setting, *International Review of Law and Economics.*

Zhao, J., and D. Zilberman. Forthcoming. Irreversibility and Restoration in Natural Resource Development, *Quarterly Journal of Economics*.

Zilberman, D., R. C. Griffin, D. D. Parker, M. J. Mazzotta, J. J. Opaluch, W. F. Hyde, D. M. Larson, and A. C. Fisher. 1994. *Review of Principles and Guidelines*. Department of Agricultural and Resource Economics, University of California, Berkeley.

6 EVOLVING LEGAL AND INSTITUTIONAL RESPONSES TO INTERJURISDICTIONAL WATER MANAGEMENT ISSUES IN THE U.S.

Robert Wayland
Christine Lewicki
U.S. Environmental Protection Agency, USA

A watershed-based approach can be an effective means for solving complex interjurisdictional aquatic habitat and water pollution issues. Federal laws that address these issues have tended to focus on particular pollutants and sources. Consequently, they have not resulted in an integrated environmental management approach, which is what is needed to protect waters from the cumulative impacts of a multitude of activities occurring across political jurisdictions. Through improved program integration, the watershed approach can reduce costly duplication of efforts and conflicting actions among various political entities that have jurisdiction over a water body.

1. INTRODUCTION

The recent revival of the concept of adopting a watershed-based approach for the protection and restoration of this nation's water resources and aquatic ecosystems can be an effective means for solving complex interjurisdictional pollution issues. This approach has been slightly ahead of its time at several points in the history of water resource management. But institutional, technological, and cultural factors are converging to bring about a watershed renaissance.

John Wesley Powell, the first director of the United States Geological Survey, is often credited for the concept of a watershed approach. Powell wanted river basins to serve as the management unit for the development of water and natural resources. In his exploration of the arid West, Powell, with remarkable insight, suggested that the political boundaries of new western states should be aligned around river basins rather than the traditional straight-lined borders (Reisner 1986). Over the years, watershed proposals were reintroduced as a means to comprehensively integrate the planning and management of water and natural resources, but were never adopted by Congress who, more often than not, disagreed with the notion of central planning for water or natural resources and preferred to maintain control by approving individual water development projects (Adler 1995).

In early U.S. history, rivers were seen as means of transportation, sources of irrigation, and the starting point for damaging floods. Much of the federal focus was on maintaining navigation with locks and dams, rerouting flows, and constructing levees to keep rivers in their channels and out of the productive farmland in their floodplains.

With the arrival of the postwar industrial and modern agricultural boom came unforeseen pollution problems in our waters and our air. Rachel Carson's *Silent Spring* (1962) brought to the nation's attention the damage to both aquatic and terrestrial ecosystems from newly developed and widely used insecticides like DDT, Dieldrin, and Parathion, or "biocides" as she liked to call them. These pollutants and their impacts traveled from upstream and upwind communities across political jurisdictions to downstream and downwind communities.

2. LEGAL RESPONSES TO TRANS-BOUNDARY POLLUTION

The Water Pollution Control Act of 1948 and the Federal Water Pollution Control Act of 1956 provided the first funds for state pollution control programs as well as funding for wastewater treatment plant construction. These laws were largely ineffective because they lacked national or facility-specific goals and did not provide any control over municipal and industrial effluent discharges into waterways (Adler et al. 1993). In addition, Congress began a long-standing practice of addressing "water pollution" and "water development" separately. The Federal Water Pollution Control Administration, which Congress created in 1965, required states to develop standards only for interstate waters. Monitoring programs were not well developed at this time, however, so enforcing against violators of these standards was difficult at best (Adler et al. 1993).

Five years before we celebrated our first Earth Day, Congress passed the Water Resources Planning Act of 1965, whose stated purpose was "to encourage the conservation, development, and utilization of water and related land resources of the United States on a comprehensive and coordinated basis" (42 U.S.C. § 1962) [Congress was out of sync with the times, passing a law that focused on development of water supplies when the public's concern for the degraded condition of lakes and rivers was growing (Adler 1995)]. Title I created the Water Resources Council (WRC), consisting of the heads of federal departments. The WRC was directed to make recommendations to the President and Congress on the efficacy of river basin plans that were mandated in Title II (Adler 1995). Yet the WRC was destined to fail because Congress did not grant it any authority to require local, state or federal governments to cede power to the river basin commissions. Consequently there could be no implementation of river basin plans on a coordinated, integrated basis (Adler 1995). In 1973, however, the WRC issued a report in which it recommended changes in national water policy to address economic inefficiency. These included (Adler 1995):

- Water users should pay the full cost for water, rather than a subsidized rate;

- Flood control should be paid for by those who benefit from it; and
- Increasing water use efficiency should be used to meet increased demand.

The year 1970, the first year that Americans celebrated Earth Day, was historic in the evolution of institutional response to interjurisdictional water issues and other environmental management issues. Congress adopted the National Environmental Policy Act, and President Nixon created the U.S. Environmental Protection Agency. Soon thereafter the Clean Water Act was enacted, and a series of other environmental statutes as well—such as the Marine Protection, Research and Sanctuaries Act; the Federal Insecticide and Rodenticide Act; the Safe Drinking Water Act; and the Resource Conservation and Recovery Act—all of which rely on regulation and enforcement to motivate good behavior rather than using economic incentives or disincentives as tools.

The Clean Water Act was a breakthrough because it established a floor of minimum water quality standards for the states. Minimum standards helped to remedy, to some extent, trans-boundary upstream-downstream pollution problems caused by non-uniform state standards. In addition, Congress mandated that all point source discharges reduce pollution to the degree afforded by "Best Available Technology" for toxic pollutants and "Best Conventional Technology" for conventional pollutants. This was intended to preclude pollution havens as well as get the first major round of pollution controls in place without the necessity of time-consuming and expensive monitoring and modeling efforts.

There is an underlying tension in the Clean Water Act. The Act calls for coordinated water quality planning in both sections 208 and 303, but implementation is not linked to these provisions. These provisions, adopted in 1972, were ahead of their time and fell into disuse when they either proved to be irrelevant to the decisions needed in the 1970s and 1980s or actually impeded these decisions.

Section 319 of the Clean Water Act calls for states to describe in their nonpoint source assessments how they will incorporate intergovernmental coordination. It also calls on states to develop and implement a nonpoint source control program on a watershed basis. Section 319 also gives states relief from interstate pollution problems in navigable waters by allowing states to petition the administrator to convene a management conference with all States that are contributing to the problem. Section 401 also provides for interjurisdictional pollution relief. It gives states certification authority and the power to veto federal permits or licenses when these activities threaten state water quality standards.

3. INSTITUTIONAL RESPONSES TO TRANSBOUNDARY POLLUTION

The U.S. Environmental Protection Agency (EPA) has had 20 years of progress in achieving best available technology. By 1994, about 52,000 National Pollution Discharge Elimination System permits were issued, covering approximately 250,000 facilities (Charlton 1997). Naturally, the system for controlling pollution by mandating a uniform technology standard for all bodies of water is not perfect since in some jurisdictions there is over-control of pollutants and in others there is too little pollution control. The good news is the country has seen some progress addressing point sources of pollution.

Much of the challenge remains in implementing a second round of controls to address common trans-boundary problems such as wet weather pollution sources like stormwater, combined sewer overflows, and nonpoint sources. Another challenge is to implement controls based upon the gap between best available technology for point sources in a watershed and the remaining reductions needed to meet water quality standards.

Because of the success the nation has had at controlling point sources of pollution, the pollution problems of the 1990s have a different focus. Today the nation is focusing more on controlling nonpoint sources of pollution, and preventing habitat degradation and wetlands deterioration. In 1994, the states and Native American tribes reported that agriculture (largely a nonpoint source of pollution) is the most widespread source of pollution in the nation's surveyed rivers and lakes, affecting 60 percent of impaired river miles and 50 percent of impaired lake acres (U.S. EPA 1995). The federal laws that address these problems have tended to focus on particular sources, pollutants, or water uses, and have not resulted in an integrated environmental management approach. Consequently, significant gaps exist in our efforts to protect waters from the cumulative impact of a multitude of activities (U.S. EPA 1996).

To address these problems, which originate in pastures, suburban lawns, and parking lots, EPA, the states, and their local partners are moving "back to basins." The EPA's watershed efforts were initially concentrated on the "jewels" like the national estuaries and the Chesapeake Bay. These served as models for place-based watershed approaches where multiple levels of government and private stakeholders integrate and tailor their efforts to the highest priority problems within hydrologically-defined geographic areas. Because watersheds' characteristics and institutions governing these watersheds are unique to specific areas, each watershed will have its own unique approach to addressing its problems according to local priorities.

As EPA's programs have evolved, the agency is focusing on expanding the tool box to help every community that wants to protect its "special place." The federal government, which faces time and financial constraints, is less likely to invest in every "special place" as it did in the cases of the national estuaries and great waters, particularly in light of government downsizing. So a framework has been developed for communities to use to save their own aquatic ecosystems—a more bottom-up approach—and one that a community can custom fit to its own needs and desires. The watershed approach has enjoyed widespread support in recent years. Water Quality 2000, a joint effort of more than 80 parties representing all levels of government, industry, academia, and environmental groups, encouraged its adoption (Adler 1995). The Western Governors Association, in collaboration with the Western States Water Council, is embracing this approach as a way to enhance local input. Even H.R. 961, introduced in the 104th Congress, and which environmentalists referred to as "The Dirty Water Act," had language encouraging a watershed framework.

A watershed or basin approach offers many benefits. People are more likely to support protecting waters and aquatic resources to which they feel a connection rather than the more abstract notion of broadly protecting the environment. A watershed approach can address overlapping and conflicting responsibilities of multiple governments with jurisdiction over a single body of water. It can also address multiple issues in a more coordinated fashion, such as water quality and quantity; ground and surface water; and land use and water impacts (Adler 1995). Through improved program integration the watershed approach can reduce costly duplication of efforts and conflicting actions (U.S. EPA 1996). Drawing upon the additional resources, tools, capabilities, and authorities of the various levels of government and private sector affords better, more holistic protection for bodies of water and aquatic ecosystems.

As EPA undergoes "government reinvention" and experiments with market mechanisms to achieve further progress, a watershed approach can help foster the establishment of pollutant trading schemes or wetlands mitigation banks. Because these actions require permits, developing them within a watershed context could enhance the predictability that future actions will be permitted, reducing costs for the private sector. As a result, the watershed approach can help enhance local and regional economic viability in ways that are environmentally sound and consistent with watershed objectives (U.S. EPA 1996).

There has recently been an emergence, or perhaps just more of an awareness, of new localized and large-scale trans-boundary issues. For example, the Pigeon River is a hotbed of interstate conflict. Its headwaters begin in the Blue Ridge Mountains in North Carolina. Champion International, which has a plant on the river, has received variances from the State of North Carolina; the last one required EPA approval, which it received. When the river flows into Tennessee, about 35 miles downstream from the Champion plant, it does not meet that state's standards. Knoxville uses the river as a source for drinking water. This is a classic example of where establishing minimum water quality standards has not been enough to solve this trans-boundary issue. A watershed approach where the multiple stakeholders and governments come together to reach consensus could provide a solution. A significantly more complex and larger trans-boundary issue that is receiving increasing attention is the hypoxia problem in the Gulf of Mexico. Twenty-eight states have watersheds that drain into the Mississippi River, thus contributing to this problem. Research shows that a significant contributor to the hypoxia problem is nutrient runoff from agriculture (Turner et al. 1991). Neither the current regulatory system nor the economic incentives in the Farm Bill have adequately addressed the issue of nutrient runoff from agriculture.

4. CONCLUSION

Legal and institutional responses to the nation's water quality and aquatic ecosystem problems have evolved slowly. The idea of managing natural resources via a watershed approach has surfaced and resurfaced over the years. The progress to date, including the desire for consensus-based solutions, the availability of desktop Geographic Information System (GIS) tools, together with models of progress like the Chesapeake Bay Program, all contribute to EPA confidence that the time is right for an idea that has for so long been slightly ahead of its time.

REFERENCES

Adler, R. 1995. Addressing Barriers to Watershed Protection, *Environmental Law* 25:974-1106.

Adler, R., J. Landman, and D. Cameron. 1993. *The Clean Water Act, 20 Years Later.* Washington, D.C.: Island Press.

Carson, R. 1962. *Silent Spring.* Boston, MA: Houghton Mifflin Co.

Charlton, T. 1997. Personal communication. Office of Wastewater Managment, U.S. Environmental Protection Agency.

Reisner, M. 1986. *Cadillac Desert.* New York: Viking Penguin, Inc.

Turner, E., and N. Rabalais. 1991. Changes in Mississippi River Water Quality this Century: Implications for Coastal Food Webs, *BioScience* 41:140-147.

U.S. Environmental Protection Agency. December 1995. National Water Quality Inventory 1994 Report to Congress, EPA841-R-95-005. Washington, D.C.: U.S. Government Printing Office.

U.S. Environmental Protection Agency. July 1996. Watershed Approach Framework, EPA84-96-001. Washington, D.C.: Government Printing Office.

7 DEALING WITH INTERSTATE WATER ISSUES: THE FEDERAL INTERSTATE COMPACT EXPERIENCE

Richard A. Cairo
Susquehanna River Basin Commission, USA

The resolution of interstate water conflicts has been a difficult problem in the United States. A reluctant U.S. Supreme Court has urged Congress to legislate solutions and has urged the states to make use of the compact clause of the U.S. Constitution. The Delaware and Susquehanna River Basin Commissions are federal-interstate compact agencies with comprehensive powers to deal, not only with interstate conflicts, but with all other water resource management issues. These commissions are the first interstate compact agencies to include the federal government as a full-fledged signatory member. This chapter traces the evolution of the federal-interstate compact commissions, reviews their most important powers, and cites several case studies of their actions.

1. INTRODUCTION

This chapter provides some insights into how two federal-interstate compact agencies in the eastern United States are avoiding conflicts and promoting cooperation in the management of two large river systems—the Susquehanna and the Delaware River basins. Before delving into specific examples of their work, we need to first look at the conditions which set the stage for the establishment of the compact agencies. This gives us the opportunity to review the various legal and institutional devices that have been applied to the problem of interstate water management.

2. DEFINING THE PROBLEM: FEDERAL VERSUS UNIFIED NATIONAL GOVERNMENT

Most national governments throughout the world are unified, i.e., the national government is the only sovereign entity operating in the governmental system. The national government is responsible for every level of administration from the smallest local matter like the probate of a will to matters of great national significance like the regulation of commerce and control of natural resources. The national government may use local administrative compartments to carry out this national administration; however, such compartments are mere creations of the government, with no sovereign powers of their own. Thus, few, if any, interjurisdictional problems can be expected to arise in the management of water and other natural resources confined to the boundaries of that nation.

By contrast, several large nations, including the United States of America, Canada, and Australia, operate as federations of sovereign or quasi-sovereign states. The United States, in particular, has a storied tradition of preserving and protecting the rights of its states. The U.S. federal government is itself a creation of its thirteen original states. The states reserved to themselves all sovereign power not specifically granted to the central government (U.S. Constitution, Amendment X). This includes the power to control and manage their own natural resources under inherent police powers (*N.W. Pipeline v. State Corp. Commission of Kansas*, pp. 493, 512). As Robert Clark (1967) points out in his treatise on water rights, "State boundaries have been fixed by man without much evident regard for hydrologic features created by nature" (p. 294). The stage is thus set for the development of classic interjurisdictional conflicts among the sovereign states.

3. SOLUTIONS TO THE PROBLEMS OF INTERJURISDICTIONAL CONFLICTS

This section considers two approaches that have been used to the solution of interjurisdictional conflict: judicial solutions and legislative solutions.

3.1 Judicial Solutions

Interjurisdictional conflicts over the waters of interstate rivers have very often resulted in litigation before the United States Supreme Court. The High Court, under its original jurisdiction (U.S. Constitution, Article III, Section 2, Clause 2), has adjudicated disputes on the Laramie, the North Platte, the Connecticut, the Delaware, and the Colorado Rivers, to name just a few (*Wyoming v. Colorado; Connecticut v. Massachusetts; New Jersey v. New York;* and *Arizona v. California*, 1963). Original jurisdiction means that the Supreme Court acts as a trial court, first taking evidence, then making determinations of both fact and law.

The federal common law principle employed by the Supreme Court in the resolution of water right suits between states is known as "equitable apportionment." As Clark (1967) stated, equitable apportionment is a label, not an analysis (p. 324). The label describes the exercise that the Court goes through to render a fair and just judgment as between two co-equal sovereigns. The Court may draw on any number of water right doctrines or laws in helping it reach a fair apportionment, including riparian rights, prior appropriation, and federal or state statutes.

Though its jurisdiction is clear under the Constitution, the Supreme Court is generally reluctant to exercise its original jurisdiction. As the Court has said in *Colorado v. Kansas* (p. 393):

> *The reasons for judicial caution in adjudicating the relative rights of states in such cases is that, while we have jurisdiction of such disputes, they involve the interests of quasi-sovereigns, present complicated and delicate questions, and, due to the possibility of future change of conditions, necessitate expert administration rather than judicial imposition of a hard and fast rule ... Not every matter which would warrant resort to equity by one citizen against another would justify our interference with the action of a state...*

It is clear that the Court regards litigating sovereigns in a different light from ordinary litigants. In such cases, the Court sees itself more in a diplomatic than a judicial role where caution must be exercised to give due regard to the rights of co-equal sovereigns.[1]

Neither is the Court well equipped to act as a trial court. It must appoint a master to hear the evidence and make recommendations as to findings of fact and law. Invariably, multiple objections are filed to the master's report that the Court must ultimately sort out. In the end, the whole process can turn out to be an administrative nightmare for the Court.

These difficulties led the Court to strongly suggest the use of non-judicial forums for the resolution of interstate water right controversies and "negotiation and agreement pursuant to the compact clause of the constitution (*Colorado v. Kansas,* p. 393; *Arizona v. California*, 1963).

3.2 Legislative Solutions

Legislative solutions to interjurisdictional conflicts have involved three approaches: congressional edict, interstate compacts, and, more recently, federal-interstate compacts.

3.2.1 *Congressional Edict*

The Supreme Court has clearly recognized that the power of Congress to apportion interstate waters is superior to the Court's use of the doctrine of equitable apportionment.[2] Congress has several legislative options available to it for doing so.

---

[1] See *North Dakota v. Minnesota* where the Court stated that, "The jurisdiction and procedure of this court in controversies between states of the Union differ from those which it pursues in suits between private parties. This grows out of the history of the creation of the power, in that it was conferred by the Constitution as a substitute for the diplomatic settlement of controversies between sovereigns and a possible resort to force. The jurisdiction is therefore limited generally to disputes which, between states entirely independent, might be properly the subject of diplomatic adjustment."

[2] In *Arizona v. California* (1963, pp. 565 and 566), Justice Black stated, "It is true that the Court has used the doctrine of equitable apportionment to decide river controversies between states. But in those cases, Congress had not made any *statutory* apportionment...Where Congress has so exercised its constitutional power over waters, courts have no power to

The first option is by a direct act. An example is the *Boulder Canyon Project Act* (P.L. 70-642, 45 Stat. 1057; 43 U.S.C.A. §617), which authorizes the Secretary of the Interior to impound 30 million acre-feet of water and deliver it pursuant to contract on federal terms irrespective of state law. Among the contracts upheld in an earlier case of *Arizona v. California* (1931) was that between the Secretary and the Metropolitan Water District of Southern California providing for an interbasin transfer of Colorado River water (see Report of the National Water Commission 1972, p. 318). Based on both *Arizona v. California* rulings, the National Water Commission Report (1972) went on to assume that a Congressional authorization of "a transfer from a river basin in one state to a different basin in another state would be upheld by the Supreme Court even where the source river does not cross or form the border of the two states."[3]

In addition to making direct allocations by law, water law scholars agree that Congress may delegate that power to an executive official such as the Secretary of the Interior. In fact, Congress has done both in the Boulder Canyon Project Act (Corker 1967; Clark 1967, pp. 366-367). Certain quantities are allocated to the basin states, with discretion left in the Secretary to allocate any shortages.

Rather than have the allocation authority vested in a single executive branch official, Congress can provide an administrative mechanism through which water allocation determinations can be made with notice and hearing prior to action (Clark 1967, p. 372). Interstate compacts with agencies for continuing administration are particularly suited to this task.

3.2.2 *Interstate Compacts*

Compacts are agreements among sovereigns. The power of states to enter into compacts is an inherent right of sovereignty not derived from the Constitution of the United States; however, the Constitution qualifies this right by requiring that Congress consent to such agreements (U.S. Constitution, Article I, Section 10, Clause 3).[4] Hence, compacts are a third legislative option available to Congress for the apportionment of interstate waters, though state legislative action is also involved.

Interstate compacts, even those with only state signatories, possess many attributes of federal law. While interstate compacts are not federal law in the conventional sense, "Congressional consent transforms an interstate compact…into a law of the United States" (*Texas v. New Mexico* which quotes *Cuyler v. Adams*). Interstate compacts therefore pre-empt any conflicting state laws under the

---

substitute their own notions of an 'equitable apportionment' for the apportionment chosen by Congress" (emphasis added).

[3] Note that, while the Water Commission Report seems to allude to both the 1931 and 1963 cases, the 1963 case contains the strongest support for Congressional authority in these matters.

[4] See also Clark (1967, p. 361).

supremacy clause of the Constitution (*West Virginia ex rel. Dyer v. Sims; Jacobsen v. Tahoe Regional Planning Agency*).

Having the attributes of federal law, and affecting the rights, obligations and relationships of its signatory state parties, the ultimate construction of the meaning and effect of an interstate compact lies with the Supreme Court of the United States. As the High Court has stated (*West Virginia ex rel. Dyer v. Sims,* p. 28. *Hinderlider v. LaPlata River & Cherry Creek Ditch Co.*),

> *Just as this Court has power to settle disputes between states where there is no compact, it must have final power to pass upon the meaning and validity of compacts. It requires no elaborate argument to reject the suggestion that an agreement solemnly entered into between states by those who alone have political authority to speak for a state can be unilaterally nullified or given final meaning by an organ of one of the constructing states. A state cannot be its own ultimate judge in a controversy with a sister state. To determine the nature and scope of obligations as between states, whether they arise through the legislative means of compact or the "federal common law" governing interstate controversies, is the function and duty of the Supreme Court of the nation*

This passage from the Court's decision in *West Virginia ex rel. Dyer v. Sims* establishes not only the Court's power to construe interstate compacts, but the limitations on the power of any single state party to do so. In some ways, a state gives up or contracts away a piece of its sovereignty when it enters into a compact. It is bound by the compact provisions and cannot unilaterally withdraw from the compact or change the compact's provisions unless the terms of the compact, agreed to by *all* the parties, so provide. The state signatories do receive compensating benefits in that the authority they confer upon the compact or the compact agency extends into the other compact states where they would normally be powerless.

Since the compact clause of the Constitution went into effect, states have entered into over 200 interstate compacts (Voigt 1972, p. 53). Up to the time of enactment of the Delaware River Basin Compact (P.L. 87-328; 75 Stat. 688, 1961), interstate compacts dealing with water resources were single purpose in nature, relating wholly to such matters as water quality, interstate allocations of water, interstate bridges or interstate port authorities. Some did not provide for continuing administration, but included procedures for later adjustments.[5]

Where continuing administration was included, some compact agencies were granted significant regulatory powers, while others were granted only planning or coordinating powers. Some had federal representation, but none had federal signatory membership. Only with the enactment of the Delaware River Basin Compact (DRBC) did a multi-purpose, administrative-regulatory, federal-interstate compact agency come into existence (Voigt 1972, p. 54). The Susquehanna River Basin Compact (SRBC), with a similar administrative agency, followed a decade later (P.L. 91-575; 84 Stat. 1509, et seq. 1970). These two remain the only federal-

[5] For example, see the Colorado River Compact found in Appendix 2, p. 371, Special Master's Report filed January 16, 1961, 364 U.S. 940 (1961).

interstate compact commissions in the nation, though such compacts are being considered elsewhere in the U.S.[6]

3.2.3 *Federal-Interstate Compacts—The Case for Federal Participation*

As noted, interstate compacts have many of the attributes of federal law. Federal involvement in their formation, scope, and purpose is inevitable. Not only must Congress consent to them, but Congress may also impose any conditions on a compact that are constitutional. The signatory states, by accepting and acting under it, assume the conditions that Congress attaches (*Petty v. Tennessee-Missouri Bridge Commission)*.

Interstate compacts are all about promoting and achieving harmony or comity among the several states. Harmony among the several states which form the union is one of the reasons advanced by *The Federalist Papers* for the adoption of the federal constitution some 208 years ago (Hamilton 1787). In this sense, interstate compacts, and the agencies which administer them, deal with issues that are closely related to the constitutional functions and interests of the United States.

One conclusion that can be drawn from the *Arizona v. California* declarations regarding the power of Congress to apportion interstate waters is that interstate compact agencies are the agents of Congress. Therefore, it is not a great leap of faith to advocate the direct and full participation of the federal government in interstate compact administration, especially compacts dealing with rivers flowing across or forming state boundary lines.

The idea of a federal-interstate compact agency to manage the water resources of a big, interstate river is not new. Writing in a 1925 issue of the *Yale Law Journal*, future Supreme Court Justice Frankfurter and Landis (1925, p. 701) observed that

> *Whenever the pressure is felt, one answer is clear: No one state can control the power to feed or to starve, possessed by a river flowing through several states. A great number of our streams have this potency. Moreover, there cannot be a definitive settlement. Population, engineering, irrigation conditions constantly change; they cannot be cast into a stable mold by adjudication or isolated acts of administration. The whole economic region must be the unit of adjustment; continuity of supervision the technique. Agreement among the affected states and the United States, with an administrative agency for continuous study and continuing action, is the legal institution alone adequate and adapted to the task.*

[6] Based on inquiries received by the offices of the DRBC and SRBC, other jurisdictions considering federal-interstate agencies include the states of Alabama, Georgia, and Florida, which share the waters of the Apalachicola-Chattahoochee Flint River Basin. The American Water Resources Association has suggested a federal-interstate compact commission for the management of the Colorado River Basin. Finally, Upper Mississippi and Missouri Basin interests are considering the DRBC/SRBC models, along with a group from the Sacramento River Delta in California.

Federal involvement in the management of water resources is another factor favoring federal membership in interstate compacts. The House Judiciary Committee, in its report on the proposed Delaware River Basin Compact in 1961 (the Susquehanna Compact is almost identical), found a

> *...need for integrating the program activities of the many different federal agencies concerned with river basin development. In no realistic sense can an exclusively interstate (as opposed to federal-interstate) basin agency plan for, or integrate the activities of the federal government. The federal government can be controlled only by itself. Having the federal government as one of the primary parties to the compact is the best way to enable the basin agency to effectively coordinate and integrate the programs of the federal agencies. Without integration, comprehensive development plans cannot be efficiently implemented. If the basin agency were to be non-federal, it would mean that coordination and integration would rest solely on voluntary cooperation (and that) national water resources study commissions show conclusively that voluntary cooperation has never yet worked in a satisfactory manner.*

There is no denying that the federal government is an active player in the management of the water resources of the Delaware and Susquehanna basins. The Corps of Engineers has authority provided by §10 Rivers and Harbors Act and §404 CWA Dredge and Fill in both rivers. The Corps also builds and operates water supply storage. Its civil flood control works in each basin have a tremendous influence on river flows and potential flooding.

The National Weather Service operates a river forecast center that makes forecasts of river and stream stages for several river basins (the Susquehanna, Delaware, Potomac, Raritan, and James River Basins) that are critical during flood events. The U.S. Geological Survey operates the stream gauges that measure stream flows and is involved in many other water quantity- and quality-related programs. It also serves as the Delaware River Master to fulfill requirements of the 1954 Supreme Court Decree. The U.S. Environmental Protection Agency (EPA) has extensive authority relating to water quality. The Fish and Wildlife Service and National Marine Fisheries Service have active programs relating to each basin.

In the Delaware, the National Park Service manages three segments of the National Wild and Scenic River system. In the Susquehanna, the Federal Energy Regulatory Commission (FERC) licenses four major hydroelectric facilities in the lower river; and in the Delaware FERC licenses two major hydroelectric facilities. The wide extent of federal water resource-related programs in the Delaware and Susquehanna basins point to the need for a single federal player empowered to represent the interests of all, resolve disputes, and bind all agencies via one federal vote.

When the Delaware Compact was before the Congress for approval in 1961, a question was raised on the constitutionality of federal participation in the Compact. This was the first time that full federal membership in an interstate compact had ever been proposed. The issue was whether federal participation would somehow interfere with the constitutional powers exercised by the federal government vis-à-vis the states.

An April 25, 1961, Justice Department opinion addressed to Presidential Special Assistant Frederick G. Dutton from Deputy Attorney General Nicholas Katzenbach cites a Columbia Law School opinion (Gelhorn and Grad 1960) on the subject. The Columbia opinion concluded that federal participation is not constitutionally prohibited. The government had previously entered into a variety of contractual-like arrangements with the states to address all kinds of problems. Added to this was the desirability of federal-state cooperation and judicial support for interstate compacts as a device for dealing with intergovernmental problems.

Nevertheless, the opinion recognized that formal participation in a compact could not be allowed to bind the government on the free exercise of those enumerated powers delegated to the federal government by the Constitution. For this reason a provision was added to both compacts preserving the right of Congress to withdraw the federal government or to stipulate the conditions under which it would remain a member of the compact.[7]

Also, a series of federal reservations were added, not to the body of the compact where they would have to be re-approved by the states, but to the enabling legislation enacted by Congress approving the compact.[8] These reservations set out a number of other conditions which would safeguard federal authority and prerogatives; e.g., the President can override any requirements of the SRBC Comprehensive Plan if he determines that the national interest so requires.[9] These reservations are in keeping with the doctrine announced in *Petty v. Tennessee-Missouri Bridge Commission* allowing the regulation of interstate compacts by Congress.

4. FEDERAL-INTERSTATE COMPACTS IN THE DELAWARE AND SUSQUEHANNA RIVER BASINS

Federal-interstate compacts went into effect in the Delaware River Basin on October 27, 1961, and in the Susquehanna River Basin on January 23, 1971. It is not the purpose of this chapter to offer a complete history of these compacts. It suffices to say that the framers of the Delaware Compact intended to deal with a host of water resource problems and conflicts which had already beset that basin, while the Susquehanna framers sought to prevent such problems and conflicts in their relatively undeveloped basin (Voigt 1972, p. 69).

The Delaware and the Susquehanna compacts are almost identical in their contents and, as noted above, are the only two of their kind in the nation. They contain some very special powers for the comprehensive management of water resources that are worth reviewing here.

---

[7] See P.L. 87-328; 75 Stat. 688, §1.4; P.L. 91-575; 89 Stat. 1509 et seq., §1.4.

[8] See P.L. 87-328; 75 Stat. 688, §15.1, Reservations; P.L. 91-575; 89 Stat. 1509 et seq., §2, Reservations.

[9] See P.L. 87-328; 75 Stat. 688, §15.1 (s), Reservations; P.L. 91-575; 89 Stat. 1509 et seq., §2 r, (2) (ii), Reservations.

4.1 Scope of Authority

As Voigt (1972) pointed out, unlike many other interstate compacts, the Susquehanna and Delaware compacts are multi-purpose in scope. Their authority extends to any and all matters that relate to the water resources of the basin from flooding to fisheries. There are a few qualifiers to this authority, but overall it is clear that the commissions can extend their authority to almost any matter involving water.

The 1954 Supreme Court Decree apportioning the Delaware's waters is the main qualifier in the Delaware Compact. The compact preserves the rights and prerogatives of the parties to that Decree. No provision of the compact may be construed so as to impair or diminish those rights (*New Jersey v. New York;* Delaware River Basin Compact, P.L. 87-328, §3.5). Beyond this, it appears that the framers of the Delaware Compact wished to vest the Commission with the very broadest and most comprehensive authority.

The Susquehanna Compact also confers broad and comprehensive powers but with a bit of a twist. Section 3.5 of the compact states that the Commission shall

> *Assume jurisdiction in any matter affecting water resources whenever it determines... that the effectuation of the comprehensive plan or the implementation of the compact so requires. If the Commission finds, upon subsequent hearing requested by an affected signatory party that the party will take the necessary action, the Commission may relinquish jurisdiction.*

This ties in with the policy announced in Section 3.2 of the Susquehanna Compact "to preserve and utilize the functions, powers, and duties of the existing offices and agencies of government."[10] Thus, while the Susquehanna Compact framers were conferring broad powers upon the Commission to manage the water resources of the basin, they were also directing the Commission to work, wherever possible, through the existing system.[11] The signatory parties are to be partners in the implementation of the Comprehensive Plan for management of the basin's resources.

---

[10] The Delaware Compact contains a similar provision at §1.5. This is not referred to as a policy of the signatory parties, only a purpose. Also, the Delaware provision states that the Commission may use these existing agencies to the "fullest intent it finds feasible and advantageous." The Susquehanna provision grants no such discretion. Combined with the history of the Susquehanna Compact provided by Voigt, this language makes a clear inference that the Susquehanna framers wished their creation to be more deferential to existing authority.

[11] Referring to the proceedings of the Interstate Advisory Committee on the Susquehanna River Basin, which drafted the Susquehanna Compact, author William Voigt noted, "None of us had any desire to create a superagency. On the contrary, we made every effort to give existing agencies opportunity to do all in their statutory power to achieve optimum conservation and appropriate use of the basin's resources" (Voigt 1972, p. 105).

4.2 Special Powers

The Delaware and the Susquehanna River Basin Commissions were given many different tools to carry out basinwide water resource planning and management. Space does not permit a complete treatment of each, but the most important are as follows:

4.2.1 *Comprehensive Planning*

Both commissions are directed to "formulate and adopt a comprehensive plan … for the immediate and long range development and uses of the water resources of the basin ... The plan shall include all public and private projects and facilities which are required, in the judgment of the commission, for the optimum planning, development, conservation, utilization, management, and control of the water resources of the basin to meet present and future needs" (Delaware Compact, §§3.2 & 13.1; Susquehanna Compact, §§3.3 & 14.1). The commissions have also included goals and objectives they wish to achieve and guidelines and principles that should apply to all water resource development. This conforms to the preamble and purpose sections of the compacts that proclaim the benefits of comprehensive, multi-purpose planning, cooperation, coordination, and uniform treatment of all water users.[12]

The plans are more than just wishful expressions of ivory tower bound planners. As noted below, they carry real legal clout. They are the blueprints for water resource development in both basins. They cover not just the management actions that the commissions should take, but the actions of the signatories as well.

A good example of this is provided in Articles 11 and 12 of the Delaware and Susquehanna compacts. The identical language in §§11.1 and 12.1 states that

> *For the purpose of avoiding conflicts of jurisdiction and of giving full effect to the commission as a regional agency of the signatory parties, the following rules shall govern federal projects affecting the water resources of the basin, subject in each case to the provisions of §1.4 of this compact: (b) No expenditure or commitment shall be made for or on account of the construction, acquisition or operation of any project or facility nor shall it be deemed authorized, unless it shall have first been included by the commission in the comprehensive plan.*

Sections 11.2 and 12.2 make similar provisions for the projects of the signatory states or their political subdivisions. Clearly, the comprehensive plan is the linchpin of Commission authority.

[12] See Article I of both compacts.

4.2.2 *Project Review and Approval*

The Delaware Compact grants the very broadest of powers to the Commission in the review and approval of water resource projects. No project having a substantial effect on the water resources of the basin shall be undertaken without prior Commission approval (Delaware Compact, §3.8).

In keeping with its view that the Commission's authority should only be invoked in circumstances where the signatory parties either failed to act or could not act, the Susquehanna framers limited the Commission's project approval authority to four situations (Voigt 1972, p. 105):

- All projects on or crossing the boundary between any two signatory states;
- Any project involving the diversion of water;
- Any project found by the Commission or a signatory water agency to have a significant effect on water resources within another signatory state;
- Any project included in the comprehensive plan or having a significant effect thereon (Susquehanna Compact, §3.10).

Under both compacts, the standard for approval is whether the project will impair or conflict with the comprehensive plan.

4.2.3 *Allocation of Interstate Waters*

Probably the most basic power is the power to do what Congress or the U.S. Supreme Court would otherwise have to do—allocate the interstate waters of the basin. This is all part of the commissions' purpose of removing causes of controversy and promoting interstate comity.[13]

The Delaware Commission is directed to use its power to allocate water in accordance with the doctrine of equitable apportionment. It may not allocate in any way that disturbs or impairs the rights awarded to the parties under the Supreme Court decree without the unanimous consent of the parties. More is given below on how a modification of the decree was engineered through the Commission.

No Supreme Court decree applies in the Susquehanna Basin, so the Commission is free to make an unhindered allocation of water. Interestingly, the doctrine of equitable apportionment is not mentioned, though allocations are not to be considered a prior appropriation of the waters of the basin.[14]

---

[13] See Delaware Compact, §1.3 (e); Susquehanna Compact, §1.3 (5).

[14] Prior appropriation refers to the water law concept that prevails in most of the Western U.S. states whereby he who is first in time is first in right.

4.2.4 *Project Construction*

Each commission may

> *Plan, design, acquire, construct, reconstruct, complete, own, improve, extend, develop, operate, and maintain any and all projects, facilities, properties, activities, and services which are determined by the commission to be necessary, convenient, or useful for the purpose of this compact.*[15]

Again, the more restraint-conscious Susquehanna Compact directs that the Commission should only engage in the construction, operation, and maintenance of a project when no other competent agency can act or fails to act.[16] Both compacts allow the commissions to issue bonds to finance such projects.[17]

4.2.5 *Standards for Projects Affecting Water Resources*

Each commission may "establish standards of planning, design, and operation of all projects and facilities...to the extent they affect water resources..."[18] Lacking the broad project review mandate of the DRBC, the SRBC has relied on this power to promulgate regulations relating to consumptive use of water and withdrawal from surface and groundwater. The DRBC has used this authority to establish water conservation plumbing standards, water quality standards, and flood plain land use standards.

4.2.6 *Delineation of Specially Protected Areas/Emergency Declarations*

Through Article 10 of the Delaware Compact and Article 11 of the Susquehanna Compact, the commissions may delineate areas where the demands for water may outstrip the supply. Within such areas, the commissions may prescribe special withdrawal and diversion regulations. If a drought affects the basin or any portion thereof, these same articles authorize the commissions to declare a drought emergency and limit water usage or withdrawals for the duration of the emergency.

---

[15] See Delaware Compact, §3.6 (a); Susquehanna Compact, §3.4 (1).

[16] See Susquehanna Compact, §1.3 (6).

[17] See Delaware Compact, Art. 12; Susquehanna Compact, Art. 13. The commissions may not pledge the full faith and credit of its signatory parties on such bonds without their consent.

[18] See Delaware Compact, §3.6 (b); Susquehanna Compact, §3.4 (2). See also sections of compact on water quality and flood plain management, Delaware Articles 5 and 6; Susquehanna Articles 5 and 6.

5. SELECTED CASE STUDIES ON RESOLUTION OF INTERSTATE CONFLICTS

The primary mission of the compact commissions is the prevention and resolution of conflicts over the waters of interstate rivers. There are several noteworthy examples of how this purpose is being carried out in the Delaware and Susquehanna basins.

5.1 A Delaware Case Study: The Good Faith Negotiations

As mentioned above, the waters of the Delaware River Basin have been parceled out among the states of New York, New Jersey, and Pennsylvania under a Supreme Court Decree (*New Jersey v. New York*). If the need arises to modify such a decree, the usual method is a petition to the Supreme Court to reopen the case and take additional evidence on the terms of a modified decree.

Changing conditions in the Delaware Basin resulted in a call for renewed negotiations on the requirements of the decree. Rather than return to the Supreme Court for such a modification, the compact members turned to the administrative forum provided by the DRBC.

After four years of negotiations (known as the good faith negotiations), the parties agreed to a set of interrelated management steps designed to respond to changed conditions. Some of the recommendations developed in the negotiations process involve voluntary modifications in the release requirements of the 1954 decree during certain low flow conditions. The Supreme Court, even in the face of some doubts expressed by the Court's appointed river master, has not disturbed these negotiated adjustments to the decree.[19]

5.2 A Susquehanna Case Study: The City of Baltimore

The City of Baltimore, while obtaining most of its water from a system of reservoirs ringing the City, installed an intake on the Susquehanna River just above the Conowingo Hydroelectric Dam in the mid-1960s. Up to the present, this intake had been used mainly in drought emergencies. The City now proposes to expand the use of its Susquehanna intake, starting with the sale of up to 30 million gallons per day to Harford County, Maryland.

Upon learning of the proposed sale, the SRBC notified the City that Commission approval would be required because (1) the sale constituted a diversion of the waters of the basin; (2) the withdrawal would come from the Conowingo Pool which extends from the Conowingo Dam back to the Pennsylvania state line and therefore might have an interstate impact (this was particularly true because of the

[19] Information on the good faith negotiations conducted through the Delaware River Basin Commission supplied to the author by David Goldberg, General Counsel to the DRBC.

large power production users just on the other side of the state line);[20] and (3) as an out-of-basin diversion, the withdrawal would constitute a consumptive use of the basin's waters.

The City's first reaction to the Commission's invocation of authority was to cite a Maryland Reservation to the Compact alleging that nothing in the Compact should be construed as impairing any rights the City has to the waters of the Susquehanna River Basin (Maryland Act of 1967, Chapter No. 391, Section 2-Reservations). The Commission countered that this reservation had not been agreed to by any of the other parties to the Compact and therefore its validity was highly questionable.[21] The Commission and the City nearly came to legal blows at a July 14, 1994, meeting of the Commission at Easton, Maryland, where legal counsel for the Commission and the City argued their respective cases on Commission jurisdiction. Following counsels' arguments, the Commission agreed to hold its invocation of authority in abeyance pending a possible negotiated settlement.

The Commission proposed an operating procedure whereby the City would cease withdrawals when the flows through the tailrace of the Conowingo Dam reached the minimum flows specified by the Federal Energy Regulatory Commission (FERC). This flow has been dubbed the Q-FERC flow. The City has commissioned a technical study of whether it can operate its water system in this manner. As of this writing, the technical staffs of the Commission and the City are discussing the study findings with the hope of reaching a modus operandi satisfactory to the City, the State of Maryland, the Commission, and all other users of the Conowingo Pool.

5.3 Delaware and Susquehanna Case Studies: Dispute Prevention

If the water needs of all users are satisfied, the possibility of conflict is automatically avoided. The regulations of the Delaware and Susquehanna River Basin Commissions pertaining to the withdrawal and consumptive use of water help remove causes of controversy, as intended by the compacts.[22] Downstream users and the receiving waters of the Chesapeake Bay and Delaware Estuary are protected.

Just getting the respective signatory parties together several times a year helps head off conflict. If trouble is brewing between the parties, they can talk it out and explore options for resolving the problem before it turns into a major controversy.

---

[20] These are the Peach Bottom Nuclear Power Plant and the Muddy Run Pumped Storage Plant. PECO Energy, which runs both of these facilities, expressed concern about the impacts of the withdrawals on the Conowingo Pool and asked that the Commission give the proposed withdrawal close scrutiny.

[21] The Commission's arguments were based mainly on the principle established in *West Virginia ex rel. Dyer v. Sims* that no state party to an interstate compact can unilaterally place restrictions upon the operation of that compact.

[22] See generally 18 CFR Parts 401 and 803 for the project review regulations of the Delaware and Susquehanna River Basin Commissions.

6. FUTURE OF FEDERAL-INTERSTATE COMPACTS

It was a challenging task in the 1960s to steer federal-interstate compact agencies like the SRBC and DRBC through the Congress and state legislatures. Today it might well be impossible.

Despite the federal-like functions of the commissions in resolving and preventing interstate disputes, the mood in Washington, D.C., these days is to shrink the role of the federal government. After years of uninterrupted support, the U.S. House Appropriations Subcommittee on Energy and Water Development recommended termination of federal contributions to the Delaware and Susquehanna Basin Commissions in FY 1997. The Subcommittee, influenced by a January 1995 report of the Heritage Foundation that described the commissions as providing only regional benefits, recommended that the states assume the full burden of funding.

The full Congress adopted the subcommittee's recommendation and the funding cut-off became effective in FY 1997. Taking a cue from the action of Congress, the White House has not included funding for the two commissions in the FY 1998 budget. The commissions now face the formidable task of persuading Congress to write them back into the budget.

While excluding the commissions from the budget, the White House is proposing to continue federal participation through the appointment of a high-staff-level person in an existing federal agency to serve as the federal member and/or alternate member.[23] Meanwhile, Congress has made no effort to withdraw the federal government from the compacts as it could under Section 1.4 of both compacts. It appears then that, at least for the immediate future, federal participation will continue even without funding. Meaningful long-term federal participation is not as certain.

Ultimately, policy makers will need to be re-persuaded that there is a legitimate federal role; that there is a real federal interest in the management of waters that flow across state lines and affect nationally significant estuaries; that there is a real federal interest in having administrative mechanisms available to quell interstate disputes; and that there is a real federal interest in having an agency that can coordinate the water resource management programs of the state and federal governments under one watershed management plan. What seemed so obvious 25 years ago, must now be reproven.

REFERENCES

Clark, R.E., ed. 1967. *Waters and Water Rights, Vol. 2.* Indianapolis, IN: The Allen Smith Company.

Corker, C.E. 1967. Water Rights in Interstate Streams. In R.E. Clark, ed., *Waters and Water Rights, Vol. 2.* Indianapolis, Indiania, The Allen Smith Company.

Corpus Juris Secondum, 81A, Section 16.

Delaware River Basin Compact, P.L. 87-328, 75 Stat 688.

Frankfurter, F., J.M. Landis. 1925. The Compact Clause of the Constitution—A Study in Interstate Adjustments, *Yale Law Journal* 34: 701.

---

[23] See language in the proposed FY 1997 Supplemental Appropriations Bill.

Gelhorn and Grad. 1960. Opinion on the Constitutionality of the Proposed Delaware River Basin Compact, *Columbia Law School.*

Goldberg, D. Delaware River Basin Commission, Information on the Good Faith Negotiations among the parties to the 1954 U.S. Supreme Court Decree apportioning the waters of the Delaware River among the states of New York, New Jersey and Pennsylvania.

Hamilton, A. 1787-88. Concerning Dangers from Dissensions Between the States. From *The Federalist,* No. 7.

National Water Commission. 1972. Water Policies for the Future, Chapter 8, in *Interbasin Transfers.*

Susquehanna River Basin Compact, P.L. 91-575, 89 Stat 1509 et seq.

Voigt, W. Jr. 1972. *The Susquehanna Compact: Guardian of the River's Future*. Rutgers University Press.

TABLE OF CASES

Arizona v. California, 283 U.S. 423 (1931)
Arizona v. California, 373 U.S. 546 (1963)
Colorado v. Kansas, 320 U.S. 383, 393 (1943)
Connecticut v. Massachusetts, 282 U.S. 789 (1931)
Cuyler v. Adams, 449 U.S. 433 (1981)
Hinderlinder v. LaPlata River & Cherry Creek Ditch Co., 304 U.S. 110 (1938)
Jacobsen v. Tahoe Regional Planning Agency, 566 F2d 1353 (1977); affirmed at 440 U.S. 391
New Jersey v. New York, 347 U.S. 995 (1954)
North Dakota v. Minnesota, 263 U.S. 365, 372 (1923)
N.W. Pipeline v. State Corp. Commission of Kansas, 489 U.S. 493, 512 (1989)
Petty v. Tennessee-Missouri Bridge Commission, 359 U.S. 275 (1959)
Texas v. New Mexico, 462 U.S. 554 (1983)
West Virginia ex rel. Dyer v. Sims, 341 U.S. 22 (1951)
Wyoming v. Colorado, 353 U.S. 953 (1957)

8 SHARING THE WATERS OF THE MURRAY-DARLING BASIN: COOPERATIVE FEDERALISM UNDER TEST IN AUSTRALIA

John J. Pigram
University of New England, Australia
Warren F. Musgrave
New South Wales Premier's Department, Australia

The Murray-Darling Basin is a complex, somewhat stressed, river system extending across a number of states in the Australian federation. Aspects of its management have been the subject of successful interstate collaboration for over 80 years. The initial focus on the provision of water services (supply and navigation) has been widened by increasing concern for the integrity of the system as a result of evidence of unacceptable pressure on supplies for consumptive use, and of environmental deterioration. Responses to this concern are reviewed and discussed. They include regulatory reform, increased charges for water, and development of an effective market-based property rights system for resource allocation and trade. Though much remains to be achieved, the Basin continues to provide an example of productive, cooperative federalism in river basin management.

1. INTRODUCTION

Australia is a large island continent with an area of some 4.7 million square kilometers (almost 3 million square miles), making it the sixth largest country in the world. It is also the world's lowest continent, and this, combined with a highly variable climate, contributes to its overall status as the driest inhabited continental land mass.

Australia has the least river water, the lowest runoff, and the smallest area of permanent wetlands. Few permanent, freshwater lakes exist, and over much of the inland, groundwater is the only practical source of water. These circumstances help explain the prominence given to water resources development since the earliest days of European settlement. Australia stores more water per capita than any other country in the world, and irrigation agriculture is the largest user of water, accounting for 70 percent of all water used. The bulk of this irrigation is concentrated in the Murray-Darling Basin in southeastern Australia.

2. IRRIGATION IN THE MURRAY-DARLING BASIN

The Murray-Darling river system drains an area roughly the size of France and Spain combined (Figure 1). Although it is the fourth largest river system in the world, its yield is low. The Amazon River would carry the river's annual flow in less than one day. As with much of the rest of Australia, the Basin is characterized by highly variable precipitation and a marked contrast between the predominantly winter rainfall zone in the south, and the predominantly uniform and summer rainfall zones further north. These differences are reflected in varying levels of availability and reliability of water supply, and in attitudes towards water use, across parts of the region.

Figure 1. The Murray-Darling Basin
Source Murray-Darling Basin Commission

The Murray-Darling Basin straddles four states—Queensland, New South Wales, Victoria, and South Australia—each with jurisdiction over water use within its borders. However, water does not respect artificially created political boundaries, and it is simply not possible to tackle the physical and technical problems associated with water quantity and quality in the Basin effectively on a state-by-state basis.

Although the Commonwealth or federal government has no constitutional responsibility for water, it can exert considerable influence because of its dominant financial and political position, and because it is strategically placed to integrate and influence state policies. This, coupled with a growing recognition of the importance of cooperative federalism, has resulted in the federal government playing a key catalytic and leadership role in the recent development of water resource policy (Musgrave and Pigram 1995).

3. COOPERATIVE FEDERALISM

As an interstate river basin, the Murray-Darling system provides considerable scope for Commonwealth leadership and cooperative federalism. This was demonstrated by the establishment in 1985 of the Murray-Darling Basin Ministerial Council. The Council consists of the Ministers holding responsibility for the environmental, land, and water portfolios in each of the governments involved, including the federal government. It has a broad mandate to coordinate management of the whole Basin, with specific responsibility concerning the management of abstractions and of water quality in the Murray River and the lower Darling. In general, the policies of the Council are executed by the relevant agencies of the component state jurisdictions.

The Council is serviced by the Murray-Darling Basin Commission, replacing the River Murray Commission; and there is a Community Advisory Committee, consisting of both regional community and special interest groups, which has direct access to the Ministerial Council. Through its policy responsibilities and the mainly Commonwealth funds it has to dispense, the Council plays a major role in reforming and restructuring the irrigation industry of the Basin.

The Ministerial Council is well placed to provide an overview of the status of the Basin's irrigation industry. Its position in this respect is that (Blackmore and Keyworth 1995)

- natural resource degradation, mainly waterlogging and salinization, threaten the long-term viability of the industry,
- the large investment in irrigation infrastructure earns a very low rate of return,
- the relatively low incomes are being earned by a significant number of farm businesses,
- the restricted ability of poor farmers to invest threatens the sustainability of the industry, and
- sustainability of the industry requires ongoing maintenance and renewal of the infrastructure, which should be funded by the industry.

To attack these problems the Council has introduced an Irrigation Management Strategy with the objective of achieving an economically and environmentally sustainable industry in the region by the year 2010 (Blackmore and Keyworth 1995).

Reflecting their constitutional responsibilities for land and water management, the four Basin states have implemented their own programs of reform with varying degrees and rates of change. Common elements have been price reform, the development of water markets, and hesitant steps towards regional planning. All Basin states are moving, to some extent or other, to adopt cost recovery approaches to the regulation and management of water. In particular, water supply services are being separated from regulatory functions. In some cases, this is being done within single agencies, while in others (e.g., Victoria), the functions have been placed in separate agencies. New South Wales is well advanced in separating water retailing from bulk supply by privatizing or corporatizing its irrigation areas or districts.

4. AGENDA FOR REFORM

These measures are symptomatic of an industry-wide trend towards water reform in the face of strong pressures for change, particularly in the past decade. The water-using sector of the economy has come increasingly under criticism as the perceived source of widespread resource degradation and extensive impairment of riverine environments. At the same time, growing demands for alternative uses of water have arisen for a range of environmental purposes and for recreation and tourism. To a large extent, these demands reflect increased environmental awareness and community desires for an improved quality of life. Satisfying these demands calls for far-reaching changes in water allocation systems and a new approach to water management. There are now encouraging signs that governments at all levels are seeking to respond positively to the challenges posed by a complex and evolving array of water issues and priorities.

The impetus for reform in Australian water policy can be linked to the switch in emphasis from resource development to resource management. Marked fluctuations in technology and economic growth in the 1980s, accompanied by emerging environmental concerns and changes in social values and priorities, underlined a need for more responsive organizational structures and institutional frameworks. The process of adjustment has been ongoing with consequent changes to administrative arrangements, water allocation and pricing policies, and to the legislation which underpins the work of water authorities.

As noted earlier, a number of measures to improve water management have been introduced in recent years, notably the adoption of tradable water entitlements and rationalization of water pricing. The process of reform accelerated following the release in 1992 of the Industry Commission Report on Water Resources and Waste Water Disposal. The Commission examined institutional, regulatory and other arrangements which contribute to inefficient and unsustainable resource use, and to emerging environmental problems, and recommended ways to revise these arrangements. A key finding was the need for an integrated approach in which policies directed at the efficient provision of water services must be tied to policies that achieve sustainable water use.

5. COAG DELIBERATIONS

The Industry Commission Report was followed in February 1994 by the Water Policy Agreement put in place by the Council of Australian Governments (COAG). The Council endorsed the findings of its Working Group on Water Resource Policy regarding deficiencies in water use and management and associated degradation of the resource base. In particular, the Working Group targeted

- allocation of water to the environment and the need for balance between environmental and developmental concerns,
- adoption of an integrated catchment management approach to water resource management,
- pricing reform, including full cost recovery, the removal of cross-subsidies, and provision for asset maintenance and refurbishment,
- adoption of tradable water entitlements,
- clarification and consistency of property rights to water,
- institutional and organizational reforms,
- structural adjustment consequences and social impact of reform, and
- community consultation and education programs.

The Agreement is seen as signaling a new urgency on the part of governments, both state and federal, to promote efficient, sustainable use of water in Australia. The strategic framework adopted by COAG for the reform of the Australian water industry was reinforced in April 1995 by the Council's endorsement of the Second Report of the Working Group on Water Resource Policy. This report documented the significant progress which had been made in implementing the 1994 Agreement. All states, along with the Murray-Darling Basin Commission, are now engaged in the reform process.

6. NATIONAL COMPETITION POLICY

At its meeting in April 1995, the Council of Australian Governments also endorsed the implementation of a National Competition Policy. The policy and associated reforms arose out of the report of the Hilmer Committee, and extend trade practices legislation to state and local government business enterprises. Competition is now encouraged in the business activities of governments and other sectors of the economy.

Essentially, the aims of the National Competition Policy are

- to remove unnecessary barriers to trade and competition,

- to ensure that no buyer or seller acts against the public interest by engaging in anti-competitive behavior,
- to require all businesses (including governments) to operate with the same market rules,
- to allow anti-competitive behavior only if it is in the public interest, and
- to reduce complex regulations and administrative duplication between governments.

The policy complements the view of COAG that many government businesses in Australia, including water authorities and management agencies, lack the necessary element of competition to drive improvements in customer service and efficiency gains. Whereas firm target dates have been set for implementation of reforms, the process also involves a commitment to proper consultation with the sectors and communities affected.

7. IMPLICATIONS FOR WATER USE

This, then, summarizes the background of water reform in Australia with reference to some of the principles underlying the agenda for reform. The concern is with promoting economic efficiency in water use in a competitive environment, set within the context of ecologically sustainable development of the water resource. In the process, urban and rural water users will be called upon to adapt to some significant changes in the allocation and management of water.

The water industry can expect to be required

- to pay more for water,
- to allocate increasing amounts of water to the environment,
- to explore opportunities for more flexible water use, including extension of tradable water entitlements,
- to achieve higher use efficiencies and "do better with less" through adoption of best management practices,
- to conform to more demanding environmental regulations,
- to fund maintenance and replacement costs of infrastructure,
- to develop effective technologies for water recycling and reuse, and
- to implement integrated environmental management of land, water, and biophysical resources of catchments to promote ecologically sustainable development.

8. ONGOING CONCERNS IN THE MURRAY-DARLING BASIN

Despite the advances achieved and their undoubted impact on the operating environment of the water industry, problems remain, many of which are identified with the Murray-Darling Basin. A major concern is overcommitment of the Basin's water resources and the consequences of this in terms of the health of the river system.

Most large-scale licenses for abstraction of water from the rivers were issued in the 1960s and 1970s when knowledge about the behavior of river systems was more limited. Economic development flourished as a result of increased irrigation activity, but with it came environmental damage, attributable at least in part to reduced flow regimes.

In 1993, the Murray-Darling Basin Ministerial Council expressed concern that river health was continuing to decline because of increasing demands for water diversions from the system. Particular issues included increasing water salinity and frequency of algal blooms, declining biodiversity in riverine ecosystems, and a decrease in the frequency of beneficial flooding and wetland replenishment (Prime Minister's Science and Engineering Council 1996).

Moreover, it was apparent that the situation could worsen because of the large amount of water allocated but not yet activated or, for a variety of reasons, underutilized (Table 1). Indeed, the historical practice had been to encourage the use of water by allocating more water than is likely to be available in the belief that not all license-holders would use their full allocation. A further concern was, therefore, that the expansion of activity in the emerging market for trade in water entitlements could activate these "sleeper" or "dozer" licenses and produce a substantial increase in the volume of water diverted.

9. WATER AUDIT

In view of the above concerns, the Council directed the Murray-Darling Basin Commission to conduct an audit of water diversions from the rivers of the Basin, comment on the effects of the increase in diversions, and assess the likely impact of these trends.

The results of the audit showed that water diversions grew by about 8 percent across the Basin between 1988 and 1994, an increase of 790 gigalitres or four times the annual water usage for the City of Adelaide in South Australia. This increase is continuing at a rate of one percent per year, and there is potential for a further 14.5 percent growth above current levels of water use (Murray-Darling Basin Ministerial Council, 1995).

Assessment of the consequences of such development indicated that security of water supplies would be reduced, forcing irrigators to reduce diversions or attempt to purchase additional water on the market. In terms of the likely impact on river health and the environment, regulation and water diversions have reduced the size of river flows, and the variability and seasonal flow regimes in parts of the Basin. These changes in the volume and pattern of flows have resulted in

- an increase in salinity in the lower Murray River,
- a reduction in the frequency of flooding and a decline in the extent of wetlands,
- a decline in native fish populations, and

Table 1. Limits to Diversion Imposed by the Allocation System

River System	Limits* to Diversion Imposed by the Allocation System (GL)	Actual Water Diverted (GL)	Water Diverted as a % of Diversion Limit (%)
New South Wales (NSW)			
Border Rivers	292	222	76
Gwydir	316	300	95
Namoi	387	248	64
Macquarie/Castlereagh/ Bogan	823	471	57
Upper Darling	549	189	34
Lower Darling	255	213	84
Murrumbidgee	4268	2443	57
Murray	3362	2053	61
Total NSW	**10252**	**6139**	**60**
Victoria			
Upper Murray/Ovens/ Kiewa	2365	1567	66
Lower Murray	460	284	62
Goulburn/Broken/Loddon	2630	1710	65
Campaspe	134	101	75
Total Victoria	**5589**	**3662**	**66**
South Australia			
Private Pumped Diversion	301	239	79
Government Pumped Diversion	295	229	78
Reclaimed Swamps	106	106	100
Total South Australia	**702**	**574**	**82**
Queensland			
Border Rivers	100	74	74
Macintyre Brook	19	10	53
Condamine/Balonne	177	162	92
Total Queensland	**296**	**246**	**83**
ACT	63	63	
Total for Basin	**16902**	**10684**	**63**

* Note that for regulated streams the diversion limit has been calculated by adding the actual diversion to the difference between the announced allocation and the on-allocation use. For unregulated streams, the diversion limit is the licensed area converted to a volume of water. Figures are the Average Actual Figures for 1988/89 to 1992/93. Key points to note from this table include:

- The difference between announced allocation and on-allocation use is large.
- Over the last five years, diversions from most systems have been constrained by factors other than the announced allocations.

Source: Murray-Darling Basin Ministerial Council (1995, p. 9).

- more frequent occurrence of river conditions conducive to the development of algal blooms with further consequences for water quality.

10. CAP ON DIVERSIONS

In response to the Water Audit, the Murray-Darling Basin Ministerial Council recognized that further increases in the level of diversions in the Basin were unsustainable. In June 1995, the Council, therefore, introduced an immediate moratorium on further increases in diversions, establishing an interim "cap" at the 1993-94 level of development. The stated aim of the cap is to achieve a balance between a healthy river system and sustainable consumptive use for irrigation and industrial use, and for domestic purposes. A working group has been established to review longer-term management of the cap and to consider arrangements for implementation of a final cap in June 1997. The group is also examining equity issues and other special circumstances affecting water use in the Basin states.

11. IMPLEMENTATION PROCESS

Whereas all parties appear to be in agreement in principle that the current and projected increases in diversions were unsustainable, and that an interim moratorium was necessary, implementation of the final cap on diversions raises some complex issues. In this context, it is important to recall that the Murray-Darling Basin spans four states, with the federal government also exercising significant influence.

Water policies and philosophies, the circumstances in which the water is used, the manner of its allocation, and the way it is managed, certainly do differ across the Basin. The situation is made even more complex because of legal requirements under the Murray-Daring Basin Agreement for New South Wales and Victoria to supply a specified annual water entitlement to South Australia. Apart from that provision, each state has taken a different approach to water management, reflecting particular needs of their agricultural systems.

In South Australia, where the state's water entitlement is assured and where horticultural crops predominate, water allocations typically have 100 percent security. In Queensland, although there is no general security policy, full allocations are available between 75 and 99 percent of years (Wahlquist 1996). A Water Allocation and Management Planning (WAMP) process is in place in that state and it is considered that full use of existing entitlements is not likely to exceed sustainable levels of development. It is even suggested that the final cap in Queensland could be higher than current levels of use.

In Victoria, irrigation is focused primarily on horticulture and on permanent pastures for dairying. Climatic conditions ensure that water rights have a high level of security (met in full 97 years in 100). After providing for these entitlements and leaving enough water in storage for the following season, the remainder is offered as water sales. "Sleeper" licenses are less of a problem in Victoria, where irrigators must pay for the water whether they use it or not.

New South Wales, which has the highest share (57.4 percent) of total Basin diversions, is the state where the cap is likely to have the greatest impact. Irrigators have a high security license for permanent plantings and a general license for annual irrigation. The reliability objective for the latter allocations is to supply full entitlement in 70 of 100 years. However, reliability of supply varies across the state, particularly in the north where rainfall and runoff are less predictable. In general, the tendency has been, after providing for high security supplies for the following year, to maximize use of available water on annual crops, e.g., cotton. In many instances, a significant proportion of water needed has been obtained by diversion of "off-allocation" flows resulting from freshes in tributaries downstream from storages. However, access to these supplementary supplies has now been curtailed in the interests of the riverine environments. Environmental objectives are discussed further below.

It is clear from this brief overview that significant differences exist between the four Basin states, in water policies and attitudes towards the resource, in water allocation procedures and security of supply, in water uses and crop requirements, in maturity of water resources development, and in measures already in place to manage water use in the interests of efficiency and sustainability. These differences are brought out when ongoing initiatives in water reform across the Basin are examined. Changes under way at various stages cover property rights in water, provision of environmental water requirements, water markets and tradable water entitlements, water pricing reform, and organizational restructuring.

12. PROPERTY RIGHTS

Clarification of property rights in water is a key element in the package of reform measures adopted by the Council of Australian Governments for the water industry (COAG 1994). Property rights are fundamental to addressing many of the water management issues which arise in the Murray-Darling Basin. One logical way to respond to increasing scarcity of water resources is to adopt a market-orientated property rights system for resource allocation, whether this be for consumptive or instream uses.

In Australia, bureaucratic arrangements historically adopted for regulation of water use were designed for the development of the resource under conditions of relatively abundant supply, rather than the reallocation of scarce resources, as is the situation in the Murray-Darling Basin today. Such a regulatory, "command-and-control" approach to water management is unlikely to provide for socially optimal use under conditions of increasing relative scarcity. Self-regulation, driven by market mechanisms linked to specified and enforceable property rights, is seen as preferable to meet changed circumstances of water demand and supply and facilitate re-evaluation of water resources to match emerging needs.

Implementation of the cap on water diversions in the Murray-Darling Basin will inevitably lead to redistribution of existing water entitlements. Redistribution can be anticipated between consumptive users offstream, e.g., irrigators, and from consumptive use to instream use for environmental purposes. In either case, secure property rights in water will provide a sound basis for market forces to bring about more rational and equitable allocation of the resource.

If continuation and expansion of intrabasin water trading between irrigators, for example, are to be encouraged as a means of redistributing capped water within the Basin, then participants in the market must be sure of what they are trading, and any conditions pertaining to the entitlements transferred. A well-defined property rights framework will underpin such rationalization of water use, and enable water to move to its highest value use both within sectors of the water industry and intersectorally.

Initiatives are under way in various jurisdictions to evolve a workable system of property rights to enable water markets to operate effectively and efficiently. These moves reflect one of the important reforms endorsed by the Council of Australian Governments, to (COAG 1994, p. 28)

> *...implement comprehensive systems of water allocations or entitlements backed by separation of water property rights from land title and clear specification of entitlements in terms of ownership, volume, reliability, transferability and, if appropriate, quality.*

The Council went on to state "that trading arrangements in water allocations or entitlements should be instituted ... no later than 1998" (COAG 1994, p. 29).

It needs to be recalled that the COAG deliberations and decisions are not mere recommendations. Substantial payments to the States are contingent upon the implementation of the reform agreements, and monitoring and reporting to the Council on progress on implementing the water reform package are undertaken annually.

In regard to interstate trading, the Council stressed the need for consistency in approaches to property rights, pricing and cost recovery, and environmental considerations (COAG 1994, p. 29):

> *Where cross-border trading is possible ... trading arrangements [should] be consistent and facilitate cross-border sales where this is socially, physically, and ecologically sustainable.*

Lack of consistency in determining property rights regimes in the Basin states could be a real impediment to the development of a "common market" for interstate trade in water. In the absence of consistency, there is potential for outcomes that are inequitable and sub-optimal from economic, social, and environmental perspectives. Likewise, distortions in trade could occur from differentials in pricing policies or subsidies. Where such imbalances attach to water entitlements under different jurisdictions, they need to be identified and dealt with under an agreed conversion process according to specified circumstances in the state of destination, or endpoint of use.

Some of these issues are to be addressed in a pilot scheme planned for the Mallee Region which straddles the New South Wales, Victorian, and South Australian borders on both sides of the Murray River (Cummins 1996). The project is seen as an attempt to "fast track" the development of a workable system to allow market-driven trade in water property rights between states. Although the Mallee Region is seen to have a relatively high degree of commonality in conditions of water use, a number of impediments to trade will need to be overcome. Possible distorting influences include differentials in subsidies, pricing and tariff structures, regulations, land tenure arrangements, environmental water requirements, and commitments under the Murray-Darling Basin Salinity and Drainage Strategy.

The Mallee project will establish an experimental "free trade" zone to address these cross-border issues and determine conditions under which trading can take place. The pilot scheme should identify current inconsistencies in institutional and market arrangements which need to be resolved for progress towards a workable implementation of the cap on Basin diversions.

13. ENVIRONMENTAL WATER REQUIREMENTS

The decision to cap water diversions in the Murray-Darling Basin was a critical first step in establishing an appropriate balance between consumptive uses and environmental needs of a healthy river system. To make this balance operational, consistent methodologies are called for to specify allocations to the environment as a legitimate user of water (COAG 1994).

National principles for the provision of water for ecosystems have been endorsed by the Agricultural and Resource Management Council of Australia and New Zealand, and are set out in Table 2. These principles provide useful directions for policy as to how environmental water requirements should be incorporated into the management of Australia's water resources. Of note is the recognition (Principle 3) of the need for a legally specified water entitlement for the environment. This calls for environmental water provisions to be enshrined as explicit entitlements in a tradable property rights regime, as proposed for other water uses and users (Claydon 1995, pp. 17-18).

> *If water for environmental purposes [is] to be recognized as a legitimate use of water—as legitimate as any other use—the environmental water provisions will need at least as strong recognition under law as the provisions for any consumptive uses in planning frameworks or property rights regimes.*

Claydon advocates explicit rights for the provision of water for the environment and that such rights be tradable within a water market. According to Claydon (1995, p. 43), such a system would

- make the environment an equal partner in water allocation and management,
- give clear title to and protection of water provided for environmental purposes,
- improve the level of management, and accountability for environmental water provisions, and
- enable environmental managers to participate in water trading.

Table 2. National Principles for the Provision of Water for Ecosystems*

Basic Premise of Principles	
Principle 1	River regulation and/or consumptive use should be recognized as potentially impacting on ecological values
Determining Environmental Water Provisions	
Principle 2	Provision of water for the environment should be on the basis of the best scientific information available on the water regimes necessary to sustain the ecological values of water-dependent ecosystems
Provision of Water for the Environment	
Principle 3	Environmental water provisions should be legally recognized
Principle 4	In systems where there are existing users, provision of water for the environment should go as far as possible to meet the water regime necessary to sustain the ecological values of aquatic ecosystems, while recognizing the existing rights of other water users
Principle 5	Where environmental water requirements cannot be met due to existing uses, action (including reallocation) should be taken to meet environmental needs
Principle 6	Further allocation of water for any use should only be on the basis that natural ecological processes and biodiversity are sustained (i.e., ecological values are sustained)
Management of Environmental Water Allocations	
Principle 7	Accountabilities in all aspects of management of environmental water provisions should be transparent and clearly defined
Principle 8	Environmental water provisions should be responsive to monitoring and improvements in understanding of environmental water requirements
Other Uses	
Principle 9	All water uses should be managed in a manner which recognizes ecological values
Principle 10	Appropriate demand management and water-pricing strategies should be used to assist in sustaining ecological values of water resources
Further Research	
Principle 11	Strategic and applied research to improve understanding of environmental water requirements is essential
Community Involvement	
Principle 12	All relevant environmental, social, and economic stakeholders will be involved in water allocation planning and decision-making on environmental water provisions

* These are the principles endorsed by the Agricultural and Resource Management Council of Australia and New Zealand, and the Australian and New Zealand Environment and Conservation Council.

Source: Prime Minister's Science and Engineering Council 1996, p. 39.

Claydon warns that potential conflicts of interest between consumptive water users and those seeking to make provision for environmental water requirements "are likely to be stark, unavoidable and pervasive" (Claydon 1995, p. 45). However, he sees significant benefits from making environmental water entitlements tradable, given the dynamic nature of economic production systems responding to changing technology, market demands and seasonal conditions, and the varying requirements of natural ecosystems in terms of seasonality.

The extent to which market forces can be harnessed to service environmental water requirements, and the potential and the limitations of economic instruments for environmental protection, are now being actively canvassed in Australia. Some environmental issues—maintenance of aesthetic values, for example—are perceived as public goods, so the conditions for a water market to operate do not apply. A further consideration in the transition to a market-oriented water allocation system is the so-called "free-rider" problem, and the need to ensure that all beneficiaries of enhanced riverine environments pay for the maintenance of environmental quality.

Opportunities undoubtedly exist for the wider application of market-based approaches for the acquisition of property rights to meet environmental water requirements. Colby (1990) has argued that in the western states of the United States, for example, instream values are high enough to compete in the market for water rights with offstream uses. If Colby is right, it should be possible for environmental interests to become market participants in the same way as consumptive water users. How such environmental initiatives are to be financed remains controversial (Willey 1992). There is scope, for example, for the public sector to be custodian of environmental rights. Alternatively, a privately funded group could acquire water entitlements to satisfy ecological values or social purposes beyond those generally accepted by the community. This would be in line with the recommendation of the Council of Australian Governments (Report of COAG Working Group on Water Policy, 1995, 12.4.iii)

> *that where parties wish to protect environmental values at levels above that which is deemed to be necessary for sustainability ... they meet all the costs of this.*

Gaining an environmental allocation is only a first step. Management of allocations acquired to satisfy perceived environmental objectives is another matter. As Millington (1991) points out, a fixed annual allocation of water for environmental purposes is not necessarily useful to the same degree in all seasons and all situations. However, in periods when allocations are superfluous to environmental needs, water could be transferred temporarily to other uses and the returns devoted to broader objectives for management of the stream environment.

As with mandatory land and water management plans for irrigation areas in parts of the Murray-Darling Basin, there needs to be a sound management plan to ensure that water is put to optimum use for environmental purposes. Accountability for management of environmental water provisions is essential so that, after minimal discharges from storage are made for environmental and riparian purposes downstream, objectives for the sustainability of ecological values are met.

The COAG deliberations also addressed the issue of determining environmental flow requirements. At their 1994 meeting, the federal and state governments agreed that high priority be given to specifying allocations to the environment as an appropriate use of water. Despite this commitment to cooperative federalism and growing endorsement of the legitimacy of the environment's claims on the resource, the response has been hesitant and ad hoc. An example was the arbitrary and heavy-handed decision by the newly elected government in New South Wales in 1995 to limit offstream diversions and to increase environmental allocations to satisfy ill-defined and unsubstantiated environmental needs.

Environmental water needs are complex, and the 1995 Report of the COAG Working Group on Water Policy noted (i.180) "that work in relation to environmental (water) requirements is still in its infancy in Australia." Scientific knowledge of the watering requirements of wetlands, for example, remains rudimentary, while community expectations regarding the desirable state of riverine environments continue to increase.

In a more considered response, the New South Wales government has now set up working groups to establish environmental objectives for major catchments in the state. The objectives will be concerned with water quality and quantity and with the timing and variability of river flows. Options to meet the objectives set will be subject to extensive community consultation and will form the basis for state-wide catchment planning.

A comparable approach is being followed in Queensland in the Water Allocation and Management Planning (WAMP) process. The procedure seeks to identify the optimal mix and balance between water uses and the impacts of environmental flow options on water availability for other uses. This analysis is important because at the completion of the WAMP process, environmental entitlements are to be given statutory recognition in Queensland.

14. WATER ALLOCATION AND CAPACITY-SHARING

If and when agreement can be reached on environmental property rights, at least between contiguous Basin states, attention then needs to be directed to an effective system for allocating water which incorporates these rights and ensures that the security and reliability of entitlements are not eroded over time. Capacity-sharing has been put forward as an innovative approach to allocation of water property rights in highly variable climatic environments such as characterize much of the Murray-Darling Basin. As stated by Musgrave (1991, p.7):

> *Capacity-sharing is a water allocation system by which users are allocated a share of the capacity of the storage as well as inflows and seepage and evaporation losses. In effect, the storage is partitioned into substorages which are credited with a volume of water according to the hydrological behavior of the storage and its catchment. Users have non-attenuated rights in this water and can direct the manager of the storage concerning its retention or release.*

Capacity-sharing is seen to have advantages over the traditional method of water allocation, which might be termed "release sharing." By partitioning the entitlements to water at the source, rather than at the point of delivery, the conditions for an efficient water market can be satisfied and transaction costs minimized.

Capacity-sharing offers the potential for individual control of water in the substorages, as might be preferred by irrigators, or operation of a shareholding as a group. The latter arrangement might appeal to community groups acquiring shares for environmental purposes, e.g., replenishment of wetlands. Flood mitigation beneficiaries may also wish to act together in controlling a share of air space in the storage.

Proponents of capacity-sharing believe that it achieves a desirable integration of water supply and demand and is superior to traditional methods of allocation. Property rights to shares in capacity are specified and secure, and freely tradable. Transferability of stored water between the substorages facilitates exchange of water between individuals and groups according to their requirements. In the longer term, permanent transfers of shares of reservoir capacity and/or inflows would reduce uncertainty, and permit structural adjustment and more efficient use of water. In recent systems modeling, the concept has been expanded to include percentage-sharing of unregulated tributary flows downstream from the reservoir (Dudley 1996).

Although water managers have expressed reservations about the acceptability of such a novel system, water agencies have expressed interest, and the state of Victoria, in recent legislation, has provided for capacity-sharing as an optional allocation system, at the bulk water distribution level. Claydon (1995) sets out several options under which sectoral groups, e.g., environmental interests (or even individuals), could hold capacity shares and be responsible and accountable for management of water entitlements held.

Given a system of secure property rights, a water market could then become operational to permit the acquisition and transfer of water allocations for a range of environmental purposes. These include

- maintenance of stream flows for ensuring sustainability of aquatic ecosystems and other instream values, including water-related recreation,
- use of dilution flows for enhancement of water quality,
- management of water pollution through a system of tradable emission rights for discharge of irrigation drainage water,
- management of groundwater resources in conjunctive use with surface water allocations, and
- in areas, subject to waterlogging and salinity, management of depth to the water table.

15. PRICING REFORMS

Water is owned by the Crown (i.e., the State) in Australia and abstractions are made under license. Whereas marked differences in water prices and tariff structures exist throughout the Murray-Darling Basin, prices for irrigation water have generally been low. Apart, perhaps, from rice-growing, water does not figure as significant a factor of production as other inputs such as fuel, fertilizer and chemicals.

The arguments for upward adjustment of prices for water are compelling, both on grounds of promoting greater efficiency in use, and recovering costs of supply. In regard to the latter issue, questions arise as to which costs are to be recovered and how they are to be measured. In particular, a central concern is the extent to which prices would reflect the typically large capital costs invested in storage construction and water delivery. It has been argued that, since these ("sunk") costs have long since been capitalized into land and irrigation infrastructure, and the assets themselves have zero opportunity costs, adjusting water prices to recover capital costs is not appropriate (Musgrave 1996). However, there is general agreement that beneficiaries should meet future capital costs, or at least enter into joint ventures with government.

In New South Wales, the state currently meets the full costs of operating and maintaining irrigation storages and regulatory works, although these are progressively being passed to state-owned, but corporatized, irrigation areas and districts. Costs incurred in "running the rivers," i.e., water delivery from storage downriver to the point of diversion, are shared 70/30 between extractive users and the state, on the principle that there are other beneficiaries involved. Sharing of operations and maintenance costs associated with diversion and reticulation remains a point of contention. Such costs are wider than simply operating water storage and regulated rivers, and include a range of activities designed to maintain and manage river systems to achieve predetermined management plans (Independent Pricing and Regulatory Tribunal, 1996a).

Irrigation groups in New South Wales have indicated that they are prepared to accept their fair share of the efficient costs of operations and maintenance. The key issues here are "fair" and "efficient." The concern is that further cost reductions are called for in rural water delivery systems before agreement can be reached about the appropriate level of costs to be recovered. Furthermore, it is argued that ongoing costs resulting from past inefficiencies should not be charged against current water users. They maintain that the present generation of irrigators should not have to pay for refurbishment of infrastructure which previous management allowed to deteriorate.

Irrigators are also concerned about the extent to which they are being asked to pay for benefits of resource management measures where the beneficiaries are the wider public and the general community. On this issue, they find significant support from a key recommendation of the Council of Australian Governments (Second Report of Working Group to COAG, 1995, 12.1,iii):

> *the costs of public benefits/impact management which are unable to be attributed and charged to specific beneficiaries/impactors be treated as community service obligations.*

Yet, in announcing increased water charges, the new state government declared (Ministerial Media Release, September 6, 1995):

> *The new charges will be put straight back into the State's rivers ... The money will be spent on improving dam safety, improving water quality and on community action activities such as Rivercare groups ...*

Clearly, such benefits relate to the general community, and the costs associated with them should not be charged to irrigators specifically, but should be borne by the community as a government responsibility. Stream gauging might be an example.

Increases already announced in water charges and changes to tariff structures in New South Wales have encountered strong opposition. Part of the problem stems from the fact that additional cost burdens are being placed on irrigators in advance of the structural reforms necessary to facilitate adjustment at the farm level.

In this respect, the COAG Water Policy Agreement emphasizes that (Report of Working Group to COAG, 1994, 5.2)

> *The speed and extent of water industry reform and the adjustment process will also be dependent on the availability of financial resources to facilitate structural adjustment and asset refurbishment.*

Thus, in recommending increases in rural water prices, recognition needs to be paid to the structural adjustment consequences and the social impact of reform.

In regard to the issue of promoting efficiency in water use, it is of some concern that at no point were the increases in charges justified as a means of demand management in a situation of water scarcity. Indeed, a study for the Dairy Research and Development Corporation suggests strongly that the policy changes may have had more to do with revenue-raising than economic principles (Watts 1995).

That aside, it is clear that the process of water reform is ongoing and the impetus for price increases remains strong. For example, in New South Wales, a wide-ranging inquiry has recently concluded into rural and country town water-pricing principles and practice. The inquiry, undertaken by the Independent Pricing and Regulatory Tribunal (1996b), was asked to investigate

- general principles for cost allocation and pricing of water services, including bulk water supply, asset maintenance and refurbishment, and regulatory and resource management activities,
- cost-sharing between different users of services,
- alternative pricing structures for water services relative to regional differences and impacts on user groups,
- effects of regulation and the operating framework on the efficiency, benefits, and costs of service provision,
- impact of environmental and quality standards on costs and benefits of service provision, and

- appropriate rate of return on public sector assets.

The Tribunal identified a number of pricing principles that should be followed in setting changes for bulk water services (Independent Pricing and Regulatory Tribunal, 1996b, p. 2):

- water charges should be based on the most efficient way of providing water services,
- the Department of Land and Water Conservation administration of water resources through the Ministerial Corporation should achieve financial stability and deliver a sustainable level of water services,
- pricing policy should encourage the best overall outcome for the community from the use of water and the other resources used to store, manage, and deliver that water.
- the cost of water services should be paid by those who are responsible for causing, or benefit from, those services; those who cause more services to be required, or benefit more, should pay more, and
- pricing policy should promote ecologically sustainable use of water and of the resources used to store, manage, and deliver that water.

Based on these principles, the Tribunal developed a number of recommendations which should be implemented in the interests of water reform. A key recommendation was that irrigation water pricing should reflect the full cost of making the water available efficiently, including environmental costs. Other recommendations refer to the removal of cross subsidies, measures to improve operational efficiency, a charge to cover future asset refurbishment and replacement, and improvements to monitoring of extractions from unregulated rivers and from groundwater.

Interestingly, the Tribunal considered that water-pricing is likely to have a relatively minor influence on future water use, compared with non-price factors such as the Murray-Darling Basin Commission's cap on water use, environmental water allocations, water quality and river flow objectives, and the regeneration of riverine ecosystems. Finding common ground in addressing these issues will not be easy, given the divergence in the structure and organization of water administration, and in approaches to water policy across the Basin.

16. COOPERATIVE FEDERALISM UNDER TEST

The Murray-Darling Basin is marked by sharp differences in water availability, in access to the resource, in attitudes towards its use, and in procedures and institutions in place to manage it. This should not be surprising. The Basin covers a vast area of contrasting climate, resources, and land use. Each state has developed its own unique system for exploitation and management of its water resources. Differing priorities meant that irrigation proceeded at a different pace and for different purposes from state to state and within states. Again, this need not be a cause for concern; however, it does make more complex the process of achieving consensus in handling emerging water issues which have greater impact and urgency in some parts of the Basin than in others.

Thus, Queensland encompasses largely only the headwaters of Basin rivers and uses only a small percentage of their waters. New South Wales takes in the greatest part of the Basin and accounts for the largest amount of water used. It is the state hardest hit by the imposition of the cap on diversions, which may explain the firm approach it is taking to its implementation. Queensland, on the other hand, may well view the cap as an unwarranted impediment to expansion of its currently limited access to the Basin's water resources.

The state of Victoria, in normal seasons, has ample water to meet and supplement allocations. Its primary concern is with the beneficial effect it is hoped the cap will have on the level of the Menindee Lakes storages. If these storages can be replenished from more sustained flows in the Darling River, they can be used to supplement supply to South Australia, rather than at the expense of Victoria's share of waters from the Murray River system.

South Australia, at the lower end of the Basin, is not affected by the cap as far as its guaranteed share of Murray-Darling flows is concerned. It may well benefit from improved water quality as more water is left in the system.

Biophysical contrasts are matched by diversity in water management and administration and in approaches to industry restructuring across the four Basin states. It is not possible here to detail the many adjustments which have taken place, and are ongoing, as a result of changes in the political scene. It is sufficient to note that the task of establishing an agreed framework and timetable for implementation of a common set of procedures for sharing the Basin's waters is made that much more difficult by these differences.

Differing political agendas, contrasts between states in the pace and characteristics of water reform, and changing perceptions and priorities in regard to water issues mean that the prospect of a consensual approach is speculative. Agreement on water pricing, tradable water entitlements, market mechanisms, provision for environmental flows, and achievement of a common workable system of property rights in water is likely to require firm federal direction, if not financial coercion.

17. CONCLUSION

The challenge in the Murray-Darling Basin is to share the waters of this complex region in an equitable, efficient, economically rational, and environmentally compatible fashion. Contrasting conditions of water supply and demand, along with political realities linked to state's rights, inevitably will require recourse to "cooperative federalism," with the federal government exercising a decisive role in reaching workable solutions to water management problems that demand urgent attention in a coordinated Basin-wide approach.

REFERENCES

Blackmore, D., and S. Keyworth. 1995. Irrigation Management Strategy—A Component of the Natural Resources Management Strategy. In Australian Bureau of Agricultural and Resource Economics and The Centre for Water Policy Research., *Focus on Policy Development and Options for Irrigation in the Lower Murray Darling Basin.* Canberra.

Claydon, G. 1995. *Water Allocations and Entitlements. Towards a National Framework for Property Rights in Water—Review of Policy Issues and Options.* Department of Primary Industries and Energy, Canberra.

Colby, B. 1990. Enhancing Instream Flow Benefits in an Era of Water Marketing, *Water Resources Research* 26: 1113-1120.

Council of Australian Governments. 1994. *Report of the Working Group on Water Resources Policy.* Canberra.

Council of Australian Governments. 1995. *The Second Report of the Working Group on Water Resources Policy.* Canberra.

Cummins, T. 1996. *Enabling Permanent Interstate Water Trade in the Mallee.* Murray-Darling Basin Commission, Canberra.

Dudley, N. 1996. Personal communication.

Independent Pricing and Regulatory Tribunal. 1996a. *Bulk Water Prices, 1996/97.* Sydney.

Independent Pricing and Regulatory Tribunal. 1996b. *Pricing Principles for Local Water Authorities.* Sydney.

Millington, P. 1991. The Water Market: A NSW Case Study. Synopsis of a paper presented at Conference on the New Environmentalism, New South Wales Cabinet Office, Sydney, Australia.

Murray-Darling Basin Ministerial Council. 1995. *An Audit of Water Use in the Murray-Darling Basin.* MDBMC, Canberra.

Musgrave, W. 1991. Environmental and Resource Management: Exploring Some New Approaches. Paper presented at Conference on the New Environmentalism, New South Wales Cabinet Office, Sydney, Australia.

Musgrave, W. 1996. The Irrigation Industry in the Murray-Darling Basin and Aspects of its Reform. Paper to the 40th Australian Agricultural Economics Conference, Melbourne, February.

Musgrave, W., and Pigram, J. 1995. Social, Economic, and Institutional Considerations Affecting Soil and Water Conservation in Australia. In E.L. Napier, S.M. Camboni, and S.A. El-Swaigy, eds., *Adopting Conservation on the Farm: An International Perspective on the Socio-Economics of Soil and Water Conservation.* Ankeny: Soil and Water Conservation Society.

Prime Minister's Science and Engineering Council. 1996. Managing Australia's Inland Waters. Roles for Science and Technology. Paper prepared by an independent working group for consideration by the Prime Minister's Science and Engineering Council, September 1996.

Wahlquist, A. 1996. Sharing the Rivers, *The Land.* November 7, p. 21.

Watts, A. 1995. *Conceptual Issues in the Pricing of Water for Irrigation.* Dairy Research and Development Corporation, Glen Iris, Victoria.

Willey, Z. 1992. Behind Schedule and Over Budget: The Case of Markets, Water and Environment, *Harvard Journal of Law and Public Policy* 15: 391-425.

PART III

THE EXPANDING SCOPE OF TRANS-BOUNDARY WATER PROBLEMS

9 INTER-JURISDICTIONAL WATER QUALITY MANAGEMENT: EXPERIENCES FROM CHESAPEAKE BAY NUTRIENT MANAGEMENT

Kenneth E. McConnell
Ivar E. Strand
University of Maryland, College Park, USA

The evolution of strategies to improve the quality of water in the Chesapeake Bay has resulted in a set of policies known as the Tributary Strategies. These policies require a reduction of nutrients of 40 percent from a 1985 benchmark level. The 40 percent reduction is, in general, applied to all tributaries. We argue that this strategy imposes two kinds of deadweight losses on society. First it is well known that a strategy of uniform reductions does not achieve a target of pollution reduction at least cost when the sources of the pollutant are heterogeneous in costs, as they surely are for tributaries. Second, in the case of tributaries, the technical condition of equal mixing is not met, so that equal reductions in nutrients do not contribute equally in terms of benefits from improved water quality in the Bay. We develop a model to illustrate these points. The model is then extended into a political economic context to explore why the tributary strategies were "agreed" upon. Finally, we use current water quality data associated with Maryland tributaries to explore the results of the strategies to date.

1. INTRODUCTION

The evolution of strategies to improve the quality of water in the Chesapeake Bay over the past 15 years has resulted in a set of policies known as the "Tributary Strategies." These policies require a reduction of nutrient loadings (nitrogen and phosphorus) of 40 percent from a 1985 benchmark level. The 40 percent reduction is, in Maryland, applied to all tributaries. We argue that this strategy imposes two kinds of deadweight loss on society. First, it is well known that a strategy of uniform reductions does not achieve a target of pollution reduction at least cost when the sources of the pollutant are heterogeneous in costs, as they surely are for tributaries. Second, in the case of tributaries, the technical condition of equal mixing is not met, so that equal reductions in nutrients do not contribute equally in terms of benefits from the Bay's improved water quality.

Maryland's tributary strategies are the culmination of a process of understanding and controlling water quality in the Chesapeake Bay. Studies in the 1970s and early 1980s produced sufficient evidence of water quality degradation that the governors of Maryland, Virginia, Pennsylvania, and the mayor of Washington, D.C., signed an agreement in 1983 to begin addressing water pollution in the Chesapeake Bay. The agreement initiated efforts to coordinate research, place bans on phosphates in detergents and control growth in "critical areas" around the Bay.

The 1987 Conference of Governors was more specific in its proposals, which included a goal of reducing nitrogen and phosphorus loadings in the Bay by 40 percent (based on their 1985 level). It was thought that achieving this goal might be sufficient to eliminate anoxia in most areas of the Bay. As part of the 1992 Amendments to the Conference, a watershed-based approach was adopted. Each watershed or tributary was to develop a strategy to meet the 40 percent reduction. The 40 percent was applied to those sources thought to be controllable, which really represents 20 percent of all nitrogen loadings and 31 percent of all phosphorus loadings.

During the development of tributary strategies, several economists (e.g., Krupnick 1988) raised the same arguments as we raise above concerning economic efficiency. The Executive Council[1] recognized the potential for improvements in tributary strategies through trading nutrient reductions among the tributaries and possibly even states. They recognized that trading could achieve lower cost and might create more habitat improvement (benefits). Despite this recognition, Maryland has focused solely on a tributary-by-tributary approach to meeting the 40 percent mandate.

The artificial constructs called tributaries are considered by us to be boundaries as are jurisdictions defined by counties and states. Problems have arisen among them, as this quote from a tributary strategy annual report indicates:

> *Overall progress in reducing nutrient delivery to the Potomac River is hampered by the uneven pace at which jurisdictions can address nutrient reduction goals. There is no formal mechanism by which Maryland, Virginia and the District of Columbia can meet and determine mutually acceptable agreements for nutrient reductions with the Middle Potomac. As a result, Maryland runs the risk of expending vast sums of public funds and seeing its nutrient reduction efforts thwarted by inaction of other jurisdictions.* (Middle Potomac Tributary Team 1995-1996, p. 12)

This chapter is meant to follow up on the previous arguments and generate interest in inter-tributary trading within and outside of Maryland (e.g., Krupnick 1988; Camacho 1993; George 1994). We examine current status and trends in water quality data associated with Maryland tributaries to see if they illuminate obvious signs that trading might be Pareto-improving. We develop a model to organize our

[1] The Executive Council is comprised of the governors of each state, the mayor of Washington, D.C., the administrator of the U.S. Environmental Protection Agency and the chair of the Chesapeake Bay Commission.

thoughts and to elicit ways of empirically examining these points. The model is then extended into a political economic context to demonstrate why certain actions have a greater potential for implementation than other actions. The paper is thus not directed at methodology. Its purpose is more the continuation of a policy dialogue regarding the creation of a market for nutrient reduction. It is also not meant as an indictment of tributary teams or government employees (who have excelled in executing policies), but rather as a critique of a political process.

2. THE CURRENT SITUATION IN MARYLAND

Let us begin to understand the situation by looking at what has actually happened to water quality in Maryland over the last ten years (see Hennessey, 1994, for a good background on the historic development). We will focus on the ten Maryland "tributaries" which have teams designated to implement policies (Figure 1). The status and trends in these systems with regard to a variety of factors is contained in the 1995-1996 annual report of each tributary team. We focus on the nitrogen status and trends (Figure 2) for the period 1985-1995. As a crude summary, the status of nitrogen (shown in black for poor and white for good) is generally poorer the further upstream one goes. Nearly every upstream portion of a tributary in Maryland has high concentrations of nitrogen. The lowest concentration of nitrogen is found in the bays of the Eastern Shore and in the lower Patuxent River basin. The trend in nitrogen loads (shown by areas) can grossly be stated as improving on Maryland's western shore and remaining the same on the Eastern Shore. The nitrogen concentrations coming into the Bay from Pennsylvania's portion of the Susquehanna are poor, and the loads are increasing. This tells us that nitrogen concentration has improved in certain systems, but that in general the concentrations are sufficient to sustain the desired aquatic life only in limited areas.

One might also probe the degree to which the 40 percent reduction goals have been met. If the goals are met and the water quality still does not sustain appropriate aquatic life, perhaps adjustments need to be made. Figure 3 shows the 1985 baseline loads of nitrogen by tributary in black, the 1993/94 load in white, and the goal (or cap) in gray. It is clear that no area has met its goal yet and the closest to achieving the goal are the Patuxent and Patapsco/Back River tributaries. The upper western shore has actually increased its loads of nitrogen. With regard to the phosphorus loads (shown in Figure 4 using the same notation), the situation is better, with the Patuxent and Patapsco/Back River tributaries actually having surpassed the goals. These results can largely be attributed to the phosphate detergent ban of 1985.

The status and trends suggest that certain tributary teams have been more successful at meeting their goals than others. Clearly a host of factors control the success. In the following, we establish a model which may be helpful in analyzing the economic factors important to success. We then examine the little empirical work that has been done on the economic factors.

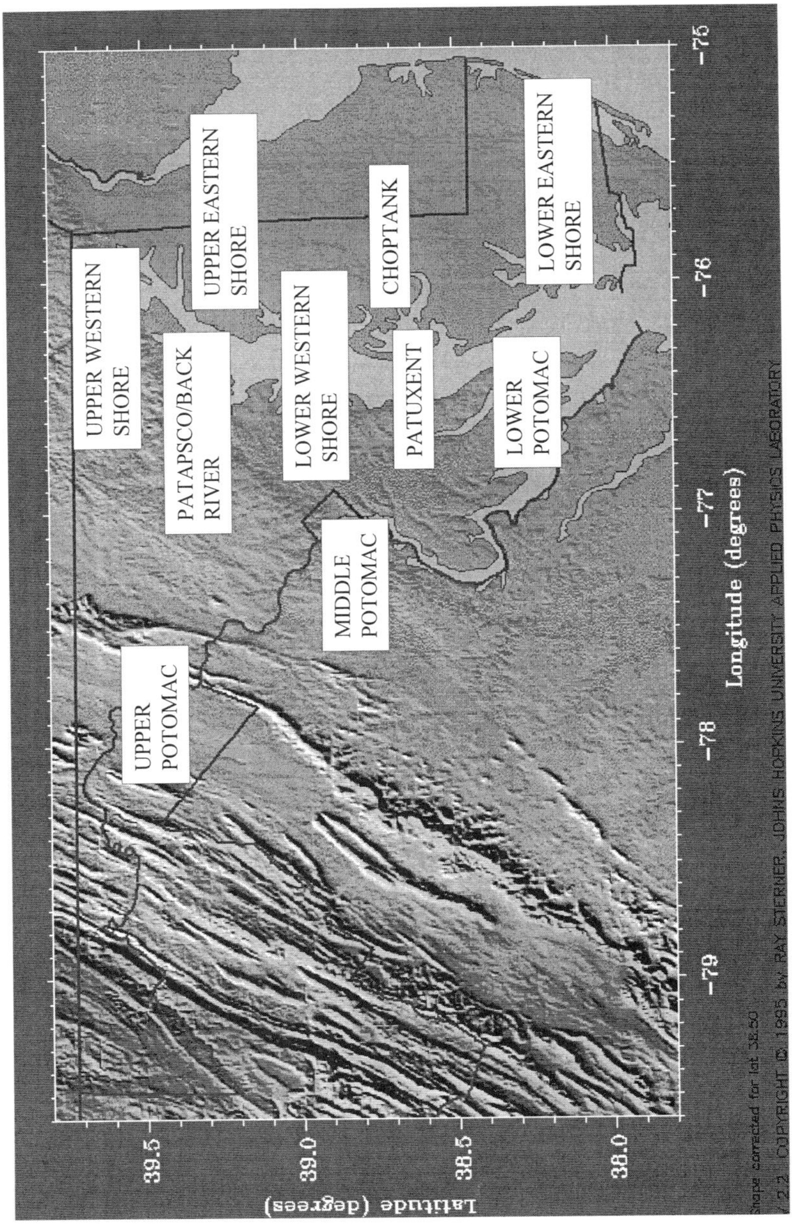

Figure 1. Maryland's Tributary Strategy Teams
Reprinted with permission by Ray Sterner, Johns Hopkins University Applied Physics Laboratory

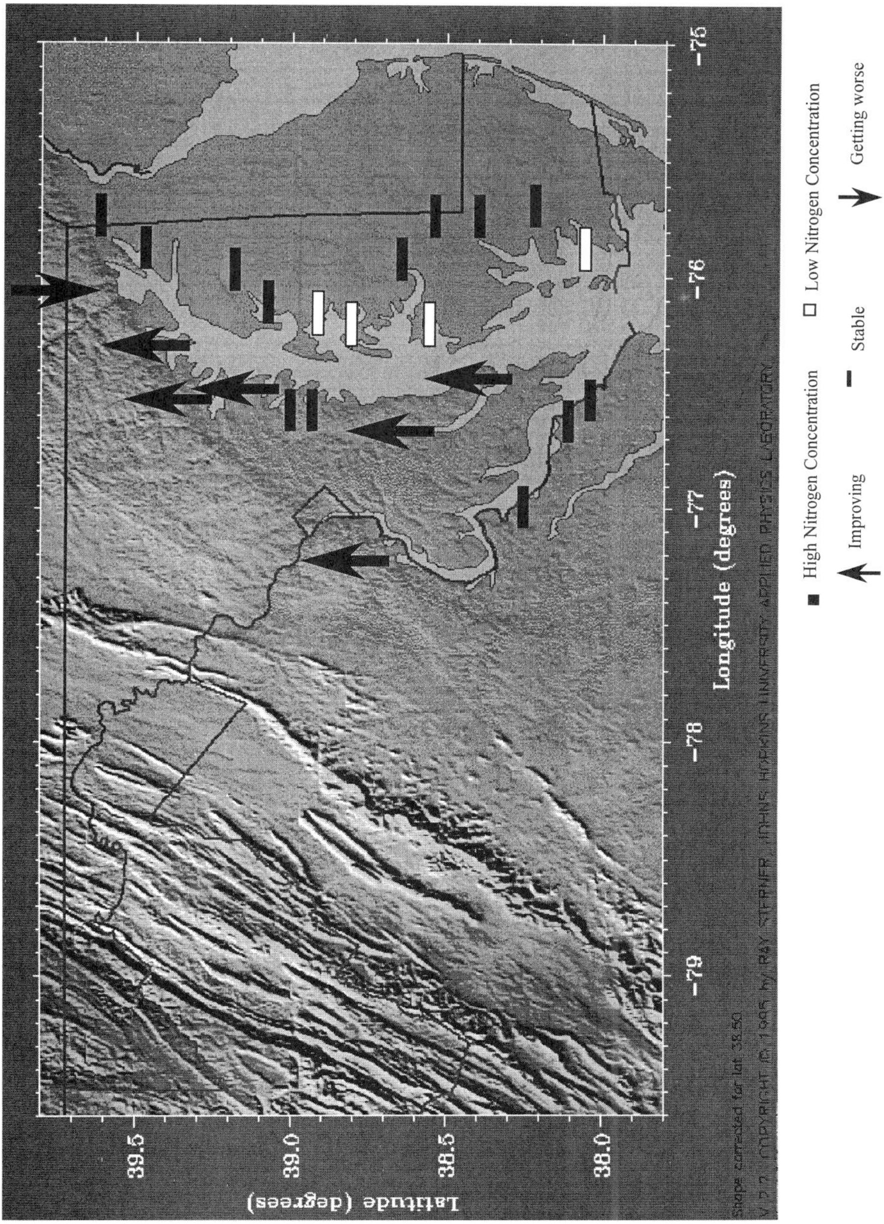

Figure 2. Status and Trends in Maryland and Susquehanna Nitrogen, by Tributary

Source: Annual Reports, Tributary Strategy Teams, 1995-1996

Reprinted with permission by Ray Sterner, Johns Hopkins University Applied Physics Laboratory

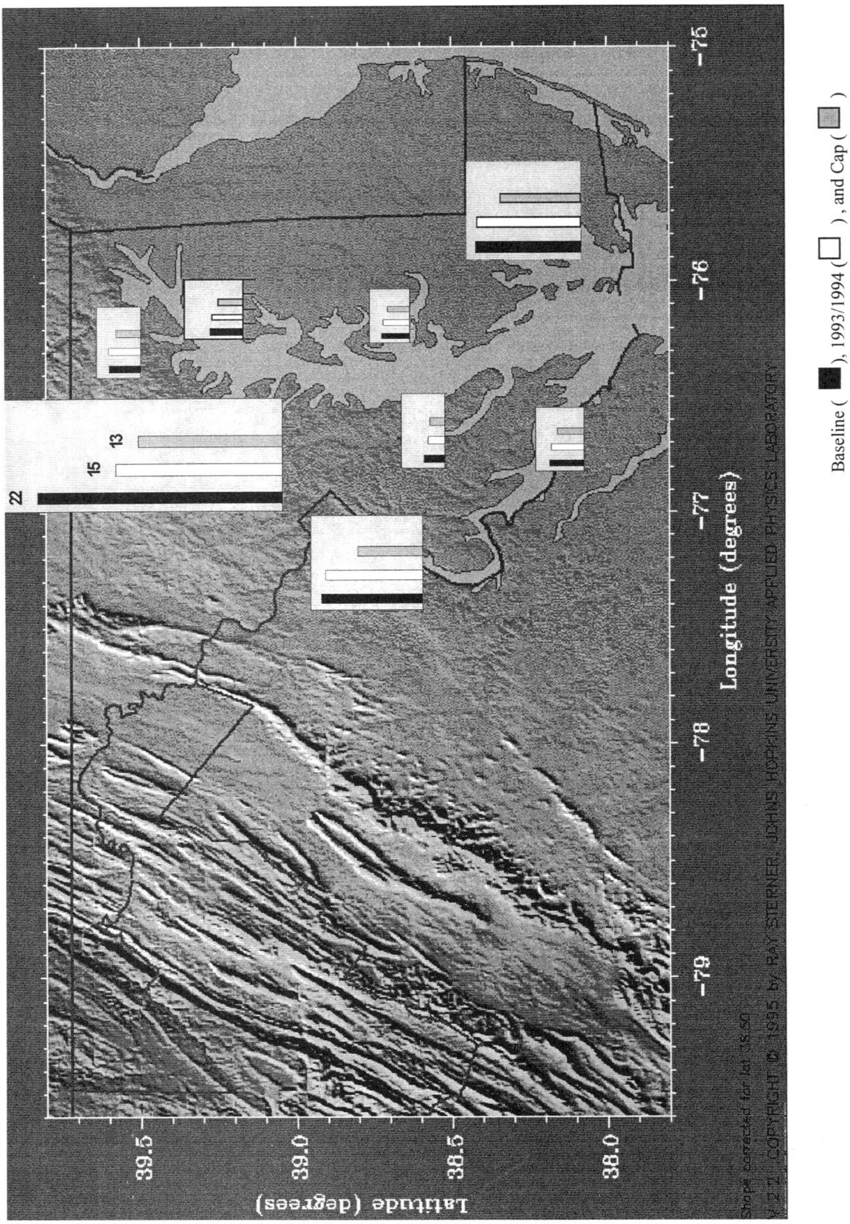

Figure 3. Nitrogen Loads by Tributary (10^6 lbs/yr)

Source: Annual Reports, Tributary Strategy Teams, 1995-1996

Reprinted with permission by Ray Sterner, Johns Hopkins University Applied Physics Laboratory

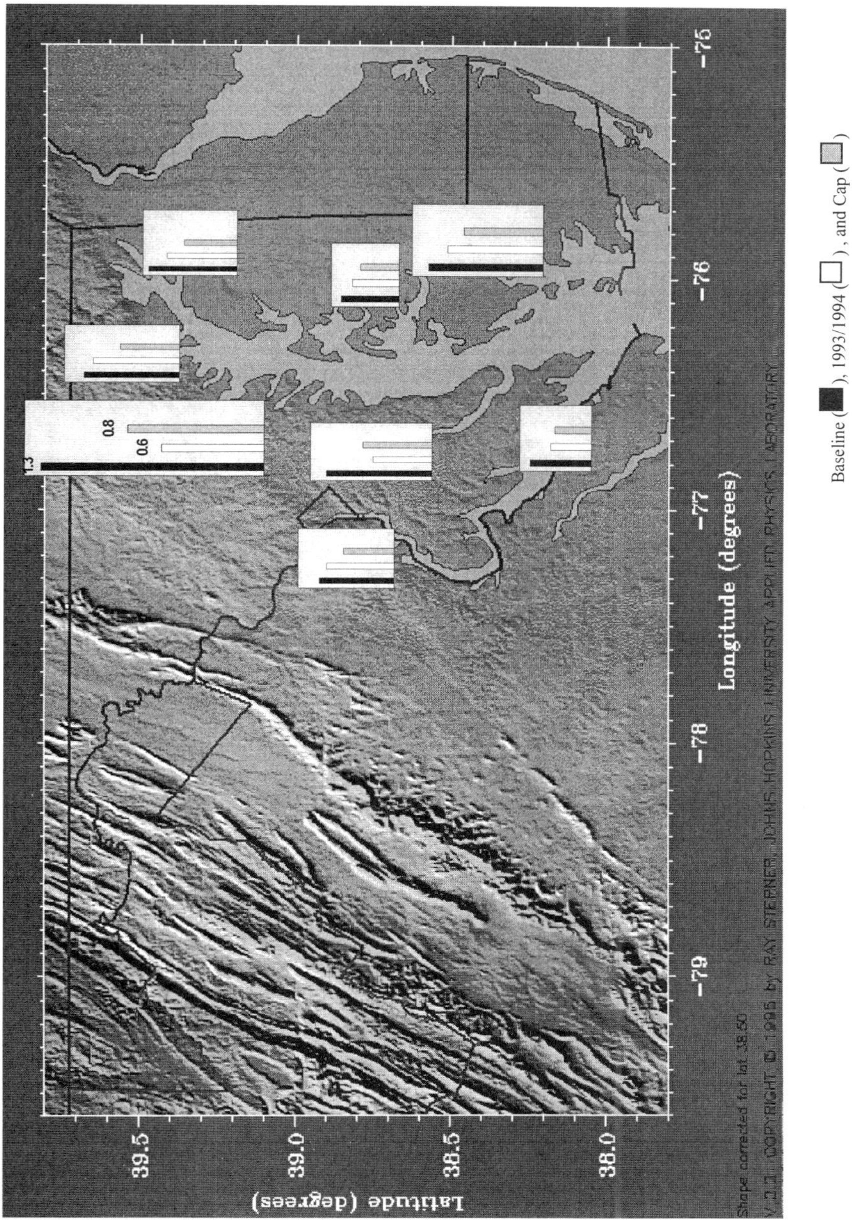

Figure 4. Phosphorus Loads by Tributary (10^6 lbs/yr)
Source: Annual Reports, Tributary Strategy Teams, 1995-1996
Reprinted with permission by Ray Sterner, Johns Hopkins University Applied Physics Laboratory

3. A MODEL OF WATER QUALITY MANAGEMENT

We organize our thoughts with a simple model of the benefits and costs of nutrient removal. Suppose that n_i represents the level of nutrients contributed to the Bay by the ith tributary. Some portion of each n_i will be subject to control, because some nutrients come from natural sources. Let T be the number of tributaries (for example, in Maryland T = 10), and $C(n_1,n_2,\ldots,n_T)$ be the social cost of nutrients in the Bay to all members of society who do not live in any of the tributaries. That is, reductions in nutrients bring benefits to people who live within the boundaries of political units in which the tributaries lie, as well as to people outside the boundaries. For example, people who live in parts of Delaware, Tidewater Virginia, western Maryland, and North Carolina gain benefits from the use of the Bay although they are not involved in programs to reduce nutrients to the Bay. Others who live farther away still may have passive use of the Bay. The damages to these beneficial uses are covered by the social cost function $C(n_1,n_2,\ldots,n_T)$ non-decreasing and convex in the n's. The 'non-decreasing' adjective is not simply an academic form of generality. In fact, evidence shows that tributaries contribute to social damages unevenly, with the upper and mid portions of the Bay contributing significantly more to water quality degradation. In some cases the nutrients from a tributary may contribute nothing to the nutrient problem of the Bay. The 1992 Amendments recognize this, noting that the 40 percent reduction goal in portions of Virginia tributaries south of the Potomac was only interim.

For benefits to people who live within the tributaries, we aggregate the individual tributaries. Let $c_i(n_1,n_2,\ldots,n_T)$ be the damages to the ith tributary from this schedule of nutrients. Like C, $c_i(n_1,n_2,\ldots,n_T)$ is non-decreasing and convex. The total damages from nutrients in the Bay are the sum of the damages for residents of the tributaries and those who live outside the tributaries,

$$\text{Total damages} = C(n_1,\ldots,n_T) + \sum_{i=1}^{T} c_i(n_1,\ldots,n_T).$$

Damages such as these have been estimated for portions of Maryland's Chesapeake Bay by Bockstael et al. (1988). The damages and their definition potentially vary between Bay managers and the public. Based on the Agreement, the damages are the loss of living resources as measured by abundance of submerged aquatic vegetation (SAV) and dissolved oxygen. According to the work of Bockstael et al. (1988), the public, however, is more likely to be concerned with the abundance of sport-fish, less algae and jellyfish in the water, and reduction in disease arising from contact with the water. It is not obvious that the public would knowingly devote resources to the improvement of SAV in the absence of more tangible rewards. Whether the two goals coincide is a matter of happenstance.

The benefits of nutrient use come from higher crop yields from fertilizer and reduced costs of disposing of human and animal wastes. Reductions in nutrients require reduced use of fertilizer and more treatment of household and animal wastes. These benefits are internal to the households and firms in a tributary's watershed or the political districts where the tributaries are located. Let $b_i(n_i)$ be an increasing and concave function representing the benefits from the use of nutrients by the ith tributary or the foregone costs of not reducing nutrients. The $b_i(n_i)$ may be thought of as the sum of individual benefits, which depend on individual use of nutrients. That is,

$$n_i = \sum_{k=1}^{N_i} n_{ik},$$

where n_{ik} is the use of nutrients by the kth person in the ith tributary, N_i is the number of people in the tributary, and $b_{ik}(n_{ik})$ is the ik[th] individual's benefits from the use of nutrients. Supposing that

$$b(n_i) = \sum_{k=1}^{N_i} b_{ik}(n_{ik})$$

is equivalent to supposing that the marginal values of nutrients are the same for each individual within a tributary. The total benefits are the sum across all tributaries,

$$\text{Total Benefits} = \sum_{i=1}^{T} b_i(n_i).$$

Reductions in nutrients can be achieved in a variety of ways. For example, typical for agriculture would be less nutrient application by using different technologies or better information and changing the applications (through timing or incorporation) of fertilizers. The adoption of more advanced technologies at wastewater treatment plants is the principal approach for household nutrient reduction. See Camacho (1993) for some analysis of the benefits of nutrient discharge.

From these definitions of benefits and damages, we can determine four kinds of choices of nutrients. The first choice of nutrients is simply to allow individuals to make their own decisions. This would occur when, for each individual k in tributary i, the net benefits of more nutrients is zero for all k,i: $b_{ik}'(n_{ik}) = 0$. This is the uncoordinated optimum. The second is the private tributary optimum (IO). It is the level of nutrients that would be chosen optimally by the individual tributaries independent of coordination. This is presumably what would have been observed prior to the implementation of programs to improve the Bay's water quality. The third is the social optimum (SO), the one that maximizes benefits minus damages. The fourth strategy is the tributary strategy (TS), that is, the 40 percent reduction of nutrient loading in all tributaries.

First, consider the actions of individual tributaries left to their own devices, the individual optimum. They would be interested in maximizing

$$\text{Net benefits} = b_i(n_i) - c_i(n_1, \ldots, n_T).$$

Since the individual tributary choices depend on what other tributaries do, it is necessary to pick a reasonable equilibrium mechanism. The most appropriate is the Nash equilibrium—individual tributary choices are optimal, given the choices of others. A tributary's choices would follow from the T joint conditions,

$$\text{(IO)} \quad \frac{\partial\{b_i(n_i) - c_i(n_1,...,n_T)\}}{\partial n_i} = 0 \ .$$

Call the n_i's that satisfy (IO) n_i^{IO}. One obvious observation is that when there is social damage from the use of nutrients within the political boundaries of the tributary, then optimal choice of nutrients would equate the marginal private gain with the social loss, at the margin. The potential here is that for some tributaries, $\partial c_i(n_1,n_2,\ldots,n_T)/\partial n_i = 0$, that is, increases in nutrients do not affect the damages that the individual members of the tributaries incur. This may arise in two ways. First, it may be that the tributary contributes its nutrients to a portion of the Bay in such a way to cause no damages. This for example would be true if the tributary emptied into Baltimore Harbor, where the social cost of disposing nutrients is negligible, due to the high level of contamination by other pollutants. The second way would be when the nutrients contribute to improved water quality in a location in the Bay that is not used by members of the tributary. It is also possible for nutrients to be flushed from the Bay, especially in its lower portion, and additional nutrients will cause no social damage.

Second, consider the social optimum. It would require the maximization of total benefits less total costs of nutrient use,

$$\text{Net benefits} = \sum_{i=1}^{T} b_i(n_i) - C(n_1,...,n_T) - \sum_{i=1}^{T} c_i(n_1,...,n_T) \,.$$

The socially optimal choice of nutrients implies that

$$\text{(SO)} \quad \frac{\partial \sum_{i=1}^{T} b_i(n_i)}{\partial n_i} - \frac{\partial C(\cdot)}{\partial n_i} - \frac{\partial \sum_{i=1}^{T} c_i(n_1,...,n_T)}{\partial n_i} = 0 \,.$$

Call the strategies that achieve the social optimum n_i^{SO}. It may be socially optimal for some tributaries to not reduce nutrients because they cause no damages. Yet it may be critical for some of the tributaries to reduce nutrients substantially, because they affect critical areas or are fundamental to the entire system. Consequently, strategies that induce optimal use of nutrients would imply nutrient reductions from zero for some localities, with little effect to large values for critical areas.

The coordinated tributary strategies require a reduction of 40 percent over the *controllable* 1985 benchmark level. For example, it is estimated that only 50 percent of the nitrogen levels are controllable, so that the tributary strategy would reduce total nitrogen by 20 percent when reducing the controllable nitrogen by 40 percent. A reasonable estimate of the 1985 benchmark level of nutrients is the individual optimum n_i^{IO}. The controllable part is the benchmark level less the uncontrollable level of nutrients: $n_i^{IO} - n_i^{UC}$, where n_i^{UC} is the uncontrollable level of nutrients for the ith tributary. Thus the coordinated solution (TS) is

$$\text{(TS)}\ n_i^{TS} = .6(n_i^{IO} - n_i^{UC}).$$

We know that in terms of the total returns from various nutrient strategies, the social optimum is higher than the individual optimum and the tributary strategy. The tributary strategy is not the optimum in two ways. First, by distributing reductions proportionately, it fails to achieve the reduction in nutrients in the cheapest way. High-valued agriculture and low-valued agriculture are both asked to reduce nutrients by 40 percent. In the case of publicly owned waste water treatment facilities (known as publicly owned treatment works, POTWs) tributaries with treatment plants and those without plants would have to achieve the same reductions. Second, the tributary strategy is inefficient, calling on all tributaries to reduce nutrients when nutrients from some tributaries do not cause social damage. It is easy to see in that case that social gains will occur by increasing nutrients for those tributaries. It is possible that the coordinated strategy will have fewer net benefits than the private optimum. This depends on the distribution of benefits and costs across tributaries.

4. PREVIOUS WORK ON THE COSTS AND BENEFITS OF NUTRIENT REDUCTION IN THE CHESAPEAKE

Research is available that demonstrates the variation in marginal costs across the various tributaries of the Chesapeake and across the various types of polluting activities (e.g., Camacho 1993). Moreover, there are studies that estimate the marginal gains to individuals from improvements in various activities across Maryland's portion of the Chesapeake Bay (Bockstael et al. 1988; Krupnick 1988). The studies demonstrate, among other things, just how difficult the accurate measurement of benefits and costs are. Thus, we rely on the qualitative nature of the information more than on the specifics of the estimates of benefits and costs.

Camacho (1993) demonstrates that there is substantial variation in the marginal costs of nutrient removal depending on the source of the nutrient and the location of the discharge. The marginal (average) cost of phosphorus removal changed by a factor of 7 depending on the wastewater treatment facility considered. In general, the costs of phosphorus removal are lower for point sources than for non-point sources. On the other hand, the marginal costs of nitrogen removal are likely to be higher for the point sources than for the non-point sources. However, there is sufficient variation and error in the measurements of these numbers to question the applicability of generalization to any specific case.

Across space, the marginal costs of a ton of soil saved (which is related to nutrient reduction, all else equal) in agriculture are highest in the upper portion of the Bay and lowest in the upper reaches of tributaries in western Maryland, Pennsylvania, and Virginia. The upstream areas are shown to be three or four times less expensive in soil savings. Unfortunately, the costs of reducing agricultural nutrients are often idiosyncratic and extremely difficult to measure.

The measurement of economic benefits is also not easy, and probably the most comprehensive analysis is contained in Bockstael et al. (1988). They found that non-use benefits were stated to be large and that benefits in general varied across the region. The benefits for water-related activities on the Chesapeake from persons in the Baltimore-Washington statistical metropolitan sampling area (SMSA) accrue mostly in an area 15 miles north and south of the Bay Bridge. It is unfortunate that the majority of activity is located in one of the relatively poorer water quality areas. The analysis also indicated that recreationists "voted with their feet" by traveling to areas with better water quality. Further, residents of the Eastern Shore have a more affordable substitute water body for water-contact activities, the Atlantic Ocean. Thus, improved water quality to residents of the Eastern Shore may not be as valuable as to residents on the western shore.

5. THE POLITICAL ECONOMY OF NUTRIENT REDUCTION

The acceptance of inefficient strategies is typically related to questions of fairness and rent-seeking. Perhaps some construction of fairness is what motivates across-the-board 40 percent reduction. A major political justification for the tributary strategies was that lawmakers of Maryland wanted to be assured that some of the benefits from water quality management would occur in their tributary. Because all were paying, it seemed "fair" that the gains from improvements should not be going to a few geographic locations, ones presumably near the Bay proper. Early efforts to address the same issue call for greater access to the Bay—more public parks and more highways leading to the Bay.

A related question in the political economy of pollution reduction concerns choice of strategies to reduce nutrients. Camacho's results, despite their shortcomings, suggest that in a typical case, reductions in nutrients can be achieved at a lower social cost from agriculture than from additional processes in waste treatment plants. Despite this we find that more progress has been made by tertiary treatment in POTWs than by reducing agricultural nutrients.

This anomaly can be explained in part by looking more carefully at the distribution of costs of reduction in nutrients. When nutrients are reduced by a rather expensive strategy of tertiary waste treatment at POTWs, the costs are typically spread across most of the population, or at least the part of the population serviced by the treatment plant. The cost can be recovered by water and sewer rates that are already in place. When nutrients are reduced by changes in agricultural practices, the marginal cost may be low, but the costs are incurred only by the farmers, who are typically few in number. Then the per capita costs incurred by farmers may be sufficient for them to engage politically.

The propensity to engage politically will likely grow as the number of affected farmers declines. Consequently, while aggregate benefit-cost analysis may favor nutrient reduction through agriculture, individual benefit-cost analysis and the median voter model suggest that POTWs would be the chosen strategy.

This result is consistent with the Olson (1971) hypothesis about small numbers and the influence of interest groups. To understand these results, let us simplify by considering two approaches: the POTW approach, which disperses the costs of controlling nutrients across all of the population who subscribe to the water supply system, and a program of farm management practices, which concentrates the costs among farmers. Suppose that these two strategies are chosen so that they yield the same reduction in damages from nutrients. Under these circumstances, benefit-cost analysis favors a program of farm management, because such a program is cheaper and provides the same reduction in nutrients as the POTW approach. A median voter model, in which households are ranked by individual net benefits of the program, would also favor the farm policy approach, because most voters, being non-farmers, would incur no costs from the farm policy, and hence would vote for that policy to maximize personal gains. The program of agricultural best management practices may have aggregate net benefits in its favor, but by concentrating costs among a relative few, it creates the incentive to organize opposition in the form of lobbies. This is a reasonable explanation of the prevalence of improvements in POTWs compared with programs of best management practices.

6. CONCLUSIONS

The rehabilitation of the Chesapeake is an ongoing process that is governed by the costs and benefits of water quality improvements, and most important, the multiple levels of federal, state, and local governments who control the methods of reducing pollutants. In the long run, taxpayers and households using the Chesapeake Bay will be best served by policies that achieve reductions in pollution, particularly nutrients, in the least cost ways. Attention to the costs of nutrient reduction is likely to be ever more important, given the well-known propensity of the costs of pollution control to rise with increasing stringency of control. This is currently hampered by the necessity of seeking cooperation from different jurisdictions of local governments as well as the state government of Maryland. This cooperation is necessary, of course, because costs and benefits fall unequally on different jurisdictions, with some counties far from the Chesapeake contributing considerable quantities of nutrients but receiving few benefits from water quality improvements.

Given the disparity among benefits and costs to different regions in the state, and the different marginal costs for different sources of nutrients, it is clear that the time is ripe for developing a state-wide nutrient trading program. A program that introduces explicit side payments is now being developed through political concessions. It would help achieve the target of 40 percent reduction in nutrients in a cost-efficient way, but because the major contributors of nutrients would be trading, it would be possible to take advantage of the ability of agriculture to reduce nutrients more cheaply.

REFERENCES

Bockstael, N.E., K.E. McConnell, and I.E. Strand. 1988. Benefits From Improvements in Chesapeake Bay Water Quality. EPA Report, Contract No. CR-811043-01-0. 123.

Camacho, R. 1993. Financial Cost Effectiveness of Point and Nonpoint Source Nutrient Reduction Technologies in the Chesapeake Bay Basin. Chesapeake Bay Program Nutrient Reduction Strategy Reevaluation, Report No. 8, CBP/TRS 84/93.

George, J.W. 1994. Cost Implications of Maryland's Tributary Strategies. In *Toward a Sustainable Coastal Watershed: The Chesapeake Experience. Proceedings of a Conference.* CRC Publication No. 149. 223-228.

Hennessey, T.M. 1994. Governance and Adaptive Management for Estuarine Systems: The Case of the Chesapeake Bay, *Coastal Management* 22: 119-145.

Krupnick, A.J. 1988. Economics and Nutrient Reduction in the Chesapeake Bay. In J.H. Cumberland, ed., *Proceedings of the Fourth Annual Conference on the Economics of Chesapeake Bay Management.* pp. 1-33.

Middle Potomac Tributary Team. 1995-1996. p. 12.

Olson, M. 1971. *The Logic of Collective Action.* Boston, MA: Harvard University Press.

Tributary Strategy Teams. Annual Reports 1995-1996 various issues, Maryland Department of Natural Resources, Annapolis.

10

LAND USE PATTERNS AND WATER QUALITY: THE EFFECT OF DIFFERENTIAL LAND MANAGEMENT CONTROLS

Nancy E. Bockstael
Kathleen Bell
University of Maryland, College Park, USA

Many environmental problems have a strong spatial component. There may also be a spatial pattern to the regulatory actions taken to solve the problem, if the resource in question either falls under the management of multiple jurisdictions or is affected by decisions of agents subject to different jurisdictions' policies. In no area is spatial pattern more important for environmental outcomes, and in no area are there more jurisdictions developing policy than in the area of land use. Land use pattern is a major determinant of water quality, and while a variety of agencies make policy that affects it, land use and land use pattern is most directly influenced by local governments. Given the relatively small size and the profusion of local governments, spillover effects are likely. Where multiple jurisdictions are attempting independently to manage land use change for reasons that include water quality protection, the results can easily be counterproductive. In fact, some have blamed jurisdictional fragmentation for at least some of the degradation of common surface and groundwater resources.

In this chapter we look at an example of the effects of differential local jurisdictional policy. We focus on the Patuxent River, a major tributary of the Chesapeake Bay, and on the land use in the lower reaches of its watershed. We model the spatial pattern of development as a function of natural and man-made features of the landscape and influences of differing regulatory policies, paying particular attention to the pattern of conversion of agricultural and forest land to residential use and its effects on the water quality of the common resource.

1. INTRODUCTION

The Chesapeake Bay Agreement, signed in 1987 by the states of Maryland, Virginia, and Pennsylvania, the District of Columbia, the U.S. Environmental Protection Agency, and the Chesapeake Bay Program's Commission, is a relatively rare example of interjurisdictional cooperation over the management of a common resource. The group has formally pledged to work together to achieve, by the year

2000, a 40 percent reduction (from 1985 levels) in nutrient loadings to the Bay. In 1992, the Executive Committee of the Bay Program directed the "Bay partners" to develop tributary strategies for watersheds of each of the major tributaries of the Bay that would accelerate progress towards the 40 percent reduction goal. In Maryland, the Tributary Strategies Initiative has developed into a series of cooperative efforts including not just federal and state entities but also local government.

To understand the importance of including local government in the Bay program, it is necessary to understand the sources of pollution of the Bay. The sources initially identified as important were sewage treatment plants, which contributed a large share of the phosphorous to the Bay, and agriculture, whose run-off of fertilizers and manure was identified as accounting for a major share of nutrient enrichment. A significant amount of the nitrogen loadings in the Bay has also been identified as coming from air deposition, whose sources include automobiles and power plants within the entire airshed of the Chesapeake. A smaller but increasing share of the nitrogen loadings in the Chesapeake have more recently been recognized as emanating from residential fertilizer use and residential septic fields. As farmland and forests are converted to housing developments, this source of pollution is growing.

The amounts, types, and spatial pattern of this pollution are driven by the pattern of land use. Given any set of technologies, emissions from sewage treatment plants and septic fields will be determined by the number of people depending on each, and this will be determined by the population density and its distribution relative to public service (public water and sewer) areas. Given any level of farming technology (and best management adoption), agricultural run-off will be affected by the amount of farmland and its spatial distribution relative to modulating land covers such as forests and amplifying land covers such as impervious surfaces. Air emissions will be affected in large part by automobile usage which itself is determined by the pattern of population relative to employment centers and other locational "attractors."

Federal and state government is responsible for most of the policy whose direct intent is to alter discharges into the environment. These include point source emissions standards and incentives for agricultural and urban best management practice adoption. However, given that land use and land use pattern in its watershed is the prime determinant of water quality in an estuary, water quality is most influenced by local government policy. It is local government that establishes zoning ordinances, growth control policies, and property tax schedules; and it is local government that provides most of the public infrastructure (with the notable exception of highways) that influences those decisions.

The 1995-96 Annual Report of the Patuxent River Tributary Strategies Team (Patuxent River Tributary Team 1996) highlights the following in their executive summary of Findings and Recommendations (p. 3):

Methods are needed for tracking the utility of and need for growth management tools. Many local governments already implement a variety of these programs. Currently these

> *tools are not tracked nor credited as part of the Tributary Strategy tracking efforts. [In addition], the trade-offs between septic systems versus sewer need to be examined.*

The Findings and Recommendations go on to suggest that the Commission should track the effect on water quality of growth management efforts at the local government level and the conservation of open space through land acquisition and easement.

Recognition of the role of local land use policies, and the specific importance of growth control instruments, land preservation programs, and public utility provision, is encouraging. Means for tracking the outcomes of these policies are becoming increasingly available as the State and some localities fund major efforts in collecting and maintaining geographic information data bases. In this paper we use these newly available Geographic Information System (GIS) data and focus on the lower Patuxent River watershed, specifically the local jurisdictions of Anne Arundel, Prince George's, Charles, and Calvert Counties (Map 1). This chapter models the spatial pattern of land use change as a function of natural and man-made features of the landscape and influences of differing county regulatory policies. We are particularly interested in the pattern of conversion of agricultural and forest land to residential use because of its ultimate effect on the water quality of the common resource. We ask in this context whether growth control strategies matter. Our preliminary analysis of these data yields some interesting results that have implications for future efforts of the Tributary Strategies Teams.

2. LOCAL LAND USE POLICIES IN THE SOUTHERN PATUXENT WATERSHED

Under current policies, the 40 percent nutrient reduction goal for the Patuxent River will probably be reached by the year 2000. However, the goal is expected to be exceeded again, by a significant amount, within 10 years. This latter prediction is based on projected population growth in the region. Demographic projections suggest that there will be 74 percent more households in the Patuxent watershed in the year 2020 than there were in 1990 when the nutrient reduction strategies were first being put into place (Maryland Office of Planning 1996). This huge potential for change in land use pattern makes the role of land use and growth controls of primary importance in water quality management.

The projected growth is not a new phenomenon in the watershed counties. Over the past decade or more, most of the counties in the watershed have experienced rapid growth, especially in the form of low density sprawl. However, most of the pressure for growth controls has not come from concerns over Chesapeake water quality. Instead, rising county infrastructure costs, together with increased congestion and loss of the rural characteristics that attracted residents in the first place, have provided most of the impetus.

The four counties in our study area are within commuting distance of Washington, D.C., and are experiencing out-migration of both employed and retired Washington, D.C., workers escaping from the city. The two northern-most counties, Prince George's and Anne Arundel, are densely developed in the sections

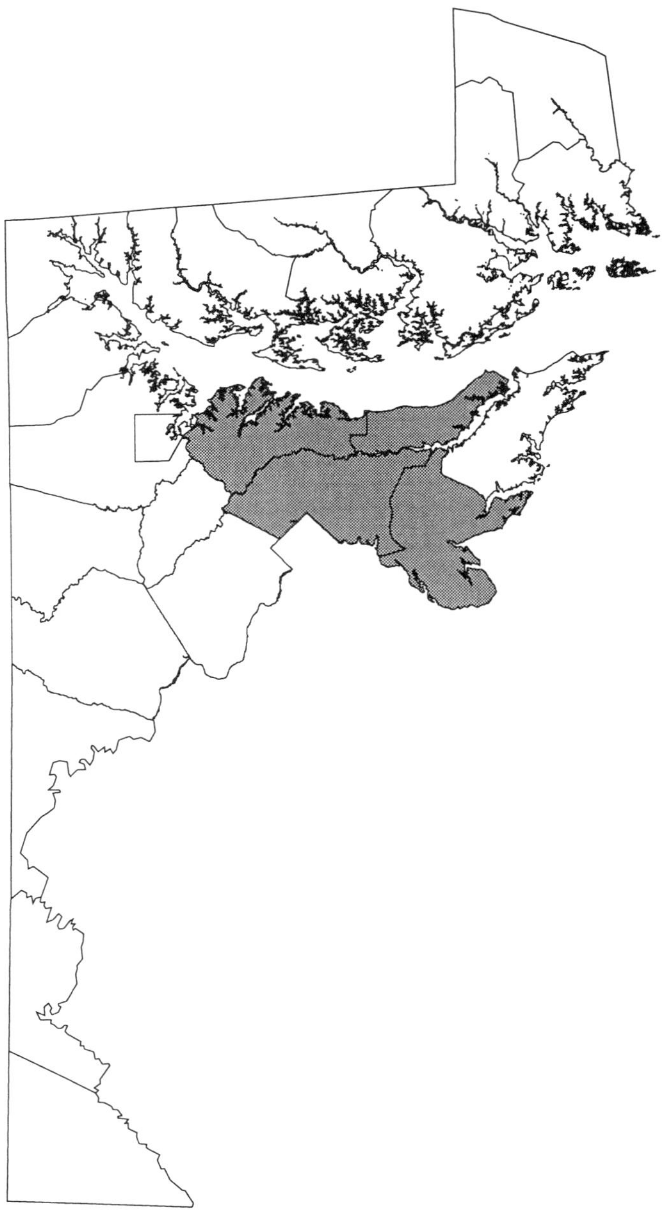

Figure 1. The Lower Patuxent River Watershed: Anne Arundel, Calvert, Charles, and Prince George's Counties

nearest the Washington-Baltimore-Annapolis corridors. But each still contains a large, relatively rural area in its southern extremity in which the primary land use/land cover is agriculture or forest. Each of the two more southern counties has pockets of dense development, but a larger share of these counties is still in rural use. The rural areas in these four counties share three characteristics of importance. Very little of the area is served by public utilities, most is zoned for "rural" use, and all are coming under intense development pressure. The "rural" zoning designation cannot preclude residential use, but it can restrict its density.

In reaction to intense growth pressure, many Maryland counties have been national leaders in adopting growth control or containment measures. All counties in the State are required to produce comprehensive plans which describe their intended land use plans and many of these have, in the last decade, focused attention on the problems of growth in the form of low density sprawl. The most common instruments used by the counties to restrict development in relatively rural areas have been: to require large minimum lot sizes or, equivalently, low allowable densities in these areas; to refuse provision of public services (specifically public water and sewer) to these areas; and to establish farmland preservation programs that encourage sale of development rights in these areas.

Large minimum lot size restrictions are a feasible, but potentially counterproductive, means of containing sprawl. To the extent that they discourage development in the rural areas of a county by making development unprofitable, they serve their intended ends. However, what development *does* occur is of the low density sprawl variety that counties are attempting to prevent. The refusal to provide public utilities has slightly different, but also occasionally counter-productive, effects. Where no public services exist, site conversion costs are higher, and profitability lower. This discourages development, *ceteris paribus*, and to the extent that development exists, counties can avoid the infrastructure costs of sewer and water services. However, the ultimate effect is often increasing problems with drinking water quality and quantity and increasing impacts on surface water quality, as well. In addition, lots dependent on well and septic cannot be zoned for intensive use and thus housing cannot be as densely clustered, should cluster zoning be invoked.

Agricultural land preservation would appear to be a completely different means for controlling rural low density sprawl, dependent on completely different considerations. It would also appear to be a control that is uniformly applied over the state, since the state of Maryland has its own agricultural land preservation program, and most of the protected land in the lower Patuxent watershed is sheltered under the Maryland Agricultural Land Preservation Program (rather than the smaller county-funded programs). However, given the mechanism by which farms come to participate in the easement program, county land use policies have a potentially large effect on outcomes. In purchasing the development rights to a farm, the state is prevented by law from paying any more than the minimum of (a) the farmer's offer price and (b) the difference between the appraised market value of the land and the value of the land in agricultural use. The actual sale is dependent on the farmer's acceptance of the price and the availability of program funds. Since the market value of the land will be affected by the zoned development density allowed on the land, county zoning policies will have a direct effect on the market value of the land. The potential exists for county policies to affect both participation in easement programs and the size of the state's outlay of funds.

We are ultimately interested in the pattern of land use conversion. The above suggests that that pattern will be significantly affected by the pattern of allowable development densities, both because density affects profitability and because the likelihood of easement sales will be affected. Therefore, county zoning policies will play a central role in determining the changing land use pattern. Maps 2 and 3 illustrate zoning and public utilities provision in the undeveloped lands of the four counties in our study area. As can be seen from these maps, policies vary considerably over counties—especially with regard to allowable development density in rural areas.

The effect of differential zoning policies is interesting to consider. A county that reduces its allowable density in an area will probably reduce the likelihood of development. But because adjacent counties provide developable land that is closely substitutable, the action of the first county can be expected to affect development probabilities in neighboring counties by deflecting development pressure.

3. A SIMPLE MODEL OF LAND USE CONVERSION

We begin by identifying all land that is, at a certain point in time, in an open space use (either agriculture or forest land) but could be converted to a residential use at the density prescribed by zoning law. The variable, W_j, is defined as the present value of the land in the jth use. Where j is the residential use (j = D), this will be the market value of the land once it is developed and sold on the residential market (minus the costs of development). Where j represents agricultural or forests (j = U), this is the present value of discounted future returns to those activities.

Values of W_D and W_U will vary over the landscape for a variety of reasons. Clearly the value of land in residential use is very much affected by its location. Location is defined not only by the commuting distance to employment centers and services, but also by the proximity to desirable and undesirable land uses, the topology of the land and its scenic amenities, and the services provided by the jurisdiction. It will also be affected by the costs of converting the land from its current use to the developed use. The value in agriculture is affected by natural features such as the quality of the soil. In addition, the W's must be interpreted as some function of the expected values of returns since economic agents cannot know them with certainty at the time a conversion decision is made.[1]

[1] We admit to the importance of risk but do not attempt to model behavioral response to varying risk.

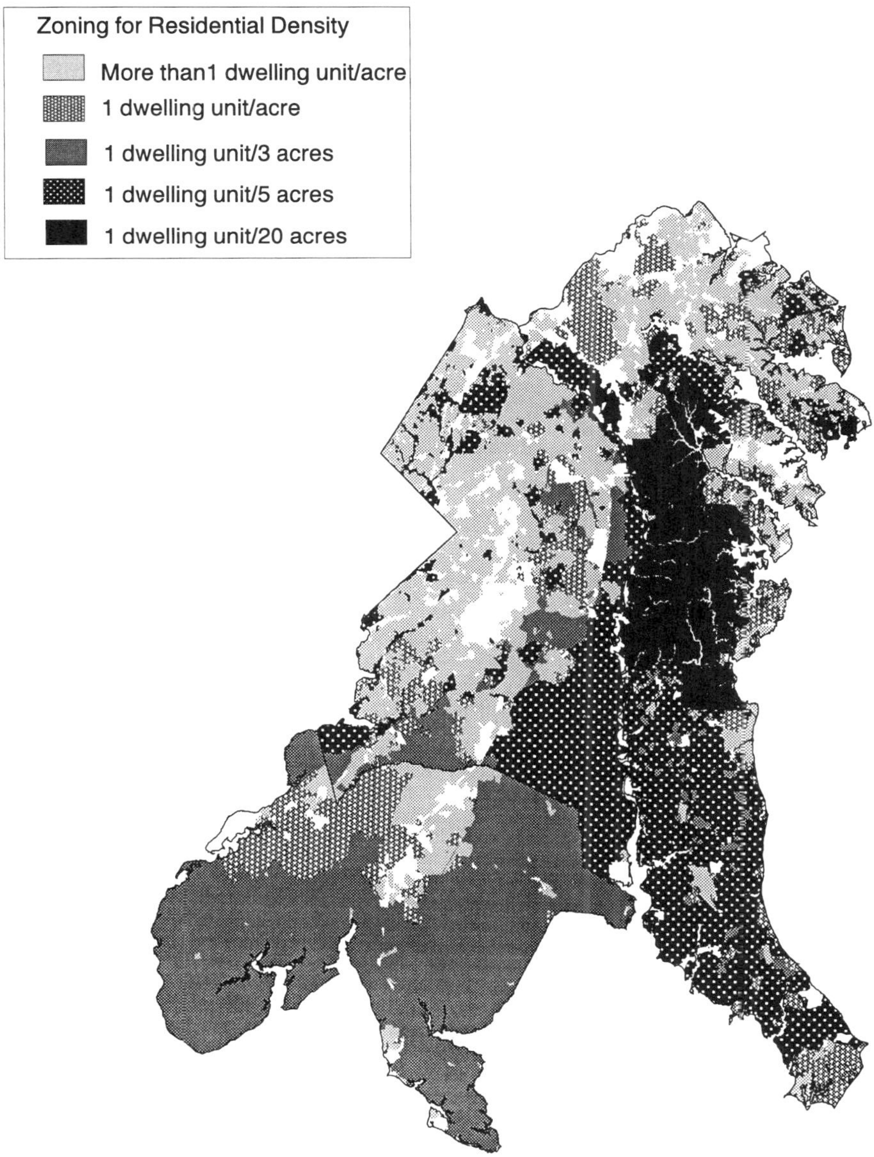

Figure 2. Differential Density of Development Zoning: Anne Arundel, Calvert, Charles, and Prince George's Counties

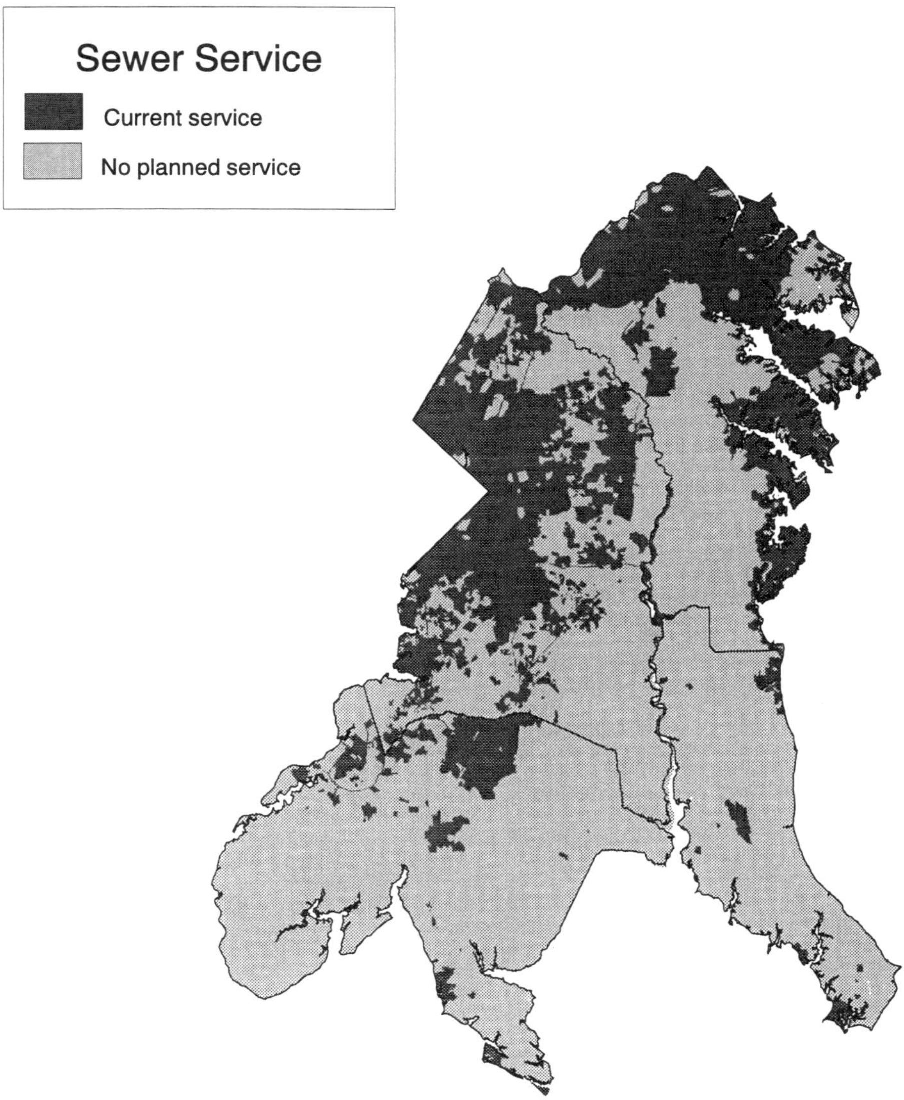

Figure 3. Differential Sewer Provision: Anne Arundel, Calvert, Charles, and Prince George's Counties

A simple model of land use conversion might turn on the relative sizes of W_D and W_U. Considering that it is unlikely that we will have information on all factors affecting these values, we might rewrite W with a systematic portion, V, which is a function of observable variables, and a random portion, η, which is unobserved to the researcher. Thus the probability that parcel i will be developed is given by

$$Pr\ (develop)_i = Pr\ (V_{iD} + \eta_{iD} > V_{iU} + \eta_{iU}) = Pr\ (\eta_{iU} - \eta_{iD} < V_{iD} - V_{iU}) \quad (1)$$

The arguments of V_{iD} will include the predicted value of parcel i in residential use net of cost of converting the parcel to residential use. The arguments of V_{iU} will be factors that affect the value in agricultural use.

The possibility of selling the development rights to the land introduces an added complexity. The easement will be sold if the State's bid exceeds the reservation price of the farmer. The maximum price the State can pay for the parcel is the appraised market value minus the value in agricultural use. This would appear to be exactly the point of indifference for the farmer, making the actual outcome dependent solely on idiosyncratic factors known only to the farmer. However, the bigger the difference between market value and agricultural value, the larger the outlay of state funds. Given a limited state budget for purchasing easements, the larger the necessary outlay the less likely the purchase will take place. Although it is an empirical question, farmers may view the likelihood as small of being able to obtain the full market value from sale of an easement of land that has a very high market value.

We restate the likelihood that a randomly drawn parcel, i, from the population of buildable land is developed at a point in time, t, as the product of the probability that the land is developed conditioned on the development rights not having been sold times the probability that the development rights were not sold:

$$Pr\ (development) = (1\text{-}Pr(easement))\ (Pr(development)|\ no\ easement). \quad (2)$$

The probability that the easement is sold is expressed as a binary discrete choice problem also:

$$Pr\ (easement)_i = Pr\ (\varepsilon_{iN} - \varepsilon_{iE} < S_{iE} - S_{iN}) \quad (3)$$

where E denotes the sale of the easement and N denotes the alternative of not selling the easement. At this point, we make the simplifying assumption that the errors in (1) and (3) are uncorrelated.

Empirically our problem is divided into three steps. Models (1) and (3) imply the estimation of discrete choice models where the decisions depend on, among other things, the expected value of the land in residential use and its value in the alternative use. We obtain estimates of value in agricultural use from a model developed by extension agents in our department which predicts agricultural value largely as a function of soil type. Forest land in the watershed has no net commercial value, although it may have considerable social value as a public good.

We estimate the expected value of land in residential use with a hedonic model of recent residential transactions.

Once in possession of a means of predicting value in residential use and some proxies for conversion costs and agricultural land values, we use this information to estimate the participation in the land preservation program. We do not yet have access to actual easement sales but we have data on agricultural district enrollment. This is a prerequisite for easement sales and precludes the farmer from selling his land for development over a five year period. Since agricultural district enrollment involves adopting nutrient management plans and water quality improvement programs, farmers are unlikely to petition to join an agricultural district unless they are serious about selling their development rights. This step in the process is important because it effectively prohibits residential development (at least temporarily) and removes that land from the pool of potential developed land. Once this land is removed from the pool, we then proceed to estimate the probability of conversion to residential use in the third step of the model.

3.1 The Hedonic Model for Residential Land in the Southern Patuxent Watershed

Taking space into account in an hedonic model is important, first, because the location of a parcel relative to amenities and disamenities is expected to have an important influence on its price. As with other hedonic applications we include in our analysis variables which measure distances to employment centers and to natural features of the landscape. However our commuting distances are more sophisticated in that they are based on actual miles traveled along road networks and not "as the crow flies." We incorporate spatial location in a second way by extending our description of location to include patterns of land use surrounding a parcel. Finally we take space into account econometrically by admitting to the likelihood of spatial autocorrelation in our data and attempting to correct for it.

The observations used to estimate the model are extracted from the Maryland Division of Tax and Assessment's data base, which includes market prices and transaction dates over the past fifteen or so years for each privately held parcel in our four counties. We estimate the hedonic model for the period from 1991 to 1993 using residential transactions on newly built houses, of which there were 5,013 with complete data in our sample. Our problem is complicated by the fact that we need to estimate a model that will predict the value of the *land* (minus the structure) in residential use. One way of attempting this is to include as many structure variables and location variables as possible in the hedonic model and hope to be able to separate out the value due to location and size of lot from the cost of the structure. This is problematic for a number of conceptual and econometric reasons and made impossible by the absence of structural characteristics for a large portion of the data base. The alternative course of action is to extract the cost of the structure from the locational value by subtracting the assessed value of the structure (a good surrogate for a hedonic estimate of structure value) from the transaction price. Assessed value of the structure is available from the tax assessment data base. We use as the dependent variable the log of this residual residential land value divided by the size of the parcel. The size of the parcel is considered exogenously determined by zoning ordinances, which is probably very close to the truth for newly developed properties such as the ones in our sample.

To measure the value of the variables associated with the location of this land, one needs to be able to locate the parcel on a map. Recently the Maryland Office of State Planning has undertaken the arduous task of geocoding the parcels in the tax assessment data base, and our analysis uses the product of their work. Being able to locate the parcels in a geographic information system means that we can then employ our geocoded maps of features of the landscape and road network to describe the parcel's characteristics more completely. In what follows we describe the explanatory variables.

Previous hedonic analyses have supported the notion that land price is a nonlinear function of lot size or, equivalently, price per acre is a function of lot size (LOTSZ). In addition to lot size, there are some obvious factors that affect land price. The first, and potentially most important for the region, is the commuting distance to Washington, D.C. (WASHDC) calculated along the roads network. To capture accessibility to shopping and other business, we include distances to the nearest large shopping center (LGSHOPD), strip mall (SMSHOPD), and town (TOWNSD). We also measure the distance to the coastline, since shore access is expected to be a desirable feature (and often a reason for relocation to these counties) and include this variable defined as inverse distance to the coastline.

By including some less common attributes in our hedonic model, we test the importance of an array of landscape pattern variables. The model reported here includes seven pattern variables, each measuring the proportion of an area surrounding the parcel that was, in 1990, in a given use. The uses included are commercial/industrial (COM), agricultural cropland (AGCROP), agricultural pasture land (AGPAST), forest (FOR), high density residential (HRES), medium density residential (MRES), and very low density residential (VRES). The actual measure is the proportion of the block of 25 (200 meter) cells dominated by each land use, where the block is defined as having the cell in which the parcel resides at its center. In addition we include dummy variables for the counties, capturing all those factors such as tax rates, education expenditures, waste removal, etc., that vary across counties but not within them.

3.1.1 *Estimation Procedures*

Several econometric problems arise in attempting to estimate this model. There are the usual ones that are perennially associated with hedonic models—those of choice of functional form and multicollinearity—but these are well known and will not be discussed here. In addition to these there are two types of problems: a potential sample selection problem and spatial autocorrelation. The problems inherent in dealing with spatial data are sufficiently challenging that we restrict our attention only to the latter. However, sample selection is still an inherently important problem, but some preliminary analysis suggests that treatment of it in our application may not dramatically alter the results.[2]

Spatial dependence among the errors is generally due to omitted variables, which are themselves spatially correlated.[3] For example, if housing values depend on neighborhood attributes and not all such attributes are included in the housing values model, the error terms will be spatially correlated. Econometric methods developed to handle problems of time-lagged dependent variables and temporal autocorrelation cannot be applied to spatial dependence problems, because unlike the time series analogue, spatial effects are two dimensional and multi-directional.[4] Time is one-dimensional and neatly orders the data. Prior years have an effect on future years. In contrast, location is two-dimensional and offers no neat ordering of data. A given observation in a spatial data set can have any number of neighboring observations at any given distance from it.

However, analogous to the treatment of temporal autocorrelation, the usual method for correcting spatial autocorrelation requires assuming a structure for the spatial dependence and estimating one or more parameters of that structure in conjunction with the parameters of the economic model. The conventional spatial autocorrelation problem is represented as a spatial autoregressive process,[5] with the standard regression model revised as:

$$Y = X\beta + \varepsilon$$

where

$$\varepsilon = \rho W\varepsilon + u,$$

which implies

$$Y = X\beta + (I - \rho W)^{-1} u,$$

where Y is an N by 1 vector of observations on the dependent variable, X is an N by K matrix of explanatory variables, W is an N by N spatial weight matrix, β is vector of parameters to be estimated, ρ is a scalar to be estimated, u is an N by 1 vector of random error terms with expected value of 0 and variance-covariance matrix $\sigma_u^2 I$, and ε is an N by 1 vector of random error terms with mean 0 and non-spherical variance-covariance matrix $\sigma_u^2(I-\rho W)^{-1}(I-\rho W')^{-1}$.

---

[2] In this problem, the hedonic model includes a non-random sample of parcels—those that were sufficiently valuable in residential use to have already been converted—leading to a potentially serious sample selection bias.

[3] They could also be due to errors in measurement which are systematically related to location. An example might be Census data which will be averaged over a larger area in rural block groups than in urban ones.

[4] See Anselin (1988) and Anselin (1995) for a review of the spatial econometrics literature.

[5] This spatial autoregressive model was suggested by Whittle (1954) and by Cliff and Ord (1973). An alternative model has been proposed by Kelejian and Robinson (1995).

The spatial weight matrix, W, contains information on the assumed spatial dependence between all pairs of errors.[6] The i,jth element of W, denoted w_{ij}, represents the relative spatial dependence between the ith and jth error, where w_{ii} is defined as 0 for all i. An observation, i, can have as many as N-1 neighbors, but the relative influence of each neighbor must be specified, *a priori*, since these elements obviously cannot be estimated independently. We use a contiguity matrix, the most common type of weight matrix in the literature, and define w_{ij} as 1 if the jth observation is within M distance of the ith observation, and 0 otherwise.[7] In keeping with usual practice, we row-standardize this matrix (divide the elements in each row by the sum of the row). As argued elsewhere, the model is not well-defined in the absence of row-standardization.

The usual procedure is to estimate the parameters (β, ρ, σ_u^2) by maximizing the likelihood function,

$$L = \ln|I - \rho W| - \frac{N}{2}\ln(2\pi) - \frac{N}{2}\ln(\sigma_u^2) - \frac{(y - \rho Wy - X\beta + \rho WX\beta)'(y - \rho Wy - X\beta + \rho WX\beta)}{2\sigma_u^2},$$

but this is problematic in large samples. The Jacobian in the above likelihood function is an NxN matrix, whose determinant must be repeatedly evaluated in searching for the maximum likelihood (ML) estimate of ρ. Alternatively, the term $\Sigma_i \ln(1-\rho\lambda_i)$ can be evaluated where λ is the vector of N eigenvalues of the spatial weight matrix. Depending on the form of W, solution for the eigenvalues may become impossible for as little as several hundred observations. In our case, we have over 5,000 observations.

Recognizing the difficulties of calculating the ML estimator for large samples, Kelejian and Prucha (1996)[8] have devised a generalized method of moments (GMM) estimator whose chief advantage is that it does not involve the calculation of the eigenvalues of W or any extensive manipulation of this potentially large matrix. The approach involves nonlinear least squares estimation of a system of equations to obtain estimates of the parameter ρ. Once an estimate for ρ is obtained, estimates for β and σ_u^2 are derived using feasible generalized least squares (FGLS).[9]

---

[6] The pattern of spatial dependence between the errors is typically presented as a maintained rather than a testable hypothesis, since it is difficult to formulate tests of alternative spatial weight constructs. Unfortunately, although arbitrarily defined, the form of the spatial weight matrix is central to estimating the spatially dependent error model.

[7] In the hedonic model, it may seem more appropriate to specify a weight matrix that allows neighbors to have differential impacts—impacts declining with distance. Unless these weight matrices are row standardized, however, they produce models with unacceptable properties. When they are row standardized, they generally lose their desirable interpretation.

[8] In Bell and Bockstael (1997) we explore a range of issues inherent in comparative estimation procedures for large samples.

[9] The parameter, σ^2, can be estimated using the GMM estimates or the FGLS estimates. Kelejian (personal communication) suggests the latter.

3.1.2 *Results of the Hedonic Model*

Both the ordinary least squares (OLS) and generalized method of moments results from the hedonic model are reported in Table 1. The estimated coefficients do not change dramatically when we correct for spatial autocorrelation, but this is not surprising since OLS produces unbiased parameter estimates and we have a rather large sample to draw on. However OLS standard errors will be biased and it is interesting to note whether significance tests based on GMM estimation are different from those based on OLS.

For most variables, the qualitative results are not changed by spatial autocorrelation correction. The price per acre declines significantly with the size of the parcel, as expected. This is empirical evidence of the common knowledge that reducing developable densities makes development less profitable. Price also declines with distance to the major employment center, Washington, D.C., as well as with distance to the nearest location for shopping and other attractions (large shopping mall, strip mall, or town center). Price, as expected, increases significantly with access to the shoreline (i.e., it increases with 1/distance to shoreline). Land in subdivisions commands a premium, perhaps because of unmeasured services provided, and other county-specific factors seem to be significant. Prices tend to be significantly lower in Charles County and significantly higher in Anne Arundel County relative to both Calvert and Prince George's Counties, *ceteris paribus*.

The significance of the coefficients on the land use pattern variables are most affected by the spatial autocorrelation correction. Two patterns emerge if one were to take the OLS results at face value. First, price in residential use seems to increase as more of the surrounding area is in open space—specifically agricultural pasture and forest (but not agricultural cropland). Price also, and seemingly contradictorily, increases with more surrounding land dedicated to medium and high density residential use. This is probably due to the high correlation between these density measures and distance to Washington, D.C., and town centers, an effect that is difficult to separate out. However, when spatial autocorrelation is considered, some of this land pattern effect disappears—specifically the positive effect of surrounding open space. In correcting for spatial autocorrelation we are inherently reweighting the errors to admit the spatial pattern in omitted variables that we believe is captured in the error term. If this spatial pattern is highly correlated with the spatial pattern included in the model, the importance of that pattern may be lost. This is an area in which more research in applied spatial econometrics is needed.

We proceed to use the results of this hedonic model to predict spatially heterogeneous values in residential use of land that is not yet developed, but could be developed as of 1994. With the dependent variable measured as the log of the residual price per acre in residential use, we explain about 53 percent of the variation according to the R^2 provided by the OLS regression. This leaves a fair amount of unexplained variation, but is a reasonably high R^2 for cross section data.

Table 1. Results of the Hedonic Model Using OLS and GMM Estimators

Variable	OLS Estimate	t-statistic	GMM Estimate	Asymptotic t-statistic
Intercept	11.988	117.596	11.636	91.257
Lotsize	-0.263	-32.383	-1.172	-23.401
Distance to D.C.	-0.011	-11.380	-0.009	-5.631
Agric cropland	-0.002	-0.771	-0.002	-0.521
Agric pasture	0.012	2.133	0.008	1.184
Commercial	-0.008	-0.841	0.013	1.294
Forest	0.005	2.398	0.004	1.553
Very low density residential	0.077	1.547	0.045	0.929
Medium density residential	0.016	5.640	0.015	4.201
High density residential	0.081	8.569	0.057	5/268
Anne Arundel Co.	1.015	22.280	1.082	14.534
Calvert Co.	0.417	5.897	0.411	3.563
Prince George's Co.	0.344	9.608	0.477	7.833
In subdivision	0.520	7.233	0.601	8.533
Distance large shopping ctr.	-0.015	-7.304	-0.012	-3.411
Distance small shopping ctr.	-0.274	-5.154	-0.041	-4.893
Distance to town	-0.290	-7.834	-0.031	-5.019
Bay access	0.006	2.073	0.008	3.171
R^2	0.5316			

3.2 Discrete Choice Land Use Conversion Model

Discrete choice models that adequately account for spatial autocorrelation are only in the development stages, and we make no attempt to incorporate them in this analysis. Consequently, both the agricultural district participation and the land use conversion problem are treated as simple binary discrete choice problems. We choose the logit specification to make subsequent scenario predictions easier to obtain. However, estimates produced by the probit specification are substantively the same.

Because our data do not allow the plotting of actual parcel boundaries and the tracking of the division of farms into buildable lots, we redefine our units of observation in terms of 200m x 200m cells of the landscape. Because the hedonic model explains price per acre in residential use and because we have spatially delineated data on land use, zoning, public utilities, road networks, etc., we can estimate the price per acre of a typical parcel in any cell using the hedonic model results. Lot size is assumed to be equal to the minimum lot size restriction, which appears to be a binding constraint on developers.

Our sample includes all those 200m cells that, as of 1994, were buildable. That is, these cells were not already predominantly developed and were not restricted by any zoning or other factors that would preclude development. We then identify those cells whose predominant use was agriculture (these are our observations for the agricultural district participation model) and further identify those that were enrolled in the agricultural district program (these are the observations for which the dependent variable is 1). For each of the cells in this sample, we predict value per acre in residential use using the hedonic model estimated coefficients. We predict the value per acre in the alternative use equal to the larger of 0 or the value in agricultural use as predicted by our extension model (based largely on soil type). Finally, we measure the proportion of the surrounding land that is in agricultural use to capture whether the cell is an isolated agricultural operation or part of an extensive area still dedicated to agricultural use.

The subsequent land conversion model uses as its sample all buildable land that was not enrolled in agricultural districts as of 1994. The explanatory variables include the predicted value in residential and alternative uses. We also create dummy variables that distinguish land that will have higher conversion costs. Specifically we identify land whose soils and slopes make excavation and construction more costly. We identify whether the land was forested and thus would need to be cleared at the time of conversion; and we identify whether sewer and water service were available or whether these utilities would have to be provided by the developer.

3.2.1 *Results of Discrete Choice Analysis*

Table 2 presents the results from these two simple logit models. As expected, the likelihood of participation in an agricultural district increases with amount of surrounding land in agriculture. That is, participation is less likely for isolated islands of agriculture amidst development. We initially hypothesized that participation in agricultural preservation programs would also be a decreasing function of the difference between market value and agricultural value of a parcel. In our model we included both values separately. We find that participation is most definitely declining with predicted value in residential use. However, we find no significant effect of agricultural value. There are two obvious explanations for this latter result. The first is that our measure for agricultural value is a relatively weak proxy for the truth, i.e., that there is a large measurement error in this variable. The second is that there is very little variation in agricultural returns over the region, but much more variation in residential values.

The same phenomenon was observed in the land use conversion model. These results are also presented in Table 2. All estimated coefficients were qualitatively as expected, except for the results for agricultural value. Likelihood of conversion increases with predicted residential value and decreases with factors that are likely to raise construction costs. However, the value in alternative use, or our measurement of it, has no significant effect in explaining conversion.

Table 2a. Results of Discrete Choice Model of Agricultural District Participation

Variable	Estimate	ChiSquare	Pr>Chi
Intercept	-3.8664	881.9268	0.0001
Residential value	-0.0217	118.6981	0.0001
Agricultural value	0.0138	0.0665	0.7964
Surrounding ag area	0.1278	372.3848	0.0001
Log Likelihood for Logistic -3388.374456			

Table 2b. Results of Discrete Choice Model of Land Use Conversion

Variable	Estimate	ChiSquare	Pr>Chi
Intercept	-2.7419	1768.819	0.0001
Residential value	0.0019	83.458	0.0001
Agricultural value	0.0320	0.805	0.3697
Forested	-0.0159	24.563	0.0001
High excavation costs	-0.9527	219.550	0.0001
Sewer provision	1.2567	611.043	0.0001
Log Likelihood for Logistic -11081.20828			

4. IMPLICATIONS OF THE RESULTS

Having made a reasonable first pass at estimating a model that explains land use conversion, we return to the original question posed by this chapter. How do growth management policies, specifically in our case—allowable densities of development, affect the water quality of the common property resource of the region? We add the second interesting question: Do inter-county conflicts in land use regulations—specifically different allowable densities of development—affect the results?

At this point we need to be specific about what we have and have not accomplished. In the hedonic analysis above we are not trying to identify the demand and supply of developable land, nor to predict future market price. The hedonic model is a mapping between characteristics of parcels and their recent transactions prices, both of which would have been observable to agents in the market. If these agents form expectations on the returns they might receive from subsequently attempting to develop any given parcel in their control, this hedonic function may approximate quite well the formation of those expectations. It may not very well approximate the ultimate price at which a parcel will sell in a subsequent time period, but that is not what we are trying to estimate We use the hedonic model merely to approximate expectations.

The discrete choice model is an attempt to estimate the importance of various explanatory variables in affecting whether a given parcel of land gets pledged to an agricultural district and, if not, whether it gets converted. The estimation is a cross-sectional one whose estimated coefficients can provide predictions of the spatial variation in the *relative* likelihoods of conversion only. As yet, we have no means of predicting total development demand which will be a function of real land prices, mortgage rates, income and employment in the region, etc.

Despite this shortcoming, the results of the model yield some interesting implications. First, one of the factors that is significant in both the agricultural district participation model and the land conversion model is the expected value in residential use, and—at least given our measure—value in agricultural use seems to have little impact on either decision. The higher the expected value in residential use, the less likely the development rights to the land will be sold to the state and the more likely the land will be converted. The second factor of importance is that minimum lot size (or, allowable density of development) is an extremely important factor in determining relative price of land in residential use.

The above two results combined suggest that the growth control measure that is likely to play the major role in determining emerging land use pattern will be county zoning ordinances. In fact, agricultural preservation programs may be having very little effect since those lands that are most likely to be enrolled are the ones that would otherwise have been least likely to be developed.

To demonstrate these apparent implications, we produce a series of scenarios in which we use the results of the model to predict the relative likelihood of agricultural district participation and land use conversion of "buildable" land cells for a subsequent round of decisions. And we do so under different assumptions about the allowable densities in the "rural" areas of our four counties. From the earlier maps it can be seen that the most rural districts of Calvert and Charles Counties are zoned for an allowable density of 1 dwelling per 3 acres. Prince George's rural zoning is divided between areas of 5 and 2 minimum acres. Anne Arundel's rural area, which makes up most of the southern part of the county, is zoned for a density of no more than 1 dwelling unit per 20 acres.

We consider three scenarios. In the first, we assume that zoning remains constant in the next "round" of decisions. In the second and third, we assume that all counties align their allowable development densities in their most rural areas. We cannot rule out the possibility that different zoning regulations will alter the total amount of development in the region and not just the spatial allocation of it, but since we do not yet have the means to predict total development, we compare the outcomes holding expected number of new dwelling units constant. In the second scenario, all rural areas are zoned at 20 acre minimums, and in the third at 3 acre minimums. The results "reported" in Maps 4 through 6 illustrate the difference in the predicted spatial pattern of development, assuming that in each case approximately 40,000 new dwelling units are added to the area. Our model predicts that differential minimum lot size zoning will deflect development from one county to another. Most obvious is the higher development pressure in Anne Arundel's rural regions when county zoning is brought into alignment.

In order to assess the effects of these different scenarios on water quality, we need a means of translating land use into nutrient loadings. Sophisticated models that predict a number of ecological effects are currently being developed by ecologists at the Maryland Center for Estuarine and Environmental Sciences, Solomons, Maryland. In the interim, we use some rules of thumb generated by the

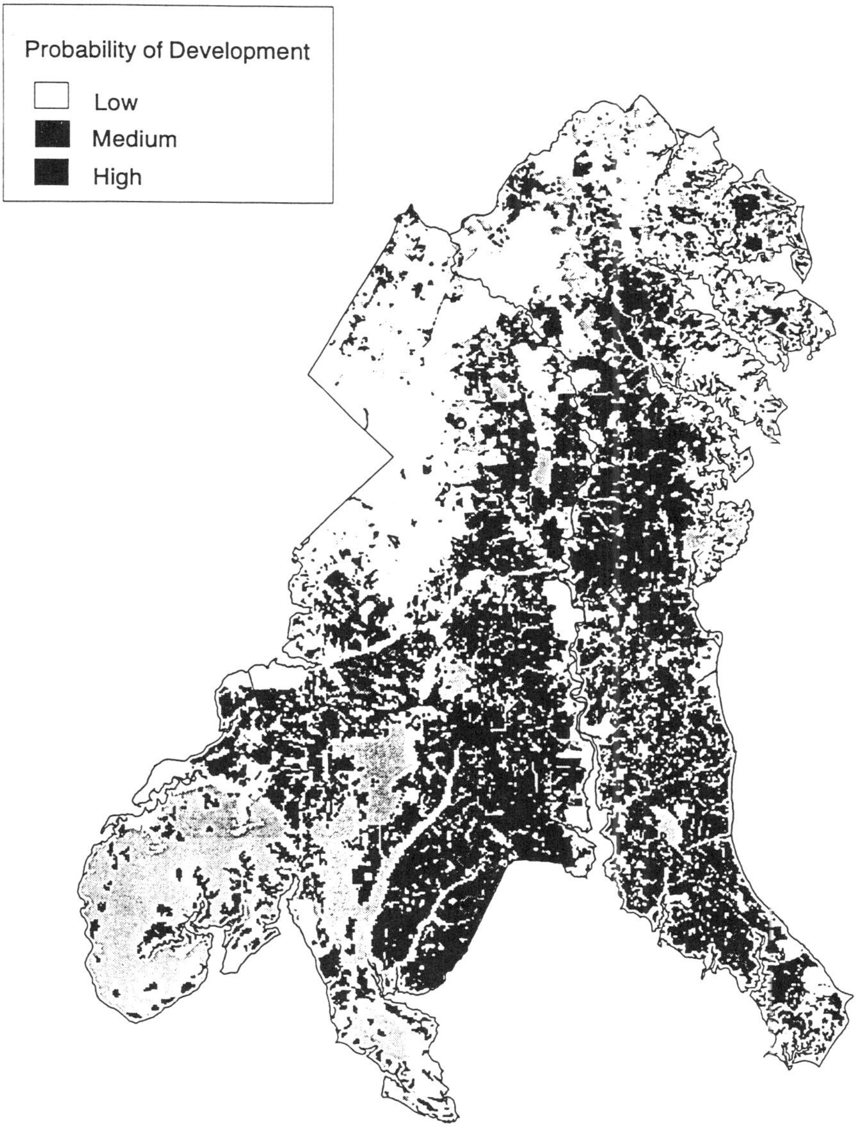

Figure 4. Predictions of Development Pressure Under Current Zoning Patterns

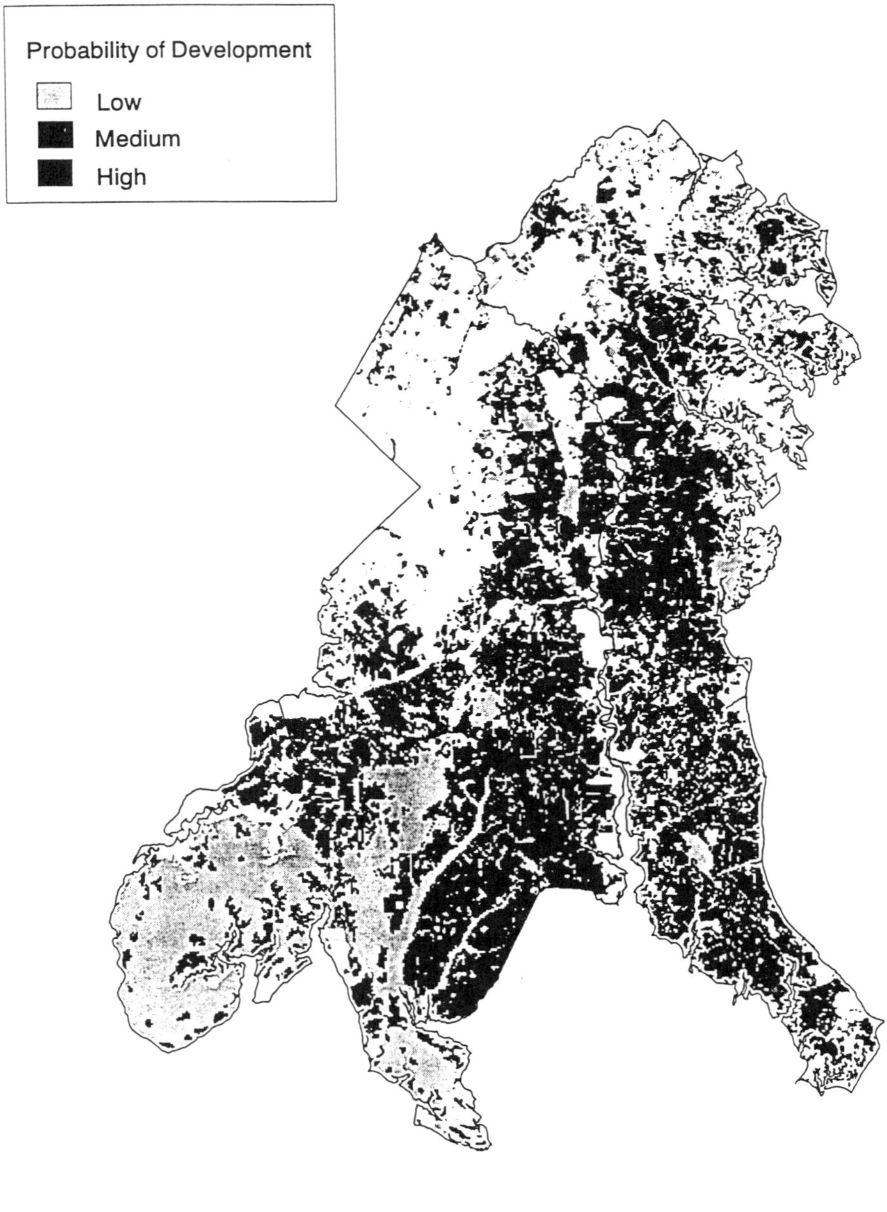

Figure 5. Predictions of Development Pressure When All Rural Areas Are Zoned for 20 Acre Minimum Lot Size

Figure 6. Predictions of Development Pressure When All Rural Areas Are Zoned for 3 Acre Minimum Lot Size

Maryland Office of Planning and the Maryland Department of the Environment. These rules of thumb provide predictions of nitrogen loadings to the Bay and its tributaries contributed by each acre in the watershed as a function of land use and soil type. Land use categories include different densities of residential use, with or without septic fields, other developed uses, agricultural cropland, pasture, forest, and brush. Since we know the predominant use of each of our cells at the current time and can use our model to predict subsequent use, we can apply these rules of thumb to our data.

Our predictions of net changes in nitrogen loadings reflect the expected value of the outcomes, given our probabilities of conversion, normalized so that under each scenario the same number of new dwelling units are built in the area. Despite the constraint that holds total number of dwelling units constant, we see some change in nitrogen loadings over scenarios. This is because the density of development, the spatial arrangement of development vis à vis public services, and the amount of agricultural land versus forest lost all affect nitrogen loadings. Note that where agricultural land is developed, a major source of nutrient loadings is being eliminated, although a new one is being created that will be more or less harmful depending on density and dependence on septic systems.

Each scenario produces an increase in the predicted nitrogen loadings into the hydrological system, but the percent increases range from 4 percent to 12 percent depending on the zoning scenario—for the same number of new housing units. More striking is the comparison of proportion of the loadings into the Patuxent Watershed relative to other watersheds. Under the current zoning pattern, we predict that 2 percent of the additional loadings generated by development will show up in the Patuxent Tributary. However, under scenario 2, where all four counties are assumed to zone their rural areas similar to Anne Arundel's current zoning, 15 percent of the additional loadings appear in the Patuxent. These two scenarios generate predictions of increases in Patuxent nitrogen loadings that vary from less than 1 percent over current levels to more than 11 percent over current levels.

5. CONCLUSIONS

The Chesapeake Bay Agreement is in some ways a model of interjurisdictional cooperation towards a common goal—the improvement of the water quality of the Chesapeake Bay and its major tributaries. Federal, state, and local governments are involved in Tributary Strategies Initiatives that seek to establish policy on a tributary-by-tributary basis that will affect the flow of nutrients to this hydrological system. The agencies involved in this endeavor are becoming increasingly aware of the impact that land use and land use pattern has on this problem, and are now recognizing the need to monitor land use change and assess the effects local government growth control policies have on the process of land conversion.

Our results suggest that at least some of these measures have an effect. The results also suggest that the resulting spatial pattern may be sensitive to the differential use of these growth control tools across counties because these counties share a common land market. Unfortunately the boundaries of the land market exceed those of any one county, and neither is aligned with the boundaries of the tributary watersheds, upon which water quality management is being based. Small changes in county growth control policies can deflect land use change and, as a consequence, deflect additional nutrient loadings from one watershed to another, making accounting for progress towards the overarching goal of the Chesapeake Bay Agreement a confusing task.

REFERENCES

Anselin, L. 1988. *Spatial Econometrics: Methods and Models*. Dordrecht: Kluwer Academic.

Anselin, L., and R.J. Florax, eds. 1995. *New Directions in Spatial Econometrics*. New York: Springer Verlag.

Bell, K., and N. Bockstael. 1997. Applying the Generalized Method of Moments Approach to Spatial Problems Involving Micro-Level Data. Working Paper 97-03, Department of Agricultural and Resource Economics, University of Maryland, College Park.

Cliff, A., and J.K. Ord. 1973. *Spatial Autocorrelation*. London: Pion.

Kelejian, H., and I.R. Prucha. 1996. A Generalized Moments Estimator for the Autoregressive Parameter in a Spatial Model. Working Paper, Department of Economics, University of Maryland, College Park.

Kelejian, H., and D. Robinson. 1995. Spatial Correlation: A Suggested Alternative to the Autoregressive Model. In L. Anselin and R.J.G.M. Florax, eds., *New Directions in Spatial Econometrics*. New York: Springer Verlag.

Maryland Office of Planning. 1996. Demographic and Socio-Economic Information. Planning Data Services.

Patuxent River Tributary Team. 1996. *Annual Report 1995-1996*. Baltimore: Maryland State Office of Planning.

Whittle, P. 1954. On Stationary Processes in the Plane, *Biometrica* 41: 434-449.

11 ESTIMATING ECONOMIC AND ENVIRONMENTAL BENEFITS OF WATER MARKETS IN A SPATIALLY DIVERSE SETTING

Susan M. Burke
Richard M. Adams
Oregon State University, USA
Andrew Draper
University of California, Davis, USA

This chapter develops a conceptual framework to measure the economic consequences of incorporating geopolitical boundaries, spatial characteristics, and water markets in the allocation of water in a trans-boundary setting. The economic model is developed in conjunction with Geographic Information System (GIS) data to assimilate the effects of spatial data on the economic analysis. The empirical focus is on the Upper Klamath Basin which straddles the California and Oregon border and hosts a variety of state and county governmental entities, as well as several Indian Nations, all with jurisdiction, and competing demands for water.

1. INTRODUCTION

In contrast to command and control allocation decisions, the adoption of a market mechanism to reallocate water, which includes the spatially heterogeneous socio-/hydro-/geo-logical nature of a water basin, has the potential to increase both stream flow and economic production. By framing the analysis in the spatial attributes of the water basin, not only is the quantification of basin-wide changes in economic product and water flow more reliable but also the basin-wide quantification can be disaggregated into political units to aid in the understanding of the political economy.

The empirical focus is on the Upper Klamath River Basin which straddles the California and Oregon border and hosts a variety of governmental entities, with overlapping jurisdictions, and competing demands for water. The primary use of water in the basin is for irrigated agriculture and environmental amenities such as several national wildlife refuges. The basin has complex hydrologic systems and varying soil types. This heterogeneity of natural characteristics implies that a spatial model is required to adequately estimate the economic and environmental effects of changes to water allocations. The trans-boundary and spatial characteristics of the basin serve to demonstrate the potential policy importance of the proposed conceptual framework.

This chapter is divided into the following four sections. The next section describes the institutions and legal arrangements common in trans-boundary issues. A brief history and description of water rights applicable to the study area is also covered in this section. The second section of the chapter focuses on the physical and hydrologic characteristics of the Upper Klamath River Basin together with the various and competing demands for water. The theoretical spatial model is described in the third section. The conclusion follows in Section four.

2. OVERVIEW OF TRANS-BOUNDARY ISSUES AND LEGAL INSTITUTIONS GOVERNING WATER RIGHTS

The use of treaties in the resolution of conflicts regarding the allocation of water resources across borders is a common practice. These treaties can be categorized into four theories: (1) absolute territorial sovereignty; (2) limited territorial sovereignty; (3) absolute territorial integrity; and (4) community (Buck et al. 1993). The theory of absolute territorial sovereignty allows for an upstream state to do as it wishes with waters flowing within its boundaries, with no regard to downstream states. Limited territorial sovereignty allows for each state to make use of the water flowing within its borders but prohibits the interference of either state in the reasonable use of water by the other. Absolute territorial integrity prohibits the restriction of water flow by upstream states to downstream states. Finally, community provides that the waters of a basin should be developed as if the basin were one political unit, with the benefits and costs of development shared among the states.

Treaty practice has shown that political entities favor the theory of limited territorial sovereignty to resolve conflicts. This is due in part to the fact that the element of compromise inherent in this form of conflict resolution strengthens the negotiating positions of each entity in other conflict resolutions, and helps define a working relationship between the entities. The adoption of limited territorial sovereignty to resolve international trans-boundary issues is exemplified in treaties between Mexico and the U.S., Brazil and Uruguay, and Sweden and Norway. The evolution of the theory as a method for resolving trans-boundary water management has its roots in the doctrines of equitable apportionment and equitable utilization. The distinction between the two is pointed out by Buck et al. (1993, p. 600).

> *Equitable apportionment suggests allocating a quantity of water, a more precise description of actual practice is allocating the right to use a quantity of water. This distinction is made because of the physical reality that it is difficult to own a unit of water within the cycle. Thus, international legal thought has rephrased equitable apportionment as equitable utilization.*

2.1 Interstate Trans-Boundary Treaties in the United States

The U.S. Supreme Court has adopted the theory of limited territorial sovereignty to resolve water disputes between individual states. Based on the doctrine of equitable utilization, three jurisdictional approaches have been employed to allocate interstate water (Johnson and DuMars 1989). The first of these approaches is made by a judge or court-appointed special water master. An example of this method of apportionment is the court-ordered decisions of *Nebraska v. Wyoming* and *Wyoming v. Colorado*, which apportioned the North Platte River and the Laramie River, respectively. The second method is so-called "congressional apportionment." This occurs when Congress authorizes the Secretary of the Interior to apportion water between states. An example of this method is the Boulder Canyon Project Act which apportioned water between the states of the Lower Colorado River Basin. The last method is by interstate compact. The study area of this chapter is ruled by an interstate compact. Under the method of interstate compacts, three variations have emerged in the United States: (1) binding with Congressional consent; (2) binding without Congressional consent; and (3) non-binding (Buck et al. 1993). The first occurs when the states have asked and received congressional assent to their arrangements. (Generally, if an interstate compact has received congressional approval it also receives federal funding.) The second arrangement is a compact among the states but lacks federal sanction. The third and last case has neither official state or federal sanction. The appropriation of water in the study area of the Upper Klamath Basin falls under the first of these categories, the federally recognized binding compact.

2.2 Western U.S. Water Law—The Doctrine of Prior Appropriation

Legal institutions that govern water use attempt to reflect social goals, and facilitate prudent, equitable, and efficient use of water resources. The legal institution that has been adopted in California and Oregon is the doctrine of prior appropriation.[1] Prior appropriation is a usufructuary right, a 'use' right, and is protected as a property right. Three attributes characterize the doctrine of prior appropriation: (1) diversion of water from its natural source, (2) the requirement of beneficial use, and (3) seniority. The federal government has recognized this doctrine and employed it in the development and implementation of irrigation works authorized by the Reclamation Act of 1902.

[1] California has a plural system of water rights which include riparian, appropriative, pueblo, prescriptive, and groundwater. For an exposition see the California Water Plan, Volume One (California Department of Water Resources 1992). The focus of this research is only on appropriative rights.

2.3 The Era of Reallocation

The appropriation doctrine, while securely established as the water law employed by California and Oregon, allows for modification and reappropriation of existing allocations. These modifications allow the doctrine to address the needs of new uses of water and the over allocation of water within a basin. Specifically, two criteria are used to determine a new prioritization of existing water rights in what has been called by Shupe et al. (1989) the 'era of reallocation'. These criteria are (1) federal reserved rights and (2) public trust doctrine. Furthermore, responding to the provisions of the Endangered Species Act (ESA) has modified the existing allocation hierarchy of water resources. Examples of each of these three methods of reappropriation follow.

One of the significant examples of reprioritization under the public trust doctrine is the recognition and development of federal reserved water rights. In 1980, in *Winters v. United States*, the Supreme Court was asked to resolve a dispute between Montana irrigators who used Milk River water and Native Americans on the Fort Belknap Indian Reservation. Native Americans had not acquired water rights under state law. However, the Court held that when Congress set aside land for the reservation it implied reserving sufficient water to carry out the purposes of the reservation. The result was to carve out an exception from the general rule that western water, even in the public domain, was fully available for appropriation under state law and that appropriators who held vested rights under state law held secure rights against all subsequent appropriators. The principle of this case became known as the reservation doctrine or the federal reserved right. The magnitude of this reserved right became clearer in the 1963 case, *Arizona v. California*, when a significant portion of the flow of the Colorado River was awarded to five Indian tribes. The implications of the reservation doctrine are not quantified in many cases because many of the Indian-reserved rights are not quantified. Most of these Indian rights predate other appropriations in the West. It is unclear how many appropriative water rights may be affected when all reserved water rights are quantified. The number will probably be large (Johnson and DuMars 1989).

The public trust doctrine is used by state officials in order to assure the protection of the public interest through the protection of fish, wildlife, and recreational resources, as well as resources which are deemed to possess aesthetic value (Walston 1989). An example of the application of the public trust doctrine is the 1983 California Supreme Court ruling in which the city of Los Angeles' rights to take water from Mono Lake were subordinated to the public interest in preserving Mono Lake (*National Audubon Society v. Superior Court of Alpine County)*. The Court held that water right licenses held by the City of Los Angeles to divert water from stream tributaries to Mono Lake are subject to state supervision under the public trust doctrine. This case was a challenge to other senior rights holders, by indicating that even presumed senior rights are no longer secure. Several western states have also established laws under the public trust doctrine that allow individuals to apply for water for stream use, or the maintenance of a minimum stream flow.

An example of reprioritization and reallocation of water under the Endangered Species Act (ESA) occurred in 1984 in the *Carson-Truckee Water Conservancy District v. Clark*, when the Secretary of the Interior, empowered through the ESA, appropriated water from a Bureau of Reclamation project for use in conserving an endangered species, rather than the municipal and industrial use for which the project had been built. As MacDonnell (1989, p. 406) puts it, "Whether these claims are characterized as property rights or just as regulation, the clear effect is to cause possible modification in the manner of exercise of state-created water rights."[2]

The allocation of water rights under the appropriative doctrine has lead to the current situation in the West where many water basins are fully allocated. Given the political, environmental, and economical costs of developing additional storage facilities, supply augmentation is not likely to be the answer to increasing demands. Increasingly, water users that need additional water are looking to existing water rights as a means for augmenting their supplies. Shupe et al. (1989) argue that we are reaching the end of the era of allocation and beginning a new era where water rights will be transferred under existing water law.

Regardless of the form that transfers take, they can be encouraged through the introduction of a market for water. Markets are becoming more common as a least cost alternative to supply augmentation. During 1991/92, California initiated an Emergency Drought Water Bank that acted as a clearing house for nearly 800,000 acre feet of water. Recent federal legislation regarding the operation of the Bureau of Reclamation's Central Valley Project in California called for easing regulations under which transfers can occur. Transfers are also occurring in an attempt to mitigate environmental and economic losses accruing as a result of a shortage of water supply. Where the transfer could be forced on the property right holder, the courts have indicated a desire to have institutions that specialize in water allocation issues negotiate settlements.

In summary, the law of prior appropriation does provide for modification of existing allocations. As the legal institutions that govern water use respond to changing social goals, and facilitate prudent, equitable, and efficient use of water resources, transfers of water rights will be required.

3. A SPATIAL MODEL

There is a well-developed literature describing potential increases in allocative efficiency to be gained by the introduction of a market for water (Burness and Quirk 1980; Hartman and Seastone 1970). Empirical estimates of gains from trade have been made for both intrastate and interstate markets (Vaux and Howitt 1984; Booker and Young 1994). Following the development of this literature, economists began considering the environmental improvements which would occur as a by-product of water markets, both through improvements in water quality (Weinberg et al. 1993; Dinar and Letey 1991) and increases in stream flows (Griffin and Hsu

[2] For a discussion of these issues see Tarlock 1985.

1993; Colby 1990). Dinar and Letey (1991) postulate that besides increasing allocative efficiency between urban and agricultural uses (the farmer's profits are increased by the sale of water and the urban sector gains by increases in available water), the environment is made better off because farmers, who now have a market to sell saved or surplus water, choose to improve irrigation efficiency. This improvement leads to lower volumes of deep percolation which reduces the quantity of agrochemicals introduced into groundwater or drainage water. Griffin and Hsu (1993) postulate that when water has in-stream value, changes in diversions which result from a water market will change the total value of in-stream use. Colby (1990) argues that markets for water allow in-stream users to purchase rights to water from off-stream users, thereby increasing stream flows.

Recently, economists have assessed the role and importance of spatial data in measuring environmental quality changes as a result of changing institutions or policies which affect production decisions (Helfand and House 1995; Fleming and Adams 1997; Bockstael 1996). These spatial analyses can by divided into two types: (1) those concerned with the relative position of modeling units, and (2) those concerned with the physical attributes possessed by the modeling unit.

An example of the first type of analysis, concerned with relative geographical position of water diversions and return flows, is the work by Griffin and Hsu (1993), who show that the value of appropriated water is intimately connected to the point of diversion. A second example of this type of spatial analysis was done by Burness and Quirk (1980) who suggest that total economic product could be maximized by granting senior appropriative rights to upstream users rather than downstream users. This allocation of water allows for the repeated reuse of the upstream user's return flow by downstream users, whereas if the water is appropriated by a downstream user, the upstream producer has less water with which to produce. An example of the second type of analysis is the work of Helfand and House (1995) who considered soil types in their analysis of changes in agricultural run-off under various efficiency control policies. Similarly, the work of Fleming and Adams (1997) considers soil type and geo-hydrology in the design of economically efficient groundwater pollution control policies.

Both of these types of spatial analysis are considered in the development of the economic and environmental impacts of a water market reported here. These spatial dimensions are captured using Geographic Information System (GIS) data. A spatial relation analysis will be performed to eliminate third party effects and quantify changes in stream flows, while the physical attribute analysis will help quantify water market participation. Each of these uses of spatial information is discussed below in turn.

3.1 The Importance of Including Relative Locational Positions in the Spatial Analysis

One concern about water markets is that their implementation may cause third party effects when return flow requirements are not considered (Gould 1989). Because water delivered to a farm which is not lost through evapotranspiration (ET) results in return flows, these return flows must be considered a source of supply to a third party. Transferring water rights may reduce return flows, which could impact a third party's water supply. The effect of diversions and return flow points can be seen by examining Figure 1.

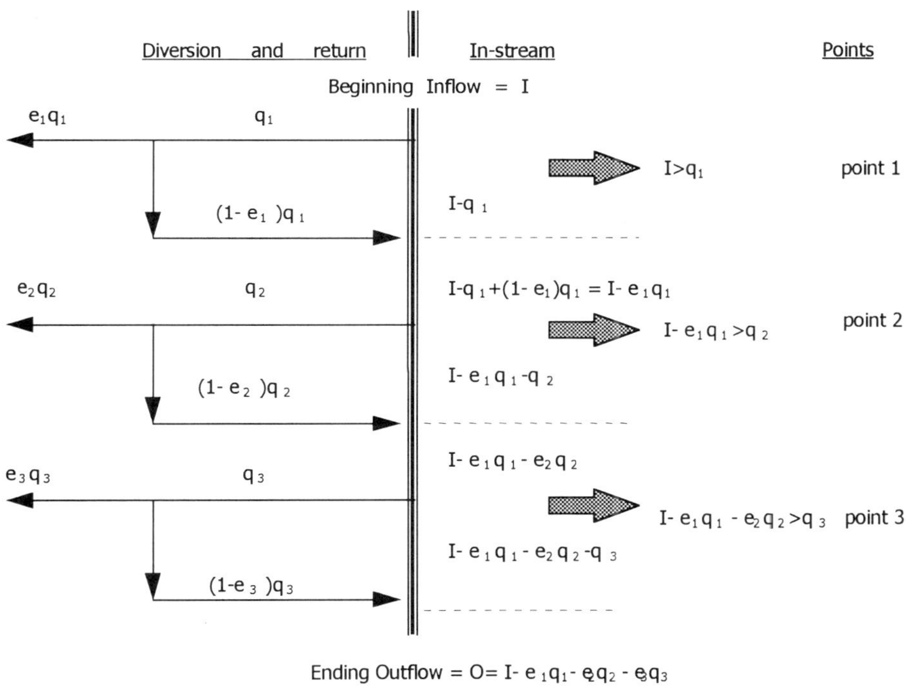

Figure 1. The Effect of Diversions and Return Flow Points on Downstream Supply and Stream Flows

In this figure the diversion and return flows are represented to the left of the center line, the stream flows are represented to the right of the center line, and the stream flow requirements implied by the diversions are listed to the right of the arrows at diversion points 1, 2, and 3, respectively. Also, e_i ($i = 1,2,3$) is the ET rate, $0 < e_i < 1$; I is the beginning inflow, and O is the remaining outflow. As can be seen in the figure, the stream flow after the first diversion point is $I - q_1$. The stream flow after the return flow of the first diversion is $I - e_1q_1$. The implication at diversion point 2 is $I - e_1q_1 > e_2q_2$. If this were not true then the stream flow would be negative, a physical impossibility.

The schematic diagram in Figure 1 conveys the third party effect which could result from the sale of q_1 downstream. Suppose that, rather than divert water for production at point 1, the quantity q_1 is sold downstream of diversion point 2. The stream flow at diversion point 2 is now, $I - q_1$, versus $I - e_1q_1$, before the sale. Since $I - e_1q_1 < I - q_1$, we are no longer assured that the stream flow at point 2 is sufficient to meet diversion. If not, then there would be a negative third party impact imposed on the second water diverter by the sale. One way to limit the occurrence of this third party effect is to restrict the seller of water to transfer only the quantity of water the seller has historically consumed, in this example e_1q_1. Thus, a third party effect will not occur.

There is, in fact, a positive externality from restricting trade to the unit of water historically consumed. This externality is exemplified by examining Figure 2, which demonstrates the effect of diversions and return flow points on stream flow/supply when the unit of water in a water market transaction is the historically consumed unit. In this figure, the same definitions apply as in Figure 1. In addition, the unit of water available for sale, s, is the consumed unit and is calculated as follows:

$$s = (a_i - q_i)e_i \tag{1}$$

where s is the unit of water traded and a_i is the allocation of water to agent i.

Now, examine the quantity of stream flow at diversion point 2 that is required to maintain a positive flow value. Here we see that the implication is $I - e_1q_1 + s > q_2$. There can be no third party impact because from Figure 1 the quantity of water which is required at point 2 to avoid a negative flow is $I - e_1q_1 > q_2$. By limiting the unit available for sale to the historically consumed unit of water, third party impacts are avoided.

To understand the positive externality this creates, compare the ending outflow (O) in Figure 1 to that in Figure 2. The ending stream flow from Figure 1 is

$$O = I - e_1q_1 - e_2q_2 - e_3q_3 . \tag{2}$$

The ending stream flow from Figure 2 is

$$O = I - e_1q_1 - e_2q_2 - e_3q_3 + s\,(1 - e_3) . \tag{3}$$

The difference between the two, an addition in stream flow, results as a consequence of restricting the quantity traded in the water market to the seller's historically consumed quantity. Generally, the quantity is

$$\Delta O = s\,(1 - e_j) \tag{4}$$

where e_j is the ET rate of the buyer of water.

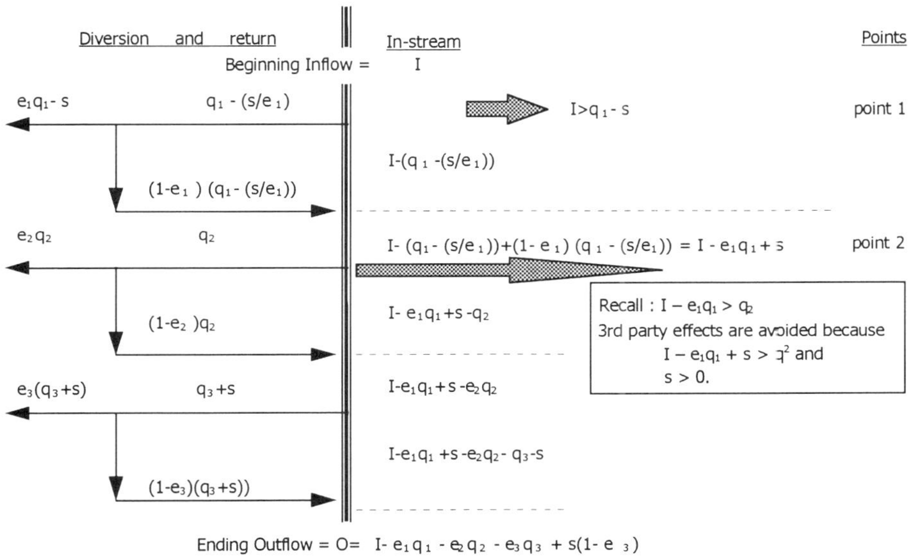

Figure 2. The Effect of Diversions and Return Flow Points to Downstream Supply and Stream Flow in a Market for Water, When the Traded Quantity Is the Historically Consumed Unit

Recall that the definition of a unit of water available for sale was made to avoid third party impacts, not to increase stream flows. Therefore, this increase in stream flow is a positive externality to users of water below the basin. Intuitively, this reduction is due to the fact that the total water available for consumption has been reduced.

In order to determine the gain to water users below the basin, the increase in stream flow must be quantified. This is accomplished by substituting equation (1) into equation (4) to reduce the change in stream flow into an expression of allocations (a_i), ET rates (e_i), and water employed in production (q_i),

$$\Delta O = (a_i - q_i)e_i(1 - e_j) \ \forall i, j; i \neq j \tag{5}$$

where subscript i represents the seller of water and subscript j represents the buyer of water.

The change in stream flow in (5) can be calculated at the optimal quantities of trade that result from solving the water market model.

3.2 The Importance of Including Physical Attributes in the Spatial Analysis

Initiating a water market between users of water in an economically and spatially homogenous setting—for example, among agricultural users within a basin—may have limited gains from trade (Zilberman et al. 1994). Gains from trade exist where there is variation in the marginal value product of water. This variation is more likely to occur in an area where there is a high degree of heterogeneity. Therefore, in order to understand the water market's impact in a basin, it is necessary to understand the degree of heterogeneity. To facilitate that understanding, GIS data is needed to identify (1) existing water use patterns in the study region, and (2) areas/farms which have a similar marginal utility of water.

Water use patterns are affected by many factors. Consumptive demand varies according to the crop and variety, planting date, planted area, and the length of the growing season. Sprinkler irrigation may be required as frost protection for some crops. Variation in ET, elevation, water table elevation, and precipitation are also included to better understand the use patterns, as are losses to deep percolation. These deep percolation losses can be affected by management practices, application methods, land topography, and soil characteristics (infiltration rate, hydraulic conductivity, and water holding capacity).

The marginal value of water is a function of crop yield, crop price, and variable input costs. These may have a spatial pattern due to differences in micro-climates, soil suitability, the presence of diseases and pests in the soil, and farm size distribution.

The goal of the GIS analysis is to determine homogenous modeling units within a basin that has a high degree of heterogeneity. These homogeneous units are delineated by overlaying GIS coverages of factors which influence water use, choice of crop, and the physical ability to trade water. The modeling unit thus defines the 'trading agent' employed by the economic model. Since the region of study is an agricultural region, the 'trading agent' is subsequently referred to as a farmer.

3.3 Economic Analysis

Using a nonlinear programming technique employing positive mathematical programming following Howitt (1995), the economic model is written to maximize farm profit from production and the trading of water.

The objective function for farmer i is

$$\max_{s,b} \Pi = \sum_{c=1}^{m} p_c f_c(q_c, l_c, e) - p_{ec} l_c - \bar{p}_w a + p_w (s - b) \tag{6}$$

subject to

$$\sum_{c=1}^{m} q_c \leq a - \left(\frac{s}{e}\right) + b \,, \tag{7}$$

$$sb = 0\,, \tag{8}$$

$$q_c, e, l_c, s, b \geq 0\,,$$

$$p_w \geq \overline{p}_w\,,$$

where

p_c = price of crop c, c=1,.....,m,
$f_c(q_c, l_c, e)$ = production function for crop c,
$\partial f_c/\partial x \geq 0$, $\partial^2 f_c/\partial x\partial y \leq 0$, x,y = q_c, l_c, e,
q_c = water ET used in production of crop c,
l_c = number of acres in production of crop c,
e = farmer's overall irrigation efficiency,
p_{ec} = per acre price of other factors of production of crop c, e.g., labor, fertilizer, capital, etc.
$\overline{p}_w$ = fixed price of farmer's appropriated quantity of water right,
p_w = market price of water,
a = farmer's initial allocation of water,
s = quantity of water sold, and
b = quantity of water bought.

Constraint (7) limits the water available for production to the remainder of the appropriated quantity less what a farmer sells on the market, multiplied by the ET rate, plus the water bought. It is derived from equation (1), the quantity of water available for sale from farmer i, rewritten as

$$q_i = a_i - \frac{s}{e_i}\,. \tag{1a}$$

Equation (1a) does not include b_i, the quantity of water purchased by farmer i. Since constraint (8) eliminates the possibility that a farmer would be both a seller and a buyer, constraint (7) can be generated from (1a) by adding b_i,

$$\sum_c q_c = a - \frac{s}{e} + b\,.$$

(Note the i subscripts have been dropped because the maximization problem represents one farmer, and the quantity of water employed in production is summed over all crops.) The first order conditions for a maximum are

$$p_c \frac{\partial f_c}{\partial q_c} \frac{\partial q_c}{\partial b} - p_w = 0 , \tag{9}$$

where $s = 0$ and $b > 0$, e.g., the farmer is a buyer of water, and

$$p_c \frac{\partial f_c}{\partial q_c} \frac{\partial q_c}{\partial s} + p_w = 0 , \tag{10}$$

where $s > 0$ and $b = 0$, e.g., the farmer is a seller of water. To form the equal marginal relationships from the first order conditions, notice that, for $b > 0$,

$$\frac{\partial q_c}{\partial b} = 1 , \tag{11}$$

and for $s > 0$,

$$\frac{\partial q_c}{\partial s} = -\frac{1}{e} . \tag{12}$$

Substitute (11) into (9) and substitute (12) into (10), and rearrange both (9) and (10) to form

$$p_c \frac{\partial f_c}{\partial q_c} = e p_w \tag{9a}$$

$$p_c \frac{\partial f_c}{\partial q_c} = p_w \tag{10a}$$

Equations (9a) and (10a) are the expected result that the value of the marginal product equals the price of water. Notice that the price of water of the seller reflects the adjustment of the market price of water by ET. This adjustment reflects the fact that only this 'consumed' quantity is available for sale under the conditions of the water market.

The objective function for the water market maximizes an additive social welfare function, summed over the farmer's profit functions,[3]

$$\max_{s_i, b_i} \Pi = \sum_{i=1}^{n} \sum_{c=1}^{m} p_c f(q_{ic}, e_i, l_{ic}) - p_e l_{ic} - \overline{p}_w a_i + p_w s_i , \tag{13}$$

[3] This objective function represents an improvement in social welfare, not a maximization of social welfare. The later function would require the valuation of inflows and outflows, which is currently beyond the scope of this research.

subject to

$$\sum_{i=1}^{n}\sum_{c=1}^{m} q_{ic} \leq a_i - \left(\frac{s_i}{e_i}\right) + b_i\,, \tag{14}$$

$$s_i b_i = 0 \;\; \forall i\,, \tag{15}$$

$$\sum_i s_i - \sum_i b_i = 0\,, \tag{16}$$

$$\sum_i a_i \leq I\,, \tag{17}$$

$$\sum_i \sum_j \left[(a_i - q_i)e_i(1 - e_j)\right] = \Delta O\,, \tag{18}$$

$$p_W \geq \overline{p}_W\,,$$

$$q_{ic}, e_i, l_{ic} \geq 0\,.$$

This social welfare function contains all the constraints and variables described in individual farmers' profit functions (6) plus constraints (16), (17), and (18). The first of these constraints, (16), is the market-clearing constraint which states that the quantity of water sold is equal to the quantity of water purchased. The next two constraints, (17) and (18), are the constraints placed on water inflow and outflow, respectively.

The solution to this maximization problem yields the increases in economic product resulting from a water market and the quantities of water sold and bought. However, high transactions costs could limit the participation in a water market. As Colby (1990, p. 1116) points out:

> *While prospects for increasing the economic returns of water use are the driving force behind markets, laws and policies affect the cost of market transactions and the attractiveness of market transfers relative to other means of transferring water. The legal and political setting determines the transactions costs associated with market transfers. Transactions costs are incurred in identifying legal and hydrologic characteristics of water rights (priority date, return flow obligations, etc.); in negotiating price, financing, and other terms of transfer; and in satisfying state laws and transfer approval procedures.*

By constraining the changes in stream flows using equation (18), we are accounting for benefits that would accrue to users within and below the basin. The out-of-basin benefit of this increased stream flow should be used to offset the transaction costs of implementing the water market. Another way to view these out-of-basin benefits would be to remove constraint (18) and simply use the results of the social welfare maximization problem to calculate the increases in stream flow [the solution to equation (5) at the optimal choice of s]. The increases in stream flow, which occur as a consequence of a water market, provide an incentive for users of water downstream from the basin to encourage the adoption of a water market for users within the basin. The accuracy of the estimates of increases in stream flow will be important to policy makers in assessing the advantages of implementing a water market.

The GIS framework allows the results of the economic analysis to be aggregated to different levels. Given the nature of the study area, which has multiple jurisdictions, the modeling unit is defined not only by hydrologic and geologic dissimilarities, but also by political boundaries. Employing GIS allows for the development of homogeneous modeling units that disaggregate the data to the level of individual choice, while allowing for the aggregation of data up to meaningful political elements for use in trans-boundary negotiations.

Another way GIS is exploited is to use the definitions of modeling units to place restrictions on trading in the market. Trades within a market may follow a continuum from most to least restrictive. For example, the most restrictive model may prohibit farmers from trading outside an irrigation district, or outside a county or state. The least restrictive model imposes no restrictions. The use of a range of trading restrictions follows the example of Booker and Young (1994), who modeled six different institutional scenarios for trading water on the Colorado River.

4. EMPIRICAL SETTING—UPPER KLAMATH BASIN

This section of the chapter describes the geographical, environmental, and hydrologic settings of the study area. The GIS data are used to demonstrate some of the heterogeneous attributes of the Basin. In addition to these physical attributes, this section also describes the current economic and political issues which are germane to this research.

The Upper Klamath River Basin (the Upper Basin) is situated on the California-Oregon border to the east of the Cascade mountains (see Figure 3). It covers approximately 5,155,000 acres. The Upper Basin is wholly contained within Klamath County in Oregon and the counties of Modoc and Siskiyou in California.

4.1 The Environmental Setting

The Upper Basin is home to a national park, a national monument, two national forests, and six wildlife refuges. The wetlands of the Upper Basin are a pinch point of the Pacific Flyway. Therefore they are essential for migratory waterfowl. The area is home to the largest wintering population of bald eagles in the lower forty-eight states. In addition to the natural resources within the Upper Basin, the

Klamath River itself is one of the Pacific Coast's most important salmon and steelhead trout rivers.

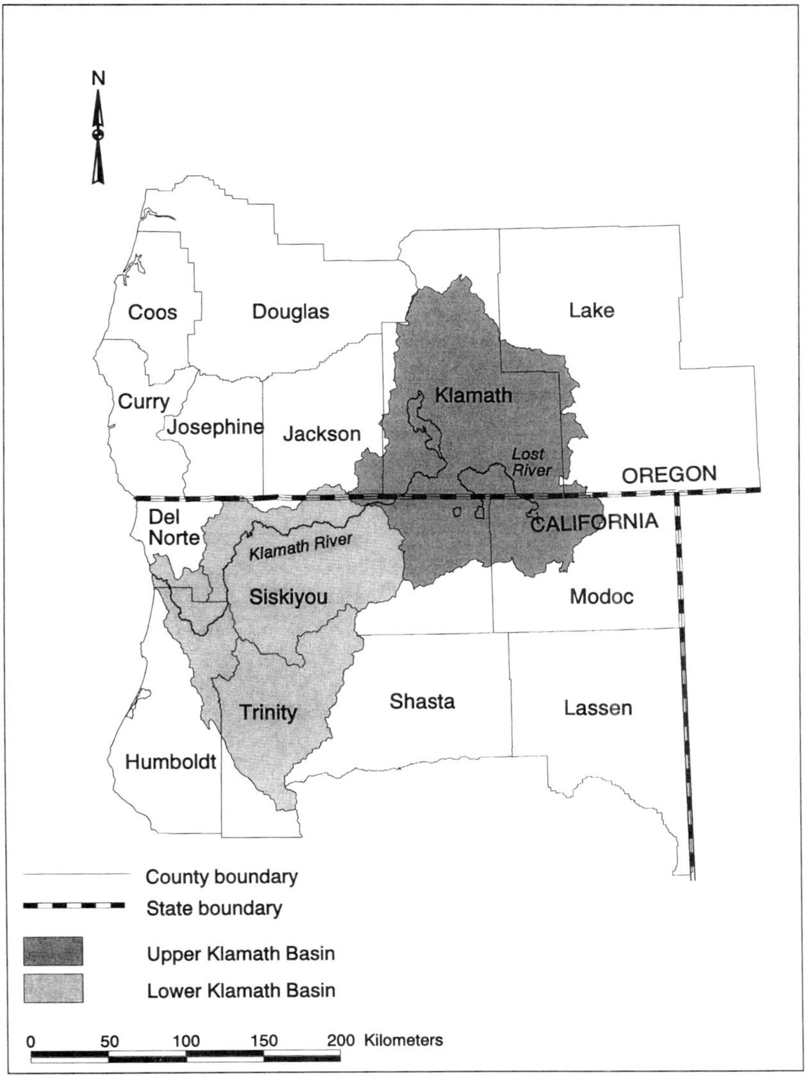

Figure 3. The Klamath Basin

4.2 Hydrology and Topology

The Upper Basin lies in the rain shadow of the Cascade mountains to the east. Precipitation is highly variable within the region. The elevation of land in the Upper Basin ranges from 700 to 3000 meters above sea level. Upper Klamath

Lake, situated near the center of the western edge of the Upper Basin, is the Basin's dominant hydrologic feature (see Figure 4). Upper Klamath Lake is approximately 90,000 acres in size with an average depth of eight feet. Two unregulated river catchments discharge into the lake: the Wood and Williamson. The outflow from Upper Klamath Lake into the head reach of the Klamath River is regulated by a dam. Historically, during periods of high runoff, this reach of the Klamath River overflowed its banks to spill into an area of marsh which included the present Lower Klamath Lake.

The Lost River system in the southeast of the Upper Basin, consisting of Clear Lake, Tulelake, and the Lost River, forms a naturally closed, internally draining basin. The Lost River originates at the outlet of Clear Lake Dam in California and flows north across the state border into Oregon. The river receives inflow from Gerber Reservoir via Miller Creek and from springs. Eventually, the river turns south and discharges into Tulelake back in California. Irrigation and drainage channels constructed in 1912 and 1941 now provide a regulated link between the Lost and Klamath Rivers. To remove excess inflow, water is pumped uphill from Tulelake to Lower Klamath Lake and then to the Klamath River.

A Southern Pacific railroad embankment across the north end of Lower Klamath Lake prevents the natural flow of water from the Klamath River, at high stage, to Lower Klamath Lake. Excess water is drained from the Lower Klamath Lake via a drain which discharges into the Klamath River.

4.3 The Agricultural Setting

River valleys and reclaimed lake beds form the main agricultural portion of the Upper Basin. They are classified as semi-arid desert. Irrigation is required for crop production. Average water use for fully watered crops grown in the area ranges from 24–36 inches per year. The length of the growing season and the susceptibility to frost are important determinants of the cropping system. The growing season varies depending on elevation and latitude from 120 to only 50 days. In the north of the basin, where the growing season is short, alfalfa, oats, hay, and pasture are the main crops. Further south, where the growing season is longer, a more diverse cropping pattern includes potatoes, sugar beets, wheat, onions, and barley.

Figure 5 shows the soil rating for agriculture. The lands immediately north and southeast of Upper Klamath Lake have the highest soil rating (derived from the State Soil Geographic [STATSGO] database, published by the U.S. Department of Agriculture).

4.4 The Klamath Project

The Klamath Project, initiated in 1905, was one of the first federal irrigation projects to be constructed by the U.S. Bureau of Reclamation (USBR). The project is located in the south central Upper Basin (see Figure 4 for location of the project). The project includes 234,000 acres of land. Water for the project is stored in Upper Klamath Lake, Clear Lake, and Gerber Reservoir.

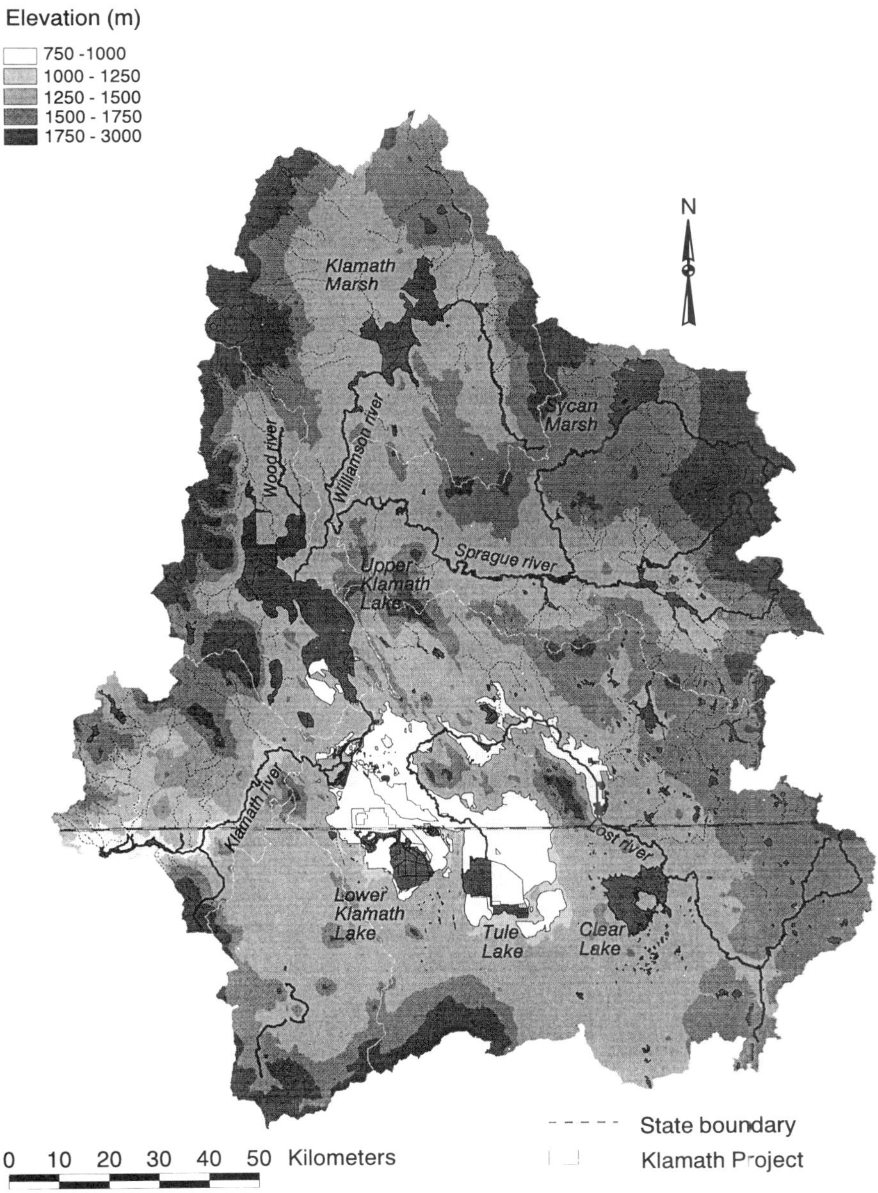

Figure 4. Topographic/Hydrologic Setting of the Klamath Project in the Upper Klamath Basin

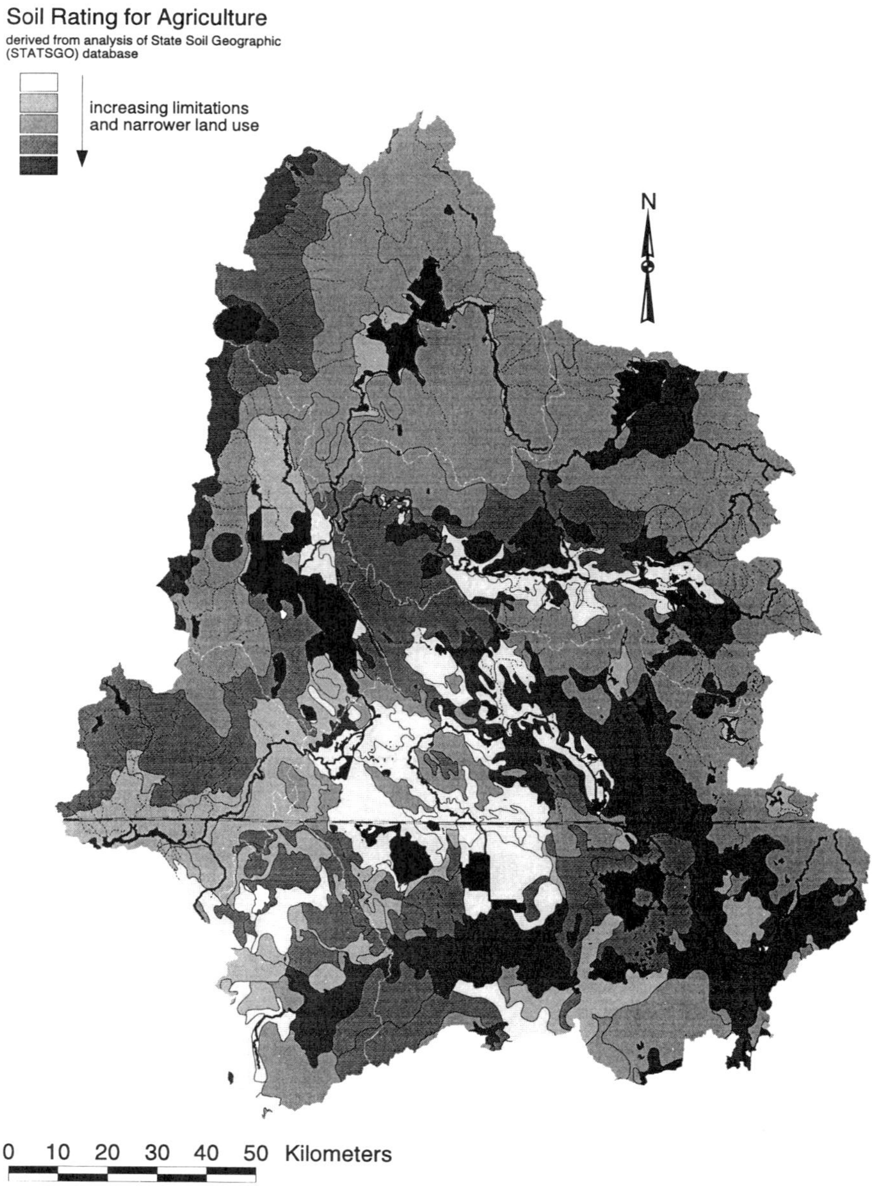

Figure 5. Agricultural Capability Class

4.5 Agricultural and Wildlife Issues

Historically, Upper Klamath Lake, its tributaries, and the Lost River have been a major habitat for a species of sucker fish. However, during this century native fish populations in the Upper Basin have declined dramatically. Following fish surveys in 1984 and 1985, two species of sucker fish were listed as endangered under the Endangered Species Act. The restoration of critical habitat area and the protection of water quality within the lake requires the USBR maintain Upper Klamath Lake at predetermined levels on a monthly basis. This constrains the regulation of lake levels for storage and thus reduces available water to the Klamath Project during critical summer months.

Within the Klamath Project the proximity of farming and wildlife is striking. Farm fields border on or are located within the Tulelake National Wildlife Refuge (TLNWR) and the Lower Klamath Lake National Wildlife Refuge (LKLNWR), and water flows directly into the marshes from irrigation drainage channels. To see how close agricultural lands are to wildlife within the Upper Basin, refer to Figure 4, and notice that both the Lower Klamath Lake and Tulelake, both national wildlife refuges, are within the boundaries of the Klamath Project. Within the Klamath Project, the distinction between canal and drain water becomes blurred as conveyance channels carry a mixture of source water and irrigation return flows. There are concerns that agriculture has impaired the water quality of the two lakes.

The Klamath River has historically supported major runs of anadromous fish species. Fish populations have been impacted by reductions in stream flows and changes in the river temperature regime caused by upstream storage. The U.S. National Marine Fisheries Service (NMFS) recently proposed to list the steelhead in the Klamath as threatened. Upper Klamath Lake is the only major reservoir which flows into the Klamath River. Thus, stream flows in Klamath River are dependent on releases out of Upper Klamath Lake.

Situated at the source of the Klamath River, the Klamath Project is currently the target for demands for reductions in water supply. To ensure the continued viability of agriculture in the Klamath Basin, efforts are underway to develop institutional solutions to respond to demands for reductions in surface water supply.

In short, the USBR's Klamath Project lies between Upper Klamath Lake and the river both geographically and politically. The Bureau must balance water level requirements in the Upper Klamath Lake with stream flow requirements in the Klamath River and wildlife requirements to the south. Meeting lake level and river flow targets is complicated by the natural hydrologic variability manifested in the basin. During a recent drought the irrigators, sandwiched between Bureau attempts to meet both the level and flow requirements, were the group that had their supply of water reduced.

As a result of this reduction in irrigation water, the farmers of the Upper Basin brought suit against the U.S. Fish and Wildlife Service for what they said was a misuse of the federal Endangered Species Act (*Bennett v. Spear* 1997). The suit ended with the 1997 Supreme Court decision which granted standing to the irrigators under the Act.

4.6 Summarizing Physical Attributes Using GIS Data

A GIS database is needed to capture the heterogeneous nature of the Upper Basin and the proximity among physical attributes. The heterogeneity of such data are demonstrated in Figures 4 and 5, which represent hydrology, topology, and agricultural capability class of the Upper Basin. These coverages can then be overlayed to define homogeneous units for use in physical and economic modeling. The process of overlaying the coverages can be best described visualizing the information on transparent paper. These coverages are sequentially overlaid one on another, during which time homogeneous units emerge.

4.7 Political Jurisdictions and Trans-Boundary Issues

In addition to the jurisdictions of California and Oregon, an Indian Nation holds treaty rights to hunt and fish on a former reservation which was located in the Upper Basin. Several other Indian Nations have been granted rights to a salmon run in the Klamath River of California, south of the Upper Basin. The federal government's jurisdiction includes the management of the Klamath Project. It also includes management of the national wildlife sanctuaries located in the Basin, as well as management of Bureau of Land Management land, a national monument, and two national forests. Due to difficulties in satisfying all of the jurisdictions represented in the Basin, management of the USBR's irrigation project has recently been delegated to four federal agencies: (1) the USBR, (2) the U.S. Fish and Wildlife Service, (3) the NMFS, and (4) the Indian Affairs Office in the Interior Department. In addition to the organization of the above federal agencies, there is also an interstate compact which has authority over water resource allocation in the upper Basin.

The Klamath River Compact Commission, formed by an act of Congress in 1957, has two stated major goals (Klamath River Compact Commission, 1957–58 and 1960–61):

(1) ... to facilitate and promote the orderly and comprehensive development and use of the waters of the Klamath River for beneficial purposes in both states.
(2) ... to further intergovernmental cooperation in developing programs for, and making good use of, the interstate waters of the Klamath River.

These goals are consistent with the doctrine of equitable utilization described earlier in this chapter. The Compact was ratified by Congress, which provides federal as well as state support of its operations. The staff of the Compact is comprised of a representative from both states and a federally appointed chairperson. The state representatives are employed by their respective state's agency which oversees water resource allocation: the Water Resources Department in Oregon and the Department of Water Resources in California. The federal appointee is a business person local to the Upper Basin.

Since the Compact's inception in 1957 it has been relatively uninvolved with resource allocation decisions in the Klamath, because there were not issues of a serious enough nature to call into use this multilateral institution. As Strand et al. (1992, p. 1154) point out, "Institutions with multilateral control are cumbersome mechanisms for management but the nature of preferences and resource may dictate their use." Because the institution represents the interests of three governmental agencies and must balance the needs of several other state and federal agencies, each with jurisdiction in the Basin, its operations can be cumbersome. In the early 1990s, facing a two-year drought and the listing of the Lost River and Tulelake sucker fish as endangered, the Compact's mechanism was reawakened to deal with water allocation issues. Currently the Compact is attempting to lead a consensus process by which solutions to these pressing problems can be examined and implemented.

Concurrently, the State of Oregon is proceeding with a process which will adjudicate the water rights of the Basin. This process of adjudication is for formalization of perfected water rights. The process documents the time and place of diversion of each right in the Basin. This adjudication is an arduous process and will not be complete for several years to come. However, the anticipation of results of the adjudication color the possible solutions to the water allocation dilemmas facing Basin agencies. The outcome of adjudication has no effect on short-term solutions to the current problems facing the Basin and will not be considered in this study.

A comprehensive analysis of the environmental and economic consequences of changes to water allocations in the Upper Basin, with its political and physical complexities, requires the integration of economic information with spatial diversity. The Upper Klamath Basin is not, however, unique in possessing these characteristics. The inclusion of spatial data in the analysis of policy changes in other regions is necessary to evaluate the efficacy of policy options. Even if, as Helfand and House (1995) and Fleming and Adams (1997) determined, the inclusion of spatial data does not play a large role in the implications of certain policy decisions, the importance of site-specific heterogeneity will vary across locations and policies.

5. CONCLUSION AND OPPORTUNITIES FOR FURTHER RESEARCH

In an era of reallocation, institutional solutions must be examined to solve problems of resource allocation. In this example, we have discussed the importance of using spatial analysis in determining some of the environmental and economic impacts of proposed institutional solutions.

The spatial element has two components. The first component is to include the physical characteristics and relative position of water. The second spatial component that is considered in this study is the political boundary. As long as the modeling units represent, among other natural attributes, their political association, the quantitative economic model can be aggregated by political unit. This flexibility will allow the model to estimate environmental and economic impacts under various scenarios of the jurisdiction of the water market.

The importance of accounting for spatial characteristics is likely to matter in the presence of heterogeneous physical characteristics where there are multiple uses for the resource in question and where there are multiple jurisdictions attempting to arrive at a consensus. Specifically, any analysis which omits these spatial dimensions would not provide the various state and federal agencies or private landholders with accurate information or tradeoffs, and may delay the consensus building process.

Several things would improve the current research. First, this work would be improved by addressing the temporal aspects to water allocation decisions. The USBR's decisions are made in real time to impact current lake levels and stream flows. Something as simple as a two- or three-stage model to address early, mid and late season would increase the informational content of this research. In addition to including seasonal aspects, including a longer time horizon would allow the model to incorporate year-to-year changes to technology as more efficient irrigation methods are adopted.

REFERENCES

Booker, J.F., and R.A. Young. 1994. Modeling Intrastate and Interstate Markets for Colorado River Water Resources, *Journal of Environmental Economics and Management* 26: 66-87.

Bockstael, N.E. 1996. Modeling Economics and Ecology: The Importance of a Spatial Perspective. Working Paper 96-23, Department of Agricultural and Resource Economics, University of Maryland, College Park.

Buck, S.J., G.W. Gleason, and M.S. Jofuku. 1993. The Institutional Imperative Resolving Transboundary Water Conflict in Arid Agricultural Regions of the United States and the Commonwealth of Independent States, *Natural Resources Journal* 33: 595-628.

Burness, S.H., and J.P Quirk. 1980. Water Law, Water Transfers, and Economic Efficiency: The Colorado River, *The Journal of Law and Economics* 23: 111-34.

California Department of Water Resources. 1992. California Water Plan, Volume One.

Colby, B.G. 1990. Enhancing Instream Flow Benefits in an Era of Water Marketing, *Water Resources Research* 26 N: 1113-20.

Dinar, A., and J. Letey. 1991. Agricultural Water Marketing, Allocative Efficiency, and Drainage Reduction, *Journal of Environmental Economics and Management* 20: 210-23.

Fleming, R.A., and R.M. Adams. 1997. The Importance of Site-Specific Information in the Design of Policies to Control Pollution. *Journal of Environmental Economics and Management* 33: 347-358.

Gould, G.A. 1989. Transfer of Water Rights, *Natural Resources Journal* 29: 457-77.

Griffin, R.C., and S. Hsu. 1993. The Potential for Water Market Efficiency when Instream Flows Have Value, *American Journal of Agricultural Economics* 75: 292-303.

Hartman L.M., and D.A. Seastone, 1970. *Water Transfers: Economic Efficiency and Alternative Institutions*. Baltimore: Johns Hopkins University Press.

Helfand, G.E., and B.W. House. 1995. Regulating Nonpoint Source Pollution Under Heterogeneous Conditions, *American Journal of Agricultural Economics* 77: 1024-32.

Howitt, R.E. 1995. Positive Mathematical Programming, *American Journal of Agricultural Economics* 77: 329-342.

Johnson, N.K., and C.T. DuMars. 1989. A Survey of the Evolution of Western Water Law in Response to Changing Economic and Public Interest Demands, *Natural Resources Journal* 29: 347-87.

Klamath River Compact Commission. Annual Reports, Fiscal Years 1957-1958 Through 1960-1691.

MacDonnell, L. 1989. Federal Interests in Western Water Resources: Conflict and Cooperation, *Natural Resources Journal* 29: 390-411.

Shupe, S.J, F.D. Weatherford, and E Checchio. 1989. Western Water Rights: The Era of Reallocation, *Natural Resources Journal* 29: 413-34.

Strand, I.E. Jr., N.E. Bockstael, and R.A. Siegel. 1992. Trade, Institutions, and Preference for Living Marine Resources, *American Journal of Agricultural Economics* 72: 1150-1154.

Tarlock. 1985. The Endangered Species Act and Western Water Rights, *Land and Water Law Review* 20.

Vaux, H.J., and R.E. Howitt. 1984. Managing Water Scarcity: An Evaluation of Interregional Transfers, *Water Resources Research* 20: 785-92.

Walston, R.R. 1989. The Public Trust Doctrine in the Water Rights Context, *Natural Resources Journal* 29: 586-592.

Weinberg, M., C.L. Kling, and J.E. Wilen. 1993. Water Markets and Quality, *American Journal of Agricultural Economics* 75: 278-91.

Zilberman, D., N. MacDougall, and F. Shah. 1994. Changes in Water Allocation Mechanisms for California Agriculture, *Contemporary Economic Policy* 12: 122-33.

TABLE OF CASES

Arizona v. California, 373 U.S. 546 (1963)
Bennett v. Spear, 117 S.Ct. 1154 (1997)
Carson-Truckee Water Conservancy District v. Clark, F.2d 257 (9th Cir. 1984)
National Audubon Society District v. Alpine County, 33 Cal. 3d 419, 445, 658 P.2d, 709, 727, 189 Cal Rptr. 346, 364 (1983)
Nebraska v. Wyoming, 325 U.S. 589 (1945)
Winters v. United States, 207 U.S. 564 (1908)
Wyoming v. Colorado, 259 U.S. 419 (1922)

12 TRANS-BOUNDARY AIRSHED MANAGEMENT AS AN APPROACH TO TRANS-BOUNDARY WATER COOPERATION: THE CASE OF THE CHESAPEAKE BAY

David H. Austin
Alan J. Krupnick
Virginia D. McConnell*
Resources for the Future, USA

This chapter examines implications for cost-effective allocation of pollution controls when preferences of coalitions organized along regional lines, or according to preferences for air versus water quality improvements, are accounted for. Results are compared to a base case in which NO_X emissions reductions must satisfy only a water quality standard, and total costs are minimized over emissions sources. Relative to a base-case result that marginal control costs must be equal across sources, stronger relative preferences for air imply shifting of control toward sources that produce greater ancillary benefits to air quality. Regional differences may require side payments to induce cooperation where benefits are low, but this will not affect how controls themselves should be allocated.

1. INTRODUCTION

It is now accepted that emissions of nitrogen oxides into the air are a major source of nutrient enrichment in the Chesapeake Bay, comprising anywhere from 10 percent to 40 percent of the Bay's nitrogen oxide (NO_X) loadings. As a result, policies to reduce air emissions of NO_X have become an important focus of attempts to reduce Bay nutrient levels. Because air emission reductions have an impact on both air and water quality, important issues arise about the efficient level of control of air sources, about the allocation of the costs of control between air and water, and about the political economy of achieving an efficient or cost-effective allocation.

* Virginia D. McConnell is also with the University of Maryland, Baltimore County.

Although there has been considerable work on the allocation of costs when there are multiple beneficiaries of improvements in a single environmental medium (see Young et al., 1982, in the case of water benefits), there has been little work examining the efficient allocation of controls across multiple sources, or when multiple media are affected by those controls, or of the allocation of the *costs* of control (see, however, McConnell 1990 in the case of vehicle inspection and maintenance programs). Unlike conventional optimization problems that are solved for the efficient outcome, and for which the distributional effects that follow are simply described, in our analysis the distributional concerns are represented in the model and can influence the efficient outcome. The analysis of cost allocation transforms the problem into one of political economy rather than of efficiency.

The Chesapeake Bay case provides a unique opportunity to examine these issues, and to do so in a unified analytical framework. Insights about appropriate ways to allocate costs will have bearing on the current policy debate about how to assign the costs of controlling airborne NO_x when both air and water media are affected. The assignment of costs may directly influence, in turn, the amount of emissions control that is instituted at each source—for it is these allocated (or assigned) costs which enter into the cost-effectiveness calculations that determine the optimal amount of emissions controls at each source.

Previous studies on the effects of air emissions controls on water quality have assigned all of the costs of control to the water benefits, ignoring the impacts of these controls on air quality through reduced ambient ozone and particulate levels. [In an appendix we present E.H. Pechan and Associates' (1996) estimated costs of nitrate reduction in the Chesapeake Bay.] This overstates the costs of reducing water pollution through air emissions controls. How control costs should be allocated between air and water as well as between disparate emissions sources, and how much control should be undertaken at each source, remain unanswered questions.

This chapter takes some first steps toward resolving some of these issues. We develop several models of optimal emissions control under various constraints. We construct a model where the only environmental standard is on the water side, but where ancillary air benefits from emissions control are recognized and accounted for. We also examine the case where NO_x emissions control must achieve both air and water quality standards. We derive as a basis for comparison the optimal control strategy for water quality improvements under a simple cost-minimization regime that ignores ancillary benefits in air quality and is constrained only by the water standard.

Because a significant number of the NO_x sources affecting the Chesapeake Bay are far upwind of it, and will enjoy fewer of the benefits from controls on their emissions, there may be more resistance from upwind regions to controlling emissions for improved Bay water quality. We address this possibility by adding political economy constraints to the problem. A different kind of political economy constraint is used to represent differing preferences for improvements in air quality versus water quality. This allows us to investigate the influence of control cost allocations within a source (over the joint air and water benefits it produces) upon optimal levels and distribution of emissions controls across sources. Cost allocations are not limited to such accounting exercises, but can also involve side payments between sources. The kind of cost allocation method used will depend on the preferences of those facing the control costs. The "rationality" and "fairness" constraints which govern cost allocation decisions are especially relevant in addressing issues arising from trans-regional air transport of pollution. Our interest here is to see how these political constraints affect the allocation of emissions controls across sources.

The models presented here will ultimately be used as the basis for an empirical analysis of cost-effective policies to reduce nutrient loadings to the Bay. The analysis will extend Pechan's (1996) earlier work to look at a broader set of policy alternatives, and to include ancillary (air) benefits in the analysis. Based on Pechan's cost estimates, and our theoretical models, optimal policies can be estimated empirically and compared to more traditional command and control policies. In addition, the "shadow prices" of the environmental standards for NO_x loadings from air sources can be estimated and compared to shadow prices of nutrient reduction due to water-based controls. If they are out of balance, future reductions of NO_x loadings should come first from the medium (air or water) which has the lower shadow price of control.

2. COST MINIMIZATION

The effect of accounting for so-called "ancillary benefits" will be to shift the control of NO_x emissions toward sources generating greater ancillary benefits for a given amount of loadings reduction in the water. This outcome may differ from one where air quality is not considered an ancillary benefit, but must also meet a standard. Beyond investigating the effects of different combinations of environmental standards, we also consider the effects of various methods of allocating control costs. Interest groups organized by location or by emissions source may credibly threaten to limit their participation to match the benefits they receive in their region or at their source. Alternatively, if there is debate over willingness to pay for water (or air) quality improvements, individual preferences may need to be accounted for, with costs within individual sources allocated to the jointly produced air and water benefits according to the relative levels produced at that source. We contrast the optimal allocation of emission controls across sources with what is implied by a base-case cost minimization model constrained only by a water quality target.

Adopting the notation used in Teitenberg (1985), in the benchmark model we seek to minimize total control costs over J sources of nitrogen oxides given that we must reduce total Bay NO_x loading to some preestablished limits $\overline{W}_i$ at each of I measurement locations.[1] (If loadings are uniformly mixed, there will be only a

[1] It is an open question how those J sources are selected. An extension of our work would be to derive the optimal extent of a regulatory region. This would involve trading off between regulator power, which presumably is greater the fewer sources that must be controlled, and costs of environmental control—because extending the boundary could lower costs by

single location and one limit $\overline{W}$, but for generality we will assume multiple locations.) Letting $c_j(r_j)$ represent the cost of reducing NO_x emissions at source j by the amount r_j —from some baseline emissions level $\bar{e}_j$ and letting the function $L_{ji}(\cdot)$ map NO_x emissions at source j into Bay loadings at site i, the objective is simply to minimize the aggregate cost of those reductions across all sources subject to achieving the water quality targets. The objective is

$$\min_{r_j} C(R) = \min_{r_j} \sum_{j=1}^{J} c_j(r_j), \tag{1}$$

subject to

$$-\overline{W}_i + \sum_{j=1}^{J} L_{ji}(\bar{e}_j - r_j) \leq 0. \tag{2}$$

That is, the regulator seeks to find the least expensive way of assuring that loadings L_{ji} of the emissions remaining after controls are installed, $(\bar{e}_j - r_j)$, are no greater than the imposed water quality standard $\cdot W_i$. Total costs, summed over all emissions sources, are given by C(R). To achieve an optimum given that reductions r_j must be non-negative—as must be the LaGrange multiplier λ_i on the environmental standard in (2)—the Kuhn-Tucker conditions (see, e.g., Varian 1992) must be satisfied. The key first-order condition for cost minimization, which we explain below, is

$$\frac{\partial c_j}{\partial r_j} - \sum_{i=1}^{I} \lambda_i L_{ji} \geq 0. \tag{3}$$

Cost minimization will not be achieved unless a "complementary slackness" condition is also satisfied, guaranteeing that for every source that optimally reduces its NO_x emissions (i.e., $r_j > 0$), condition (3) is met with equality. In other words, at the optimum point, all sources are controlled to where their marginal control costs is equal to the sum of the "shadow prices," λ_i of the environmental standards, weighted by the loadings, L_{ji}, of their NO_x emissions at each Bay receptor. For the single receptor (or uniform mixing) case, (3) implies that *sources are controlled to where the ratios of their marginal control costs, MC_j, to their Bay loading factors, L_j, are all equal,*

$$\frac{MC_j}{L_j} = \lambda = \frac{MC_{j'}}{L_{j'}},$$

for any sources j and j'. Sources with marginal costs too high to this would not reduce their emissions.

---

creating increased gains from trading over a larger market, as well as possibly bring large emissions sources under the control of the regulator.

This outcome can be achieved by regulatory fiat, but it has been shown elsewhere that a market-based, tradable permit approach, where permits for emissions produce total loadings of $\overline{W}_i$ at each receptor i, can achieve the same outcome. [See Montgomery (1972) for initial development of a trading model in this context, and Krupnick et al. (1983) for refinements.] Equilibrium permit prices of $P_i = \lambda_i$ would arise at each receptor and, subject to the qualifications in Krupnick et al. (1983), would achieve the cost-minimizing outcome without regulatory intervention (beyond issuing the correct number of permits). Separate markets would exist for permits specific to each receptor. The initial allocation of the permits need have no effect on the outcome. Under this scheme, low-cost sources would reduce emissions beyond what they are permitted to emit, and would sell their excess permits at price P_i to high-cost sources.

3. ACCOUNTING FOR ANCILLARY AIR BENEFITS

The control of NO_x sources to achieve water quality goals will inevitably also create benefits from improved air quality. These "ancillary" air benefits must be accounted for if emissions controls are to achieve the water quality standard in a socially optimal manner. Here the problem is one of social welfare maximization, rather than of cost minimization, given an imposed water quality target. That is, we seek to maximize ancillary air benefits B_a net of the costs of achieving the water standard,

$$\max_{r_j} B_a\left[\sum_{j=1}^{J}\sum_{k=1}^{K} d_{jk}\cdot r_j\right] - \sum_{j=1}^{J} c_j(r_j)\,, \tag{4}$$

subject again to constraint (2). Here d_{jk} is a source-receptor matrix mapping emissions at source j to concentrations at (air) receptor k; c_j, and r_j are as before. (Note the substitution of "concentrations" for "loadings" when the context switches from water quality to air quality.) The first-order condition for maximization of (4) which is analogous to the first order condition (3) is

$$\frac{\partial B_a}{\partial r_j}\sum_{k=1}^{K} d_{jk} - \frac{\partial c_j}{\partial r_j} + \sum_{i=1}^{I}\lambda_i L_{ji} \le 0\,. \tag{5}$$

As before, complementary slackness implies that the optimal point of control occurs when marginal control costs *net of marginal ancillary benefits* are equal to the weighted shadow prices of the constraints. This can be seen by rearranging terms in (5),

$$MC_j - MB_{a_j} = \sum_{i=1}^{I}\lambda_i L_{ji}\,,$$

where MB_{aj} is the marginal ancillary benefits, $\partial B_a/\partial r_j$, from controlling source j. If uniform mixing is assumed, this condition simplifies to

$$\frac{MC_j - MB_{a_j}}{L_j} = \lambda = \frac{MC_{j'} - MB_{a_{j'}}}{L_{j'}} \tag{6}$$

for any two sources j and j'.

Three important results are implied by this equation. First, rather than controlling all sources to where their marginal control costs are equal to $L_j\lambda$, the shadow price of the constraint weighted by their individual loading factors, here sources which produce air benefits are controlled more than they otherwise would be, as an increasing function of those air benefits. Second, accounting for the air benefits implies a different shadow price λ than before—one that reflects the true social cost of the water-quality target. Finally, and perhaps most important, the air benefits are external to the Bay clean-up. A trading market for Bay nutrients would not achieve the optimal outcome unless a mechanism were designed to reflect each source's ancillary benefits in the price of its permits. The straightforward system for the benchmark, cost-minimization model would not work here.

As a generalization of the ancillary benefits model, consider the case where both air and water targets are imposed, and neither type of benefit is considered "ancillary." With ancillary benefits out of the objective function, the problem once again becomes one of cost minimization. In the case where there is only one receptor for water quality, we can assert that only one set of constraints will be binding. Meeting the atmospheric NO_x loading target for the water either requires such stringent air emissions controls that the air quality targets are also met, or it does not. In the former case, the water targets bind. In the latter case, it is the air standards.

This statement may need some refinement for the case where there are multiple water targets (or there is non-uniform mixing in the one target). However, for the simplifications adopted in (6) and in what follows, this assertion applies. It will be invoked in working out some of the less immediate implications in the analyses which follow.

4. COST ALLOCATION ACROSS SOURCES

We next consider how political reality can require additional constraints on (5). These constraints concern how the costs of control are allocated across—or within—sources. As we show, cost allocation can alter the allocation of controls across emission sources.

There are two different cases to be considered. One concerns coalitions that are linked to individual sources and have some "hold-up" power. That is, sources cannot be compelled to install controls, but must be satisfied in some relation to the benefits they will receive from those controls. The other kind of cost allocation accounts for the different valuations people will have for air benefits versus water benefits. Allocating an individual source's control costs according to its relative levels of water and air benefits can affect the optimal amount of emissions control for each of the sources.[2]

Coalitions organized around emissions sources and possessing hold-up power require the addition to the model of constraints that assure the "participation" of these coalitions. Accounting for ancillary benefits, the objective function to be maximized remains expression (4). Two sets of constraints are added to the constraints on water quality, (2). Participation constraints assure that the amount of control at each source is individually rational for that coalition to undertake. In the extreme case that the regulators have no coercive power over the coalitions, each coalition's "threat point" will depend solely on the benefits it receives from its own actions, ignoring the benefits all others receive from its actions. If the regulators are indeed held up to this point, they will not necessarily be able to achieve a full allocation of all control costs. They may have to make a transfer from the public sector to the sources to cover costs fully in this case.

Such a transfer actually would have no real effect on the allocation of emissions control, so to make this model consistent with the others, we shall assume that the regulators possess power sufficient to impose control costs beyond individual threat points, but that they must accept some form of participation constraints.[3] Here the regulators balance the requirement of a full cost allocation against the political constraints by allocating *total* program (control) costs over *all* sources in such a way that no source incurs costs greater than the share of total program benefits (in air and water) achieved by controls at that source. This scheme appeals to some notion of fairness in that the benefits of control at a source are related to the damages for which that source is responsible.[4] With full cost allocation, the regulators' participation constraints are

$$f[B_{tot}(r_j)] - s_j \cdot C(R) \geq 0 . \qquad (7)$$

This says that the share s_j of total control costs C(R) allocated to source j cannot exceed some function f(·) of the share of total benefits, B_{tot}, of control at source j. The function f(·) is completely arbitrary, in the sense that its shape will be something the rival coalitions can determine among themselves, according to their own notions of fairness, or that the regulator can impose.[5] The actual function chosen will have real effects on the allocation of controls across sources, but there is no *a priori* "best" function in a world where the coalitions cannot be coerced.

[2] When this cost accounting, which is internal to each individual source, has real effects on the amount of control, it must also affect the allocation of controls across all sources. We do not explore the implications of this here.

[3] Since the implications for emissions reduction are the same, the requirement of a wealth transfer in the case of no coercive regulatory power is the only interesting difference between that case and the one analyzed here.

[4] Benefits will also be related to the costs of available control technologies.

[5] Young et al. (1982) show that the Shapley value approach (Shapley 1953) is superior to other well-known cost allocation methods in regards to notions of fairness, rationality, monotonicity, and other properties desirable in a cost allocation.

The requirement that a full cost allocation be achieved means that the cost shares s_j must all be non-negative, and must sum to one,

$$\sum_{j=1}^{J} s_j - 1 = 0 , \quad s_j \geq 0 \text{ for all j} . \tag{8}$$

The first-order (Kuhn-Tucker) conditions reveal that $\gamma_j = \gamma_{j'}$ for all sources j and j', where γ_j are the LaGrange multipliers associated with the participation constraints (7). This means that these political constraints are either *all* binding, or none of them are. Indicating their common value by γ, and letting ϕ stand for the LaGrange multiplier on the shares constraint (8), this means that $\gamma = \phi / C(R)$.

This condition has a very important interpretation. Recall our remarks, above, that there is no real effect of failing to allocate fully the control costs—thus violating condition (8). In this event, a transfer must be made from the public sector to cover the unallocated control costs, but this transfer does not increase the *social* cost, which comprises public and private costs. Thus ϕ, the shadow (or social) cost of the "full cost allocation" constraint (8), must be zero! The implication of this result is therefore that *side payments* (which in effect occur whenever a source pays an amount different from its own control costs) *have no effect on how much control is installed at any source.* In other words, it is immaterial in this model whether or not the regulator has any coercive power to compel sources to absorb costs beyond their threat points.[6]

This result also holds regardless of the actual cost shares s_j. These shares are arbitrary for the purposes of this analysis (even to the point where they can be negative). They will be determined by whatever balance of power exists between the sources and regulator. Of course, there is the separate issue of the regulators' ability to effect a transfer in the event they cannot fully allocate costs. The analysis of that situation is beyond the scope of this paper, although we note in passing that, were they to have insufficient power with respect to the coalitions and also the taxpayers, they would have to aim for a less stringent environmental target.

5. COST ALLOCATION WITHIN SOURCES

Finally, we consider allocating costs of control at individual sources to air and water benefits. This is somewhat akin to the problem a multi-product firm has in assigning joint production costs to its individual product lines. The firm's decision will affect the prices it charges for its individual products, and therefore the

---

[6] This is a different matter than the regulator's ability to impose controls independent of who pays. If the environmental target had been derived as part of an optimization, the environmental targets would be generating benefits in excess of costs, and the regulator would (in theory, and under the assumption that people have uniform tastes for environmental quality) need no political power to institute controls. In the case of controls on loadings to the Chesapeake Bay, the water quality target is imposed, not derived, so the regulator's political power is relevant.

quantities demanded and its profits. Here, as we shall show, only under a particular circumstance does cost allocation between air and water benefits matter. Unlike in Section 4, there are no side payments being considered in this scheme.

This problem can be set up analogously to (4)–(7), but with the allocation constraints altered to reflect allocation within, rather than across, sources. Thus each source can be thought of as an individual firm making cost allocations between its two products, air and water benefits. These allocations might depend, for example, upon the preferences of persons living in the vicinity for air quality improvements versus water quality improvements, as with the problem in (4) subject to (2) and

$$f[B_a(R_j), B_w(r_j)] - s_e \cdot c_j(r_j) \geq 0 \tag{7'}$$

where $e \in \{\text{air, water}\}$, B_w represents water quality benefits, and

$$s_{air} + s_{water} = 1,\ s_e \geq 0\ . \tag{8'}$$

By analogy with the previous analysis, it should be clear that, as long as cost allocations do not appear in the objective function (4), the way control costs are allocated to the air side and the water side at any particular source will have no effect on how much control is instituted at that source. This is because, with a single environmental target [(2) or its air equivalent], the allocation constraint (7') does not constrain the amount of control instituted at that source. The result is that controls will still be instituted using the net marginal costs rule in (6). These marginal costs (and the marginal ancillary benefits) are unaffected by whatever internal cost accounting procedure may be in force at a source.

However, allocations across benefit areas matter when they enter the objective function directly. The allocation of some of a source's control costs to air benefits effectively makes the water benefits simultaneously achieved less expensive. Even if this is considered an accounting "sleight of hand," this apparent lowering of the costs of water benefits matters when those allocated costs are what are being minimized in the objective function.

Allocating a share s_j of control costs to j's air benefits makes the effective cost of water benefits $(1 - s_j)c_j(r_j)$. It is this cost function which enters into the objective function (4) in place of full costs. Thus the problem becomes

$$\max_{r_j} B_a\left[\sum_{j=1}^{J}\sum_{k=1}^{K} d_{jk} \cdot r_j\right] - \sum_{j=1}^{J} (1 - s_j)c_j(r_j) \tag{4'}$$

subject again to (2), (7'), and (8').

The effect of this allocation of costs is to shift controls to sources which achieve more ancillary benefits—a result similar to (6) but with the added element that it is not *full* marginal costs (net of marginal ancillary benefits) which are equalized, but rather the share of marginal costs allocated to water that are equalized (net of marginal ancillary benefits). Sources which achieve more air benefits per unit of water benefits will be more tightly controlled relative to their control under the net-marginal-cost rule (6).

It is one matter to assert, as we have done, that an objective function such as (4'), with cost allocations there rather than in the constraints, is the only way those cost allocations will affect the amount of control at each source. It still remains to consider what the rationale is for putting cost shares in the objective function. Here the answer is the same one we gave when considering how much a coalition can be coerced. It depends on the distribution of bargaining power in the actual situation.

The effect of equation (6), the marginal costs net of marginal benefits rule, is to shift emissions controls toward sources that create greater ancillary air benefits (on the margin) for each unit of water benefits achieved. Since the water quality target is imposed, and will be met with equality, this is an outcome that most favors those who value air benefits most highly. The same reasoning applies to (4'). The greater is the share s_j of costs at source j that is allocated to ancillary benefits, the less the water share of those costs will diminish the objective function, and the more that source will be controlled to achieve the water target.

It should be clear that this outcome favors the "air coalition" more than the net-marginal-cost rule (6) does. This answers the question of what the rationale is for including cost allocations in the objective function. Its inclusion reflects the degree of bargaining power the air coalition has relative to the water coalition.[7] The whole of our analysis can now be seen as a description of how the outcome changes as the bargaining power of the air coalition increases continuously from zero (ancillary benefits are ignored and cost minimization implies $MC_j/L_j = \lambda$ for all sources), up to a point where ancillary air benefits are accounted for (thus shifting emissions control toward sources achieving greater ancillary benefits), to a greater point still, where preferences for air benefits relative to water benefits are strong enough that the cost shares s_j are inserted into the objective function.[8]

6. CONCLUSIONS

This chapter has identified conditions under which political economy considerations will affect the efficient allocation of controls on the air emissions of nitrogen oxides across multiple sources. By political economy, we mean the relative bargaining strengths and preferences of interest groups, coalitions, and regulators. We represent these, for the most part, by various kinds of cost

[7] There is, of course, considerable overlap between those two "coalitions." These artificial constructs are useful here for expository purposes, to represent differences in individual preferences for air and water benefits.

[8] The careful reader will have anticipated the *reductio ad absurdum* where the air coalition has "full" power. In this case, the share of costs allocated to water improvements will be zero! At that point the objective will simply be to maximize air benefits regardless of costs. This outcome, though, begs the question of why there would be an environmental constraint on water quality when nobody cares about that and everybody cares about air quality. (This is the implication of full political power for the air coalition.) The model is flexible, but setting $s_j = 1$ would break it.

allocations, both across emissions sources (as when, for instance, rate-payers in upwind states such as Ohio are reluctant to undertake emissions controls that will largely benefit persons in downwind states such as Maryland and Virginia), and within emissions sources. In the latter case, preferences over the air and water benefits achievable at individual sources govern the allocation of costs within those sources.

While we examine many levels of bargaining power, we show that only in fairly extreme circumstances, when interest groups succeed in changing the objective function rather than merely the constraints of the problem, cost allocations do not matter. We are most interested in the case where water quality standards from air emissions of NO_x must be achieved, and we wish to account for what are therefore the ancillary benefits from concomitant air quality improvements. In this case the efficient level of control is to net out ancillary benefits, on the margin, from control costs and to equate the ratio of the net marginal costs to the Bay loading factors across all sources.

Diverse individual preferences for air and water quality are represented in this outcome. Only in the case where water quality improvements are valued very little relative to air will controls have to explicitly account for cost allocations over those media. In that case, however, it seems unlikely that the water quality standards would have had the political or popular support to be implemented in the first place, and the focus would shift to control for air benefits alone.

The political economy of instituting emissions controls in upwind regions where local benefits will be low is perhaps more relevant to the problem of water quality improvements in the Chesapeake Bay. The implications of our model are that if it is economically efficient (in terms of costs and benefits) to control sources in, say, Ohio, it does not affect the optimal amount of control at those sources if they require side payments to induce them to cooperate. In other words, such distributional issues have no affect on allocation of emissions controls.

There is further work to be done before this analysis will have addressed all of the outstanding issues in this area. We must reconsider the case of joint air and water standards in the case where there are multiple receptors for each medium. We must also consider the case where coalitions form simultaneously along both media and source lines. That is, cost allocations may be required both within and across sources. However, it appears from our analysis that these issues will not matter except in extreme cases, which may call into question the environmental goals being pursued.

The most important extension of our work was noted in an early footnote. This is the issue of the optimal regulatory boundary. This is a larger project than those noted above. It involves an application of integer programming to solve for the optimal boundary when there are tradeoffs between size of regulatory domain and regulatory power. The nature of these tradeoffs will be determined by transactions costs, bargaining coalitions, and the gains from trade in large versus small markets.

Ultimately, all of this research is directed toward supporting future policy decisions affecting the allocation of emissions (or effluent) controls across both air and water sources. The empirical work we are performing in conjunction with this research will examine trading outcomes and will estimate cost-effective levels of controls across sources. Even then, however, we will not have estimated the shadow price of controlling air emissions of nitrogen oxides for water benefits. To achieve efficient levels of control, these prices must be brought to equality with the shadow price of water-based nutrient controls.

APPENDIX

E.H. Pechan's (1996) estimates of the costs of controlling NO_x emissions to improve Chesapeake Bay water quality are the starting point of an empirical investigation by the authors, and Scott Atkinson of the University of Georgia, of the costs of control taking ancillary benefits into account [using equation (6) above], and of the gains from various emissions trading schemes. Pechan produced the estimates found in Table 1 for utility and mobile source emissions in various Bay airshed states.

Table 1. Pechan Estimates of Costs of Controlling NO_x Emissions for Bay Loading Reductions by Selected Scenarios

Scenario/State	NO_x Reduction (1,000 tons)	Nitrogen (N) Load Reduction (1,000 lbs)	Totai Annual Cost ($1,000,000)	Cost Effectiveness ($/ton NO_x)	Cost Effectivenes ($/lb N)	Ratio: ($/ton/ $/lb)
Utility—(reduce 0.15lbs/MMBtu)						
Maryland	47.0	1,610	$62.7	$1,300	$39	0.33
Pennsylvania	178.2	3,510	214.0	1,200	61	0.20
Virginia	52.8	1,990	57.9	1,100	59	0.19
West Virginia	155.5	2,240	157.5	1,000	70	0.14
Kentucky	169.1	760	192.3	1,100	254	0.04
Mobile Source (Introduce LEVs)						
Maryland	13.6	410	$39.0	$2,900	$95	0.30
Pennsylvania	24.1	470	76.5	3,200	164	0.20
Northern Virginia	4.4	90	11.9	2,700	130	0.21
Virginia (all)	10.4	220	58.4	5,600	270	0.21

Source: Pechan (1996).

The first column of Table 1 shows Pechan's estimates of the reduction in NO_x emissions that would be realized by achievement of a 0.15lbs/million British thermal unit standard for utilities, and by the introduction of low-emission vehicles (LEVs) in each respective state. Gains on the utility side may differ because existing levels of control at individual sources may differ, as may current utility capacities in each state. The LEV estimates are based on Pechan's modeling of the adoption of California's LEV program in each of the states and Northern Virginia.

According to Pechan, these estimates would change "somewhat" if instead they modeled a 49-state LEV program different from California's program (Pechan 1996).

Based on these estimates, and a transport matrix mapping emissions to Chesapeake Bay loadings, Maryland and Virginia can achieve a disproportionate reduction in Bay loadings, relative to their potential NO_x emissions reductions, because of their proximity to the Bay (second column). More of their NO_x emissions end up in the Bay than do the emissions of more distant states.

The two cost-effectiveness columns take Pechan control-cost estimates (column 3) and divide them by, respectively, tons of NO_x emissions reductions (column 4), and pounds of nitrogen loadings reduced (column 5). Notice that Maryland sources are the least cost-effective to control for NO_x emissions, but for a goal of loadings reduction in the Bay, Maryland sources become the *most* cost-effective to control. Kentucky, being farthest away, has a smaller fraction of its NO_x emissions transported to the Chesapeake Bay, and so controlling these sources for Bay water quality improvements is the least cost-effective of these options.

The last column of the table tells this story in a slightly different way. The ratio of NO_x-reduction cost effectiveness to loading-reduction cost-effectiveness is greatest for Maryland, and least for Kentucky (for mobile sources, with Kentucky not represented, it is least for Pennsylvania—which is farther from the Bay than Maryland and Virginia). That is, in both collections of states the pounds of nitrogen loading reduction per ton of NO_x reduction is greatest in Maryland, least in the state farthest from the Bay. Were NO_x emissions uniformly "mixed" and transported in equal proportions to the Bay, the ratios would be equal across these states.

REFERENCES

Krupnick, A.J., W.E. Oates, and E. Van den Berg. 1983. On Marketable Air-Pollution Permits: The Case for a System of Pollution Offsets, *Journal of Environmental Economics and Management* 10: 233-247.

McConnell, V.D. 1990. Costs and Benefits of Vehicle Inspection: A Case Study of the Maryland Region, *Journal of Environmental Management* 30: 1-15.

Montgomery, D. 1972. Markets in Licenses and Efficient Pollution Control Programs, *Journal of Economic Theory* 5: 395-418.

Pechan and Associates, Inc., E.H. 1996. Costs to Reduce Airborne Nitrogen Deposition Affecting the Chesapeake Bay: Project Briefing. Prepared for U.S. Environmental Protection Agency, Springfield, Virginia.

Shapley, L.S. 1953. A Value for n-Person Games. In H. Kuhn and A.W. Tucker, eds.. *Contributions to the Theory of Games,* Vol. 2. Princeton: Princeton University Press.

Teitenberg, T.H. 1985. *Emissions Trading, an Exercise in Reforming Pollution Policy.* Washington, D.C.: Resources for the Future, Inc.

Varian, H.R. 1992. *Microeconomic Analysis.* New York: W.W. Norton & Company, Inc.

Young, H.P., N. Okada, and T. Hashimoto. 1982. Cost Allocation in Water Resources Development, *Water Resources Research* 18: 463-475.

13 ECOSYSTEMS AND SOCIAL CONFLICT: LESSONS FROM THE FLORIDA EVERGLADES

J. Walter Milon
Clyde F. Kiker
Donna J. Lee
University of Florida, USA

The relationship between social conflict and ecological processes has received relatively little attention in the economics literature. Yet recent public concern has focused on the variability of ecosystems over time and how societies adapt to environmental change. We present a heuristic framework based on Kaufmann's statistical theory of order within complex systems to consider the interplay between ecosystems and social institutions in "ecological-economic organizations." We use this framework to evaluate social conflicts over water and ecosystem management during the past century in the South Florida/Everglades region. The analysis indicates a social adaptation process leading to centralized management and a transformation of the essential attributes of the ecosystem. It remains to be determined whether new initiatives to restore the ecosystem can succeed without causing more social conflict.

1. INTRODUCTION

Conflicts over water and ecosystem management during the past century in the South Florida/Everglades region provide useful lessons about interactions between ecological and social processes. This region is characterized by one of the most unique, biological diverse ecosystems in the world yet it has also experienced explosive population growth and demands for the ecosystem's services. Currently the region is the subject of one of the most extensive and expensive ecosystem restoration projects in the U.S. (Yaffee et al. 1996).

The relationship between social conflicts and ecological processes and how societies adapt to environmental change is a topic of contemporary importance. For example, the World Commission on Environment and Development concluded that environmental change was an "important part of the web of causality" leading to social conflicts. Others cite historical evidence of conflicts emanating from environmental changes and suggest such conflicts will increase in the future (e.g., Gleick 1991;

Homer-Dixon 1991; Homer-Dixon et al. 1993).

The concern that conflict results from societies' quest to control natural resources is hardly new. Elliott (1996) points out that competition for strategic control of natural resource stocks has long been recognized by historians and political scientists as a rationale for warfare and international tensions. Moreover, economists such as Demsetz (1967) and North (1990) describe how establishing property rights helps to mitigate conflicts and foster economic growth.

What is different about this new-found interest in ecological and social interactions is an emphasis on the variability of ecosystems over time and the process of social adaptation to this variability. Despite major advances in scientific knowledge about ecological processes, social institutions may not adequately anticipate or respond to environmental change. Moreover, adaptations to these changes may have unintended social and environmental consequences. Arrow et al. (1995, p. 521) succinctly express this view,

> *Environmental damages, including loss of ecological resilience, often occur abruptly. They are frequently not reversible. But abrupt changes can seldom be anticipated from systems of signals that are typically received by decision-makers in the world today. Moreover, the signals that do exist are often not observed, or are wrongly interpreted, or are not part of the incentive structure of societies.*

In this chapter we present a heuristic framework to consider the interplay between ecological processes and social institutions in "ecological-economic organizations." We use the term organizations because we wish to describe an organic structure that spans a variety of possible ecological states and social institutions. A central premise of our analysis is that the potential for social conflict increases as the interactions between ecological and economic processes increase. Since organizations "...engage in extensive parallel and distributed processing" (Kleindorfer et al. 1993, p. 292), the flow of information becomes increasingly important as the level of complexity within the organization increases. Moreover, the traditional focus on differences in political boundaries within an ecosystem (e.g., watershed) as the fundamental cause of social conflict (e.g., Boulding 1991) can be seen as simply one source of complexity within an ecological-economic organization.

Our heuristic framework is based on Kauffman's (1993) statistical theory of order and self-organization within complex systems and Westhoff et al.'s (1996) application of this theory to explain industrial organization adaptations to economic change in the United States and Britain. We use Kauffman's (1993) approach to describe ensembles of ecological and economic attributes within an organization. We then use this framework to describe how the interaction between ecological and economic processes have created social conflicts and lead to organizational adaptation in the South Florida/Everglades region. We conclude with a discussion of issues raised by this framework and its contributions to research on conflict and cooperation in water resource management.

2. AN ATTRIBUTE MODEL OF ECOLOGIC-ECONOMIC ORGANIZATIONS

Social organizations are complex systems that evolve in response to human needs. In recent years scientists from a variety of disciplines have sought to explain the behavior of complex systems [see Waldrop (1992) for a review]. One somewhat frustrating result was the link between complexity and chaos. Dynamic, nonlinear systems that appear to be orderly may become completely disorganized over time. This may then be followed by adaptation and a new, redefined order. In economies, markets can display such behavior (Arthur 1988). Similarly, nonmarket organizations can evolve as though they were guided by an "invisible hand" (Schotter 1981, p. 139). In general there appears to be some type of anti-chaos phenomenon operating within many complex systems that inhibits flights into disarray.

Kauffman (1993) in his work on *The Origins of Order* in biological systems provides an analytical framework that is useful for studying complex organizations. Like biological systems, organizations may differ because they have dissimilar patterns of activities, not because they have different fundamental properties. Much as Kauffman considers the genome, we might consider organizations as parallel-processors, or networks, in which components regulate one another's actions either directly or through their interactions. Each organization "has what can be called local features: these characteristics describe how individual elements in the system are connected and how they may influence one another" (Kauffman 1991, p. 79). This heuristic allows us to understand how the activity of one organizational element is influenced by others and their joint interaction. And, it provides a framework to consider a variety of possible ecological-economic organizations.

For the purpose of analysis, assume there are N possible "attributes" (traits) that can be used to describe the properties of an organization. A collection of possible combinations of these attributes can be termed an "ensemble;" the ensemble contains A^N elements where A is the number of possible states for each attribute. For simplicity, it is helpful to idealize each attribute as a binary (on or off) variable ($A = 2$), which gives us a Boolean network arrangement for the ensemble.[1] Interactions between the N attributes are given by K where $K < N$. In the limit, if $K = 0$ all attributes are independent in terms of their influence on other attributes; they independently contribute to the "fitness" of an organization.[2] Alternatively, if $K = N - 1$, all attributes interact so that changes in any one attribute imply changes in all others. For the simple binary case, if an attribute has K interactions, there are 2K possible combinations of interactions. Clearly, high levels of connectedness ($K \sim N$) lead to many organizational states and near-chaotic behavior ($K = N$ implies chaos). Low

---

[1]For simplicity we assume attribute states are binary. In reality a continuum of possible states exist which could be classified as fuzzy variables (Kosko 1993) which have partial influences on other variables. Examples would be diversity that lies between narrow and broad and dynamics that are not quite fast or slow. A fuzzy neural network would be an alternative to Kauffman's Boolean representation but would add considerable mathematical sophistication to our heuristic.

[2]Fitness is used in the sense that an organization has a structure that allows it to persist and evolve through endogenous change (Witt 1993).

levels of connectedness (K is "small") allow the possibility of a "basin of attraction" in which the attributes would eventually settle into a state cycle.

This theory can be given specific content in the case of ecologic-economic organizations by considering the ensemble of 6 attributes as represented in Table 1. We use 3 attributes to represent an ecologic system: geographic scale (G), diversity (V), and dynamics (D). For simplicity, we assume 2 states for each attribute: small versus large scale, narrow versus broad biodiversity, and variable versus stable cycling dynamics. Similarly, we represent the economic system's interaction with the ecologic system by 3 attributes: the scope (S) of services derived from the ecosystem, the extent (E) of the market for these services, and the decision-making architecture (DM). Again, for simplicity, we assume 2 states for each attribute: a narrow versus broad scope of services, a small versus large extent of the market, and a polyarchic versus hierarchic decision architecture. The choice of attributes and the possible levels is not intended to be exhaustive but to illustrate the inherent complexity in ecologic-economic organizations. Even with this limited attribute set, the ensemble includes 64 possible combinations.

Consider the selection of 2 possible attribute ensembles represented in Figure 1 [where **bold** signifies an "on" (1) state for an attribute]. The ensemble on the left corresponds to the organizational structure characteristic of traditional theory: an ecosystem with small scale, narrow biodiversity, and stable dynamics; users consume a narrow scope of services within a limited market under a decentralized decision structure such as individual property rights.[3] Changing E to a large market under either a polyarchic or hierarchic decision structure typifies the organizational alternatives considered in many analyses of water allocation mechanisms (e.g., Anderson 1983).

---

[3]To illustrate the relationship between our heuristic and traditional economic models, consider an ecosystem as a set of services, $S = (s_1, \ldots, s_k)$ which have value (either from direct or "passive" use) to individuals. Assuming the extent of the market for these services is defined by $n = 1, \ldots, N$ individuals, society's objective would be to maximize the net present value of some scope of services over some finite (or infinite) planning horizon,

$$\text{Max PV}(E) = \sum_{t=1}^{T} \sum_{n=1}^{N} e^{-rt} V_{nt}(S_{nt}),$$

where V represents the indirect utility function and r is the social discount rate. This maximization can occur through a decentralized process (polyarchy) whereby each individual maximizes utility from the services subject to their budget constraint. Or, if various externalities (stock, congestion, etc.) create inefficiencies, a centralized authority (hierarchy) can manage the flow of services (through taxes, user regulations, etc) subject to the agency's budget. Typically a limited scope of services such as a single species fishery or a confined aquifer simplifies the consideration of ecosystem processes so a service can be modeled independently. Also, service flows are usually stationary over time. This allows for stable equilibrium outcomes and efficient behavioral rules such as sustainable consumption is determined by the growth rate of the service, harvesting costs, and the discount rate (e.g., Zilberman et al. 1993, pp. 99–114). Yet, once an optimal consumption path is defined, there are no endogenous sources of conflict.

Table 1. Attributes of an Ecological-Economic Organization

Attribute	Possible Levels	
Geographic scale (G)	Small (0)	Large (1)
Diversity (V)	Narrow (0)	Broad (1)
Dynamics (D)	Variable (0)	Stable (1)
Scope of services (S)	Narrow (0)	Broad (1)
Extent of market (E)	Small (0)	Large (1)
Decision-Making architecture (DM)	Polyarchic (0)	Hierarchic (1)

On the other hand, the right side ensemble indicates very different ecological and economic interactions. The large, diverse, dynamically stable ecosystem in a setting where a broad market demands a variety of ecosystem services suggests a broad array of interactions which are coordinated through a hierarchic structure. This latter configuration typifies the "multiple-use" approach to forest system management as might be practiced by a federal agency such as the U.S. Forest Service (e.g., Bowes and Krutilla 1985).

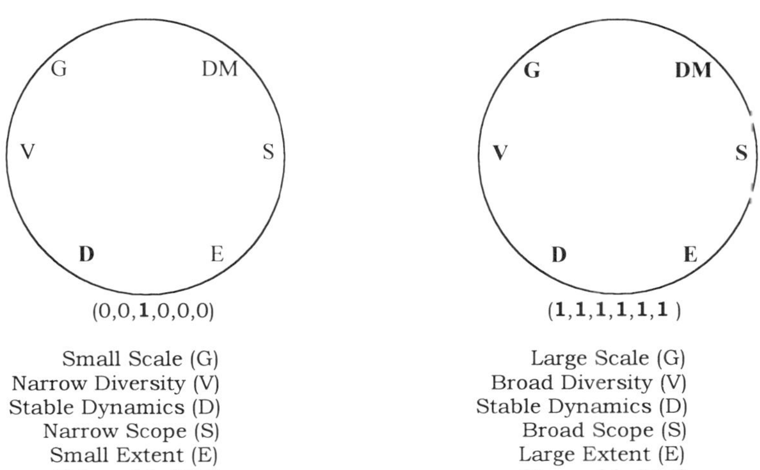

Figure 1. Possible Attribute Combinations for an Ecological-Economic Organization

Within our adaptation of Kauffman's model, changes in attribute states within an ensemble result from relations between attributes. Three functional relations are particularly useful: *Equivalence*, in the case where an attribute has only one input, implies the attribute assumes the state of its input; *Or*, in the case of an attribute with two inputs, implies the regulated attribute will assume the state of one or the other of the two regulating attributes if either attribute is active; *And* implies the regulated attribute becomes active only if both regulating attributes are active. Kauffman describes the *Or* and *And* functions as "canalyzing Boolean functions" that bring order to large complex systems (Kauffman 1993, pp. 203-204).

Systems that are sparsely connected and appear to be governed by canalyzing functions are likely to exhibit traits of parallel-processing systems where there are "frozen elements." Order emerges in these canalyzing networks because a large fraction of the attributes fall into a stable frozen state where the core of elements is nearly identical in all attractors (Kauffman 1991, pp. 83-84). Therefore, mutations of a single (or a few) attributes would have limited effects so the perturbation is "minimal." And, "avalanches of damages" caused by mutations should not propagate to most other elements in a canalized network. On the other hand, "structural" perturbations may change attribute interactions or functions resulting in an unstable state.

Change in one attribute that has limited connectedness will have a minimal effect on organizational structure and will result in little conflict. Other attributes with greater connectedness can induce structural change and the potential for major conflict. The specific relations between attributes in an ecologic-economic organization, however, cannot be known *apriori* without a great deal of understanding about the behavior of such organizations. Therefore predictions are limited by the historical context of each organization and conflicts must be explained within that historical context.[4]

In the following sections, we describe the historic changes and conflicts in the Everglades/South Florida region in the past century using the ecologic-economic organization heuristic. We then discuss possible changes and conflicts that will emerge in the future.

3. THE EVERGLADES/SOUTH FLORIDA REGION AS AN ECOLOGIC-ECONOMIC ORGANIZATION

The historic (pre-1880) Everglades ecosystem consisted of three interconnected water systems: the Kissimmee River, Lake Okeechobee, and the Everglades marsh stretching from the south shore of Lake Okeechobee to Florida Bay (see Figure 2). The spatial

[4]As Westhoff et al. (pp. 19-23) point out, the problem of predicting movement between multiple equilibria in complex systems is not unique to Kauffman's framework. The advantage of this heuristic is that it highlights the variety of organizational forms and attribute interactions which most other economic theories of organizations ignore. Our intent is to show how this framework can contribute to a richer understanding of the interplay between ecological processes and social conflicts.

scale of the region spanned nearly 11,000 square miles yet was essentially flat except for a slight decrease in elevation from Lake Okeechobee south to Florida Bay. Rainfall in the northern part of the region might have taken as long as 8 months to reach Florida Bay (Holling et al. 1994, p. 745). The primary driving force at the ecosystem scale was precipitation through highly variable cycles of dry and wet seasons and periodic hurricanes. Average rainfall across the region has been more than 50 inches per year (Fernald and Patton 1984). Yet due to the large spatial scale, the region was like a giant sponge that slowly expanded and contracted in response to rainfall events. This spatial scale coupled with a subtropical climate provided a sawgrass marsh and tree hammock dominated landscape to support one of the most biologically diverse wetland ecosystems in the world.[5]

Although Florida was colonized in the sixteenth century, few were willing to venture into the southern region. Following the Seminole Indian War, federal surveyors were sent to the region in the 1840s to map the landscape. They found land that was underwater at least three months a year and some was always underwater except during periodic droughts. The surveyors described its "utter worthlessness to civilized man, in its present condition" and as a place "fit only for the resort of reptiles" (Carter 1974, p. 62).

Over the past century this landscape has been compartmentalized into geographic subunits connected by canals, levees, and pumping stations designed to supply water for agricultural and municipal uses, provide flood control, and maintain the remaining ecosystem. The interactions between ecological and social processes, and resulting conflicts that led to this transformation, can be viewed in terms of two distinct periods.

3.1 The Wetland Frontier (1800s–1947)

At statehood in 1845 Florida's population was 55,000, all in the northern one-third of the state. More than three-quarters of the land within Florida was owned by the U.S. government. In the Swamp Lands Act of 1850, Congress established a means for the State of Florida to carve out 20 million acres of federal public domain lands. The goal of the Act was to make these wetlands productive by drainage and levee construction. Senator J.D. Westcott, Jr., of Florida, who pressed for passage of the Act, envisioned "reclaiming" the Everglades. The view at the time was that it would be easy and profitable. The image of homesteading families was promoted. Leaders recognized the growing population and affluence of the people along the Atlantic coast and saw the opportunity to market tropical agricultural products. Others such as Stephen R. Mallory (a future U.S. Senator) believed "it will be found wholly out of the question to drain all the Everglades" (Carter 1974, p. 63).

---

[5]Further details on historic conditions in the region are available in Davis and Ogden (1994); Douglas (1947); and Lodge (1994).

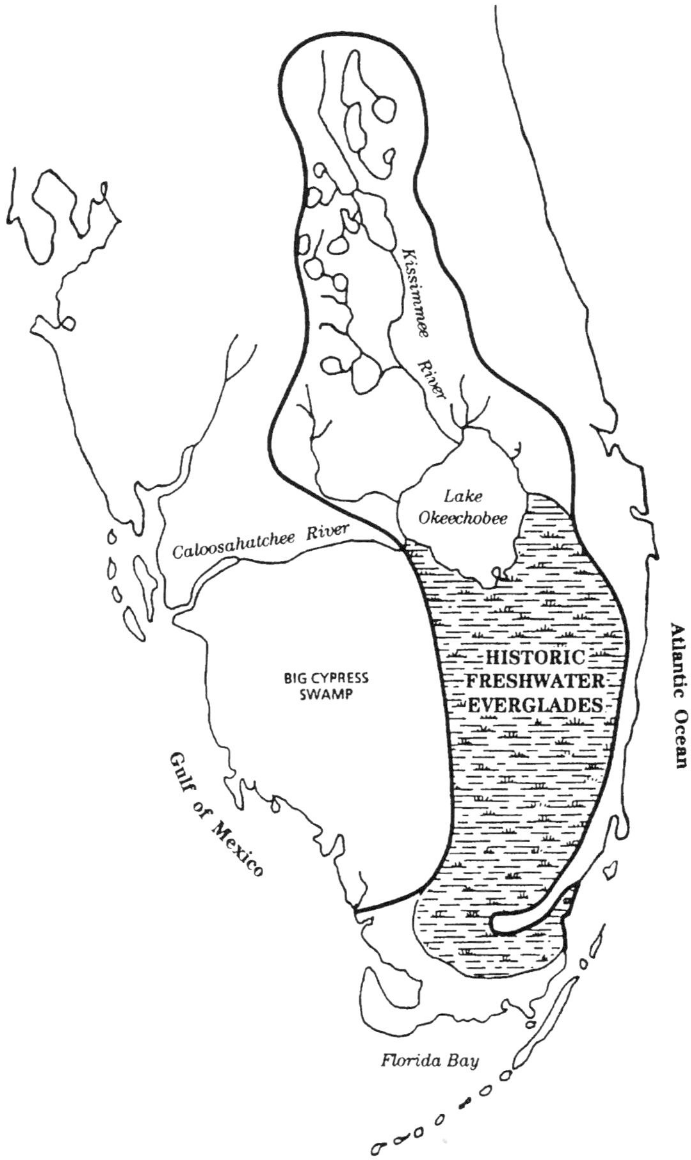

Figure 2. The Historic Everglades/South Florida Ecosystem

In reality, no drainage activities were started until the 1880s. Initial efforts to promote drainage by transferring lands to private railroad companies resulted in a financial disaster for the companies and for the Trust Board set up to administer the reclamation efforts. One of the first successful drainage efforts was completed by

Hamilton Disston, a wealthy northern industrialist, who lowered the level of Lake Okeechobee by dredging a connection to the Gulf of Mexico through the Caloosahatchee River (Figure 2). This opened parts of the area around Lake Okeechobee to agriculture. In the 1890s, railroads were completed down the east and west coasts and the railroad owners built luxury hotels to lure northern tourists. Development of the coasts began. Florida was seen as a frontier in which an enterprising person could make a fortune. By 1900, the population in South Florida grew to more than 20,000 people (Pierce 1995).

Despite these drainage efforts, most of the region remained in its natural condition. Yet the success of agriculture on the newly drained land fostered new plans by Governor Napoleon Bonaparte Broward to "tap the wealth of the fabulous muck" (Carter 1974, p. 68). By 1917, four major canals had been cut from the Atlantic to Lake Okeechobee: the Miami Canal—85 miles; the North New River Canal—58 miles; the Hillsboro Canal—51 miles; and the West Palm Beach Canal—42 miles. With the completion of the St. Lucie Canal from Lake Okeechobee due east to the Atlantic, by the mid 1920s these canals combined with Disston's's Caloosahatchee Waterway to the Gulf provided a means to drain the region. In addition, Chapter 298 was added to the Florida Statutes in 1913 which allowed owners of a majority of the acreage of any contiguous body of wet or overflowed lands to form a drainage district. A district had the power to levy a tax simply by petitioning a circuit court judge and presenting a reclamation plan. In addition to taxing authority, a district could have canals constructed both within and outside their district. The goal was to deal with the "common enemy"—flood waters (Carter 1974, p. 73) The result was numerous drainage districts acting independently to alter the flow of water throughout the region.

Any hope that natural processes had been controlled were dashed by a series of wet years during the 1920s and major hurricanes in 1926 and 1928. The preceding two decades had less rainfall than normal, so when heavy rains came the canals proved to be undersized. Moreover, the drainage districts could not coordinate flood water removal from developed areas. As a result, the hurricanes flooded millions of acres and caused thousands of deaths. State appeals to the federal government led to a U.S. Army Corps of Engineers' project to build Hoover Dike around the southern shore of Lake Okeechobee. This project stopped the flow of water from the Lake southward, but economic conditions in the 1930s forced the drainage districts into bankruptcy and ended any ongoing management of the existing canals and structures.

The 1930s were dry years but now the problem was overdrainage. Oxidation of the muck soils caused extreme subsidence in agricultural areas. In other areas, muck fires blanketed Miami and Fort Lauderdale with thick clouds of smoke for weeks at a time. The overdrainage was further exacerbated by construction of the Tamiami Trail (U.S. 41). Hydrologically, it created a barrier across the middle of the Everglades, restricting recharge of the broad Biscayne aquifer that underlies the southern portion of the region. The combination of surface water drainage coupled with increasing groundwater withdrawals resulted in saltwater intrusion along the Atlantic coast.

In addition, the seeds for recurring social conflict were sown during the 1920s when a group of horticultural and business interests in Miami pressed to preserve over two million acres of contiguous tropical habitat at the southern end of the region as a national park. While supporters saw an important tourist attraction to promote the region, others saw it as "snake swamp park on perfectly worthless land" (Carter 1974. p. 110). The proposal required the State of Florida to give land back to the Federal government, but much of this land had been deeded to railroad and drainage interests. The economic effects of the Depression increased local desire for a tourism attraction and paved the way for Congressional authorization of Everglades National Park in 1934. Continuing reluctance by the State to purchase lands for the Park delayed the dedication until 1947 and the final boundaries were not settled until 1958. These Park boundaries were little more than 60 percent of the area authorized by Congress and raised many doubts about the long-term viability of the Park in the face of ongoing drainage efforts north of the Park.

The post World War II period marks an important turning point for the Everglades/South Florida region. The war effort and improved transportation increased the population in the region to more than 700,000 by 1950 (Pierce 1995). But, heavy rains returned in the summer of 1947 followed by two hurricanes in the fall. Total rainfall for the year was 8.5 feet resulting in flooding of urban and agricultural areas throughout the region for several weeks at a time. By now it was apparent to the public and elected officials that attempts to control the natural cycles of flood and drought had failed.

3.2 Centralization and Human Domination: 1948–1994

In response to pleas from Florida public officials, in 1948 the U.S. Congress authorized (PL 80-858) the Army Corps of Engineers (Corps) to assist in resolving conflicts over flood control and water supply in the Everglades/South Florida region. The Corps orchestrated a multi-billion dollar construction effort under the Central and South Florida (C&SF) project. At the same time, the state legislature centralized water management into the Central and Southern Florida Flood Control District to operate and maintain the facilities built by the Corps (Blake 1980; DeGrove 1958). Throughout the 1950s and 1960s the Corps built hundreds of miles of levees and canals, water conservation areas (WCA), and pumping stations to control water flows (Light and Dineen 1994). When the main structural components of the C&SF project were completed in the mid-1960s, the region had been fractured into small manageable compartments (see Figure 3) that could be coordinated from a single control room.

Nevertheless, new conflicts emerged. In the natural system, water moved slowly and varied little. In the altered system, controlled water moved quickly and varied widely (Stone and Legg 1992). Plants and animals were required to adapt quickly to changing conditions or suffer the consequences. When droughts occurred in the mid-1960s, no water was delivered to Everglades National Park because District officials gave higher priority to other users (Carter 1974, p. 122). Wading bird rookeries failed throughout the Park raising concerns about the broader impacts on the ecosystem (Ogden 1994). Congressional action (PL 91-282) in 1971 was required to assure minimum flows to the Park. In 1970-71, a record setting drought led to a 750,000 acre fire, saltwater intrusion into drinking water wells, and local government efforts to restrict water usage.

In an effort to balance conflicts over flood control, water supply, and wildlife habitat, the Florida Water Resources Act of 1972 was enacted. The Act created the South Florida Water Management District to replace the Central and Southern Florida Flood Control District and added water quality and habitat protection to the District's list of responsibilities. These new responsibilities would prove to be difficult. Population in the region had surged to more than 3.5 million residents by 1980 (Pierce 1995) and increasing demands for agricultural products (especially sugarcane due to the embargo of products from Cuba (Snyder and Davidson 1994) added new competition for the ecosystem's services. A particularly serious crisis erupted in the early 1980s over the disposal of drainage water from the Everglades Agricultural Area (EAA).

During heavy periods of rainfall, drainage water was pumped from the EAA into Lake Okeechobee and the Water Conservation Areas (WCA). But, in the late 1970s when algal blooms appeared in the Lake, the District halted pumping from the EAA into Lake Okeechobee. While the action served to improve Lake water quality, it shifted the pollution problem back to the WCA. The WCA soon began sprouting noxious algae and, by 1986, algal blooms were so extensive that the water quality problem became a matter of national attention. The State responded with passage of the Surface Water Improvement and Management (SWIM) Act of 1987 to study impacts of water diversions from Lake Okeechobee, to develop and implement best management practices, to create a pollution permit process, and to monitor water quality (Stone and Legg 1992). But in 1988 the U.S. Department of Justice (joined by five environmental groups) sued the District and the State of Florida for violating state and federal water quality laws (Stone and Legg 1992; U.S. General Accounting Office, 1995). Following years of litigation and protracted negotiations, the State of Florida enacted the Everglades Forever Act in 1994. The Act reflected an agreement between the parties in the litigation which establishes a comprehensive plan to improve water quality and quantity and prevent the spread of exotic species in the "Everglades Protection Area" (see Figure 3). A central feature of the plan is the construction of man-made marshes or "stormwater treatment areas" (STAs) on former agricultural land to filter phosphorus from drainage water of the EAA. The Act envisions construction of more than 44,000 acres of STAs over the next twenty years (South Florida Water Management District 1995).

The ecological impacts of the C&SF project have been dramatic. More than 50 percent of the area covered by Everglades wetlands have been converted to agricultural and urban development (Davis et al. 1994). The water management system created "a yo-yo situation ... which oscillates wildly up and down between problems of too much water and problems of too little water. The inherent buffering capacity of natural systems has been almost extinguished or greatly weakened" (Fernald and Patton 1984, p. 126). Wading bird rookeries have been lost or disrupted resulting in large reductions in their populations (Lodge 1994; Ogden 1994). Natural vegetation has been replaced in many areas by exotic species such as Australian pine, cattails, and melaleuca (Finkl 1995). Of the native species of flora and fauna that remain, 48 are endangered, 14 are threatened, and 62 are candidates to be listed (U.S. Army Corps of Engineers 1994).

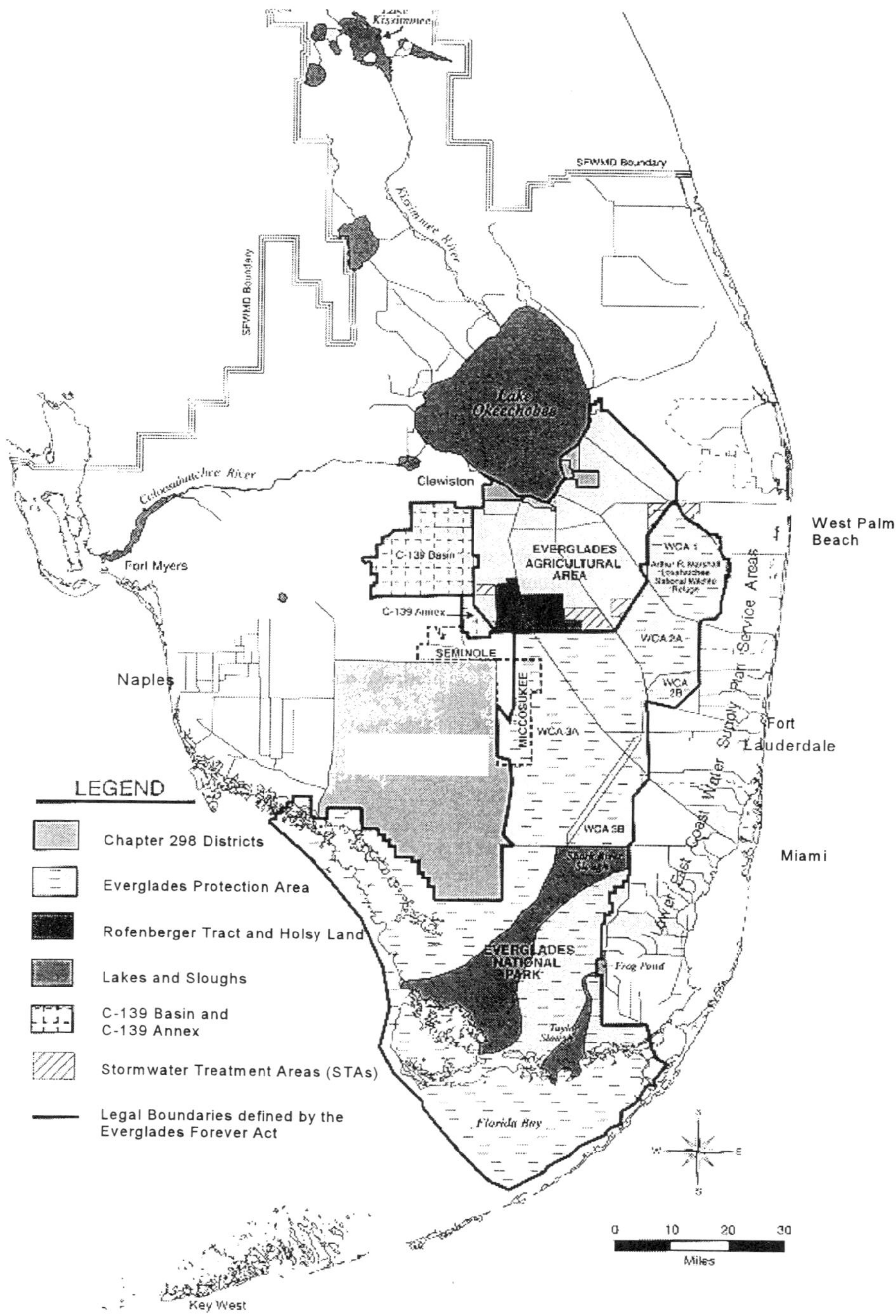

Figure 3. The Present Everglades/South Florida Ecosystem

3.3 Changes in the Ecologic-Economic Organization

These changes in the Everglades/South Florida ecological-economic organization can be represented using the attribute ensemble heuristic presented earlier. Figure 4 indicates that in the 1800s the ecological system had a large geographic scale with broad biodiversity and stable dynamics due to the continuity of the landscape. The scope of services demanded by the small population in the region prior to the 1900s was little more than water, land, and animals for local food supplies. Decisions about drainage, flood control, and water management were made by independent decisionmakers. By 1947, demands for ecosystem services and the scope of the market for these services created new sources of conflict within the organization. South Florida was becoming the focus of agricultural and urban development. Yet the flood control infrastructure could not protect these uses of the ecosystem's services. Moreover, the newly established Everglades National Park juxtaposed demands for other uses of the ecosystem. Changes in the state of these attributes set off major structural perturbations within the organization.

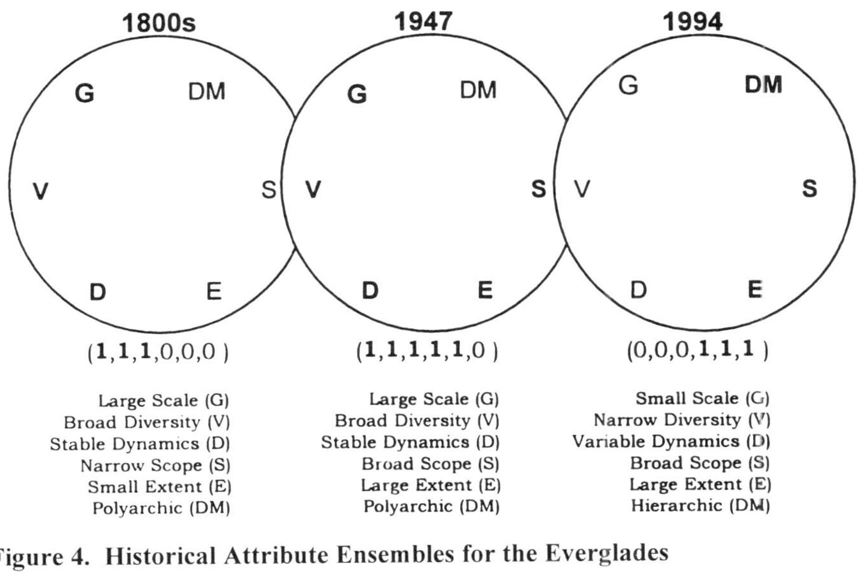

Figure 4. Historical Attribute Ensembles for the Everglades

These perturbations are reflected in the changes from 1948 to 1994 (see Figure 4) resulting in the centralization of decision-making authority within the U.S. Army Corps of Engineers and the South Florida Water Management District. This change could be characterized (using Kauffman's Boolean relations) as an *And* relation between the attributes Scope (S) and Extent (E) with the decision-making structure (DM) the regulated attribute.[6] Also, the water management infrastructure created in this period transformed the broad, continuous landscape into a checkerboard of regulated uses and activities, each operating at their own small scale under the central authority. The consequences of altering the scale and stability of the ecosystem was a reduction in the natural biological diversity.

The history of the Everglades/South Florida region reflects the conflicts between ecological processes and social institutions. Whether the changes that have evolved occurred by chance or reflect dominant ordering relationships within ecological-economic organizations cannot be determined from a single case study. The attribute heuristic helps provide a framework to consider simultaneous interactions between attributes and provides a pathway to consider whether other possible structures may improve the fitness of the organization.

4. RECONFIGURING THE ORGANIZATION

The ecological impacts of the C&SF project and the related conflicts between local, state, and federal interests have not gone unnoticed by the public and the media (e.g., Mairson 1994). In 1993, President Clinton created the South Florida Ecosystem Restoration Task Force to integrate the planning and programs of 6 federal departments and agencies [Agriculture, Commerce, Defense (Army Corps of Engineers), Interior, Environmental Protection, and Justice]. Under the direction of Interior Secretary Babbitt, the Task Force sought to initiate and coordinate environmental restoration activities in South Florida . The membership of the Task Force was expanded in 1995 to include relevant state agencies and the Miccosukee and Seminole Tribes of Florida. Congress institutionalized this entity as the official decision-making authority for federal funding of Everglades/South Florida restoration projects in the Water Resources Development Act of 1996. The District remains the local sponsor and manager of facilities and infrastructure. Under the Act, the Corps (in cooperation with other members of the Task Force) is responsible for developing a final plan "for the purpose of restoring, preserving, and protecting the South Florida ecosystem" [Sec 528 (5)(A)(I)] no later than July 1, 1999.

This approach to coordinate federal and state interests represents a significant departure from traditional decision-making authority in ecological-economic organizations, particularly for federal agencies such as the Corps (U.S. General Accounting Office 1995). Described as "ecosystem management," the goal of this approach is "to restore and sustain the health, productivity, and biological diversity of ecosystems and the overall quality of life through a natural resource management approach that is fully integrated with social and economic goals" (Interagency Ecosystem Management Task Force, Vol. 1, 1995, p. 17). The difficulty, however, is reaching agreements on ecological and economic goals across diverse stakeholders (U.S. General Accounting Office 1994).

These events raise some important questions about the further evolution of the

---

[6]This example is intended to be illustrative of the possible attribute relations; obviously other relations could be described.

Everglades/South Florida organization. Of paramount concern is whether the ecosystem management approach can change the centralized decision-making that has prevailed since 1947. The emergence of broad markets for a diverse array of services from the regional ecosystem has also created powerful stakeholders in agricultural, environmental, and urban development groups. Conflicts between these interest groups could be balanced and controlled through appointments to the District Governing Board. What remains to be determined is whether the collective efforts of these interest groups will act to block changes, performing the role of "canalyzing functions," so that the current organizational structure is frozen in place.

A closely related concern is whether some of the original ecological attributes of the Everglades/South Florida region can be restored. Certainly the original scale of the ecosystem will not return, unless nearly 3,000 square miles of former wetlands now occupied by agricultural operations, homes, and businesses were suddenly abandoned by their owners. But is it possible to restore the biological diversity and stability that once existed as if one were simply making a "photocopy reduction" of the original? It seems likely that, at best, some partial restoration of these ecological attributes can be achieved if some consensus emerges on desired endpoints as part of the ecosystem management process (Milon et al. 1997). This may be a fragile consensus leading to new rounds of conflicts if the ecological system produces "surprises" in response to restoration efforts.

REFERENCES

Anderson, T., ed. 1983. *Water Rights: Scarce Resource Allocation, Bureaucracy, and the Environment.* San Francisco: Pacific Institute for Policy Research.

Arrow, K., B. Bolin, R. Costanza, P. Dasgupta, C. Folke, C.S. Holling, B-O. Jansson, S. Levin, K-G. Mäler, C. Perrings, and D. Pimentell. 1995. Economic Growth, Carrying Capacity, and the Environment, *Science* 268: 520-521.

Arthur, W. 1988. Self-Reinforcing Mechanisms in Economics. In P. Anderson, K. Arrow, and D. Pines, eds., *The Economy as an Evolving Complex System.* Reading, MA: Addison-Wesley.

Blake, N. 1980. *Land into Water—Water into Land.* Tallahassee, FL: University Presses of Florida.

Boulding , E. 1991. States, Boundaries and Environmental Security in Global and Regional Conflicts, *Interdisciplinary Peace Research* 3: 78-93.

Bowes, M., and J. Krutilla. 1985. Multiple Use Management of Public Forestlands. In A. Kneese and J. Sweeney, eds., *Handbook of Natural Resource and Energy Economics, Vol. II.* Amsterdam: Elsevier Science Publishers.

Carter, L. 1974. *The Florida Experience: Land and Water Policy in a Growth State.* Washington, D.C.: Resources for the Future.

Davis, S., L. Gunderson, W. Park, J. Richardson, and J. Mattson. 1994. Landscape Dimension, Composition, and Function in a Changing Everglades Ecosystem. In S. Davis and J. Ogden, eds., *Everglades: The Ecosystem and Its Restoration.* Delray Beach, FL: St. Lucie Press.

Davis, S., and J. Ogden, eds. 1994. *Everglades: The Ecosystem and Its Restoration.* Delray Beach, FL: St. Lucie Press.

DeGrove, J. 1958. The Central and Southern Florida Flood Control Project: A Study in Intergovernmental Cooperation and Public Administration. Ph.D. dissertation, University of North Carolina, Chapel Hill.

Demsetz, H. 1967. Toward a Theory of Property Rights, *American Economic Review* 57: 347-373.

Douglas, M.S. 1947. *The Everglades: River of Grass.* Miami: Banyan Books (reprinted 1986).

Elliott, L. 1996. Environmental Conflict: Reviewing the Arguments, *Journal of Environment and Development* 5: 149-167.

Fernald, E., and D. Patton. 1984. *Water Resources Atlas of Florida.* Tallahassee, FL: Florida State University.

Finkl, C. 1995. Water Resource Management in the Florida Everglades: Are 'Lessons from Experience' a Prognosis for Conservation in the Future?, *Journal of Soil and Water Conservation* (11-12): 592-600.
Gleick, P. 1991. Environment and Security: The Clear Connections, *Bulletin of the Atomic Scientists* April: 17-21.
Holling, C.S., L.G. Gunderson, and C.J. Walters. The Structure and Dynamics of the Everglades System: Guidelines for Ecosystem Restoration. In Davis, S. and J. Ogden, eds. *Everglades: The Ecosystem and Its Restoration*. Delray Beach, FL: St. Lucie Press,
South Florida Water Management District. 1995. Everglades: 1995 Annual Report. West Palm Beach, FL: South Florida Water Management District.
Homer-Dixon, T. 1991. On the Threshold: Environmental Changes as Causes of Acute Conflict, *International Security* 16:76-116.
Homer-Dixon, T., J. Boutwell, and G. Rathjens.. 1993. Environmental Change and Violent Conflict, *Scientific American* 268: 16-23.
Interagency Ecosystem Management Task Force. 1995. The Ecosystem Approach: Healthy Ecosystems and Sustainable Economies, Vol. I: Overview. Washington, D.C.
Kauffman, S.A. 1991. Antichaos and Adaptation, *Scientific American* pp. 78-84.
Kauffman, S.A. 1993. *The Origins of Order: Self-Organization and Selection in Evolution*. New York: Oxford University Press.
Kleindorfer, P.R., H.C. Kunreuther, and P.J. Schoemaker. 1993. *Decision Sciences*. Cambridge: Cambridge University Press.
Kosko, B. 1993. *Fuzzy Thinking: The New Science of Fuzzy Logic*. New York: Hyperion.
Light, S.S., L.H. Gunderson, and C.S. Holling. 1995. The Everglades: Evolution of Management in a Turbulent Ecosystem. In L.H. Gunderson, C.S. Holling, and S.S. Light, eds, *Barriers and Bridges to the Renewal of Ecosystems and Institutions*. New York: Columbia University Press.
Light, S.S., and J.W. Dineen. 1994. Water Control in the Everglades: A Historical Perspective. In S.M. Davis and J.C. Ogden, edsd., *Everglades: The Ecosystem and Its Restoration*. Delray Beach, FL: St. Lucie Press, 1994.
Lodge, T.E. 1994. *The Everglades Handbook: Understanding the Ecosystem*. Delray Beach, FL: St. Lucie Press.
Mairson, A. 1994. The Everglades: Dying for Help, *National Geographic* 185: 2-35.
Milon, J., C. Kiker, and D. Lee. 1997. Ecosystem Management and the Florida Everglades: The Role of Social Scientists, *Journal of Agricultural and Applied Economics* 29.
North, D.C. 1990. *Institutions, Institutional Change, and Economic Performance*. Cambridge: Cambridge University Press.
Ogden, J.C. 1994. A Comparison of Wading Bird Nesting Colony Dynamics (1931–1946 and 1974–1989) as an Indication of Ecosystem Conditions in the Southern Everglades. In S,N, David and J.C. Ogden, eds., *Everglades: The Ecosystem and Its Restoration*. Delray Beach, FL: St. Lucie Press, 1994.
Pierce, A., ed. 1995. *Florida Statistical Abstract 1995*. Gainesville, FL: University Press of Florida.
Rosser, J.B. 1995. Systemic Crises in Hierarchical Ecological Economies, *Land Economics* 71: 163-72.
Schotter, A. 1981. *The Economic Theory of Social Institutions*. Cambridge: Cambridge University Press.
Snyder, G., and J. Davidson. 1994. Everglades Agriculture: Past, Present, and Future. In S.M. Davis and J.C. Ogden, eds., *Everglades: The Ecosystem and Its Restoration*. Delray Beach, FL: St. Lucie Press.
South Florida Water Management District. 1995. Everglades: 1995 Annual Report. West Palm Beach, FL: South Florida Water Management District.
Stone, J., and D. Legg. 1992. Agriculture and the Everglades, *Journal of Soil and Water Conservation* (5-6): 207-215.
U.S. Army Corps of Engineers. 1994. Central and Southern Florida Project Comprehensive Review Study. Jacksonville, FL.
U.S. General Accounting Office. 1994. *Ecosystem Management: Additional Actions Needed to Adequately Test a Promising Approach*. Washington, D.C.
U.S. General Accounting Office. 1995. *Restoring the Everglades: Public Participation in Federal Efforts*. Washington, D.C.
Waldrop, M. 1992. *Complexity: The Emerging Science at the Edge of Order and Chaos*. New York: Simon and Schuster.
Westhoff, F.H., B.V. Yarbrough, and R.M. Yarbrough. 1996. Complexity, Organization, and Stuart Kauffman's "The Origins of Order," *Journal of Economic Behavior and Organization* 29: 1-25.

Witt, U. 1993. Emergence and Dissemination of Innovations: Some Principles of Evolutionary Economics. In R.H. Day and P. Chen, eds., *Nonlinear Dynamics and Evolutionary Economics*. London: Oxford University Press.

Yaffie, S., A. Phillips, I. Frentz, P. Hardy, S. Maleki, and B. Thorpe, eds. 1996. *Ecosystem Management in the United States*. Washington, D.C.: Island Press.

Zilberman, D., M. Wetzstein, and M. Marra. 1993. The Economics of Renewable and Nonrenewable Resources. In G.A. Carlson, D. Zilberman, and J.A. Miranowski, eds., *Agricultural and Environmental Resource Economics*. New York: Oxford University Press.

14

PRIVATIZATION AND REGULATION OF MULTI-SOURCE WATER USAGE

Eyal Brill*
Hebrew University of Jerusalem, Israel
Gahl Hochman
University of Tel-Aviv, Israel
Eithan Hochman
Hebrew University of Jerusalem, Israel

The chapter deals with public utility firms which supply two intermediary inputs, one with positive externality, and the other with a negative one. Both inputs have an intertemporal impact on the environment. It is proven that the trajectory path is unique, and of a saddle-point nature. Based on the optimal trajectory, the chapter addresses the privatization of the public utility firm, so the tax policy changes over time, and management is separated from the owners. Finally, the optimal tax policy is compared with a specific form of standard around the steady state, when there is an error in the coefficient of return. The main finding in this section is that under imperfect information the special form of standard is superior (inferior) if the demand for water is elastic (inelastic).

1. INTRODUCTION

Water is extracted from different sources. So if water from different sources damages or benefits an area in different degrees, then conjunctive usage of water has to be considered in order to maximize social benefit. An example is water used from fresh water versus recycled water. With little doubt, each of these sources influences the environment in different degrees. The West Yezreel Valley in Israel serves as an example. Available data indicate that utilizing groundwater over time creates salinity of both soil and groundwater and deteriorates maximal potential yield. Therefore, welfare-maximizing policy must consider long-term effects of using recycled water and weigh them against using fresh water.

---

* The authors are grateful to the Department of Agricultural and Resource Economics, University of Maryland, for support that led to the completion of this paper. At the time of this study Eyal Brill held a post doctoral appointment and Eithan Hochman was visiting profess, both at the University of Maryland, College Park.

This chapter deals with a public utility supplying water from two intermediary inputs, one with a positive externality, and the other with a negative one.[1] One example, which is modeled in this chapter, is a water district that supplies regional water demand from two sources—imported surface water and groundwater pumped from a local aquifer (see Figure 1). Both inputs have an intertemporal impact on the environment, i.e., today's utilization of the damaging water input influences tomorrow's fertility. A dynamic model is used to structure the relationship and to derive the optimal allocation policy of such firms. Using a dynamic framework, it is shown that optimal policy is unique and leads to a steady-state equilibrium which is a saddle-point.

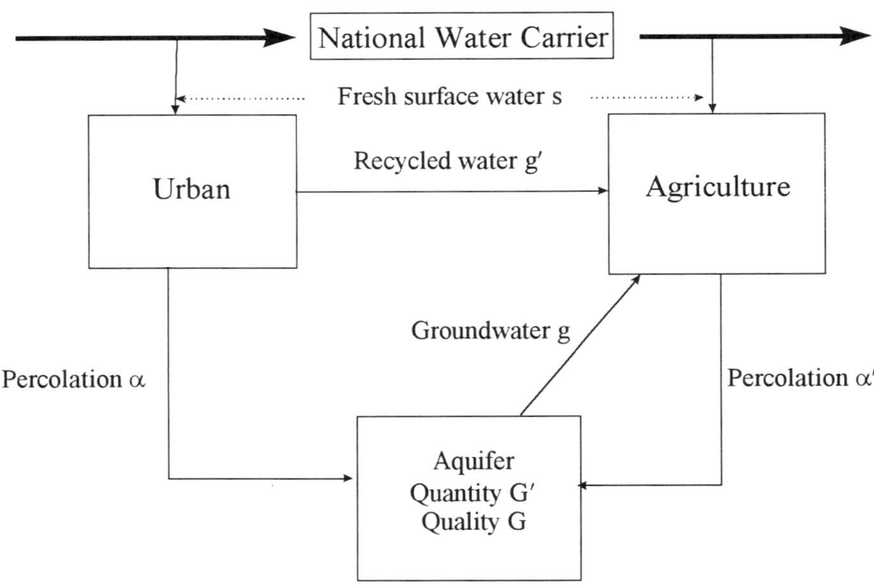

Figure 1. Schematic Flow Diagram of Regional Water Balance

Public utilities may suffer inefficiencies when compared to privately owned firms, which may favor privatization [see Vikers and Yarrow (1988) for details]. If privatization occurs, however, the newly privatized firm should be regulated in order to prevent it from exploiting monopoly power and damaging the environment by over-using the undesirable input. We address these two issues (monopoly power and externality regulation) by examining the case of a newly privatized regional water agency. We assume that the agency buys surface water from outside the region (e.g., from a national water aqueduct) and generates water from a local

[1] See the discussion on desirable and undesirable public goods in Hochman and Hochman (1980).

aquifer. Its monopoly position enables the firm to charge a price which exceeds the marginal cost of generating the water. Simultaneously, over-pumping of cheap water from the local aquifer enables the firm to increase its profits even more. Hence, regulation of the privatized firm is required (the Assertion is that the firm acts myopically or its objective function is different from the objective function of the region's social planner).

To assess the regulation policy around the steady state, a regional welfare function is examined. Using the optimal policy derived from the dynamic framework, we examine a specific form of standard and compare it with the traditional tax policy around the steady state. The main finding is that, under imperfect information (an error in the coefficient of return), a special form of standard is superior (inferior) if the demand for water is elastic (inelastic). This finding is investigated using data from the West Yezreel Valley in Israel.

Section 2 computes and describes the dynamic system, while Section 3 addresses the privatization of the public utility firm. Section 4 investigates two policy tools around the steady state and shows under what criteria one policy is superior to the other when there is an error in assessment of the coefficient of return. Section 5 summarizes the results and points out further possibilities for research.

2. DYNAMIC CONJUNCTIVE USE OF ABATING AND POLLUTING INPUTS

Assume the case of a regional water district that uses s_t cubic meters of fresh imported water each time period, and generates g_t cubic meters of water from a local aquifer. Both fresh and local water is used for irrigation. The price of imported fresh water, w, is fixed, while the average cost of generating local water is $z(G_t)$, which depends on the level of water in the aquifer, G_t, at time t. The yield response to the water salinity level affected by the water level of the aquifer is $p(G_t)$, with $\partial p_t/\partial G_t \le 0$, i.e., a decline in the groundwater level causes salinity intrusion which causes yields to decline. Let $f[g_t, s_t, p(G_t)]$ be a (stationary over time) regional benefit or utility function, with the following properties:

$$\frac{\partial f}{\partial g_t} \ge 0, \quad \frac{\partial f}{\partial s_t} \ge 0, \quad \frac{\partial f}{\partial p_t} \le 0, \quad \frac{\partial^2 f}{\partial g_t^2} \le 0, \quad \text{and} \quad \frac{\partial^2 f}{\partial s_t^2} \le 0.$$

Regional welfare maximization is attained by solving the dynamic maximization problem,

$$\underset{g_t, s_t}{\text{Max}} \int_0^\infty \{f[g_t, s_t, p(G_t)] - z(G_t)g_t - w_t s_t\} e^{-\beta t} dt \tag{1}$$

subject to

$$\dot{G} = -g_t + \alpha(g_t + s_t) \tag{2}$$

$$G(0) = G_0 \tag{3}$$

where G_0 is the state of the aquifer at time 0, and β is the subjective time preference. The equation of motion in (2) measures the change in level of the aquifer caused by pumping groundwater, g_t, given the return coefficient of water into the aquifer, α, with $0 < \alpha < 1$. Tsur (1991) gives a detailed description of such a system. It should be noted that the water rights system used by at least two states (Arizona and New Mexico) is based on the idea of a return coefficient (see, for example, Gisser 1983). Such a return coefficient was introduced also by Zilberman and Caswell (1985) in the case of drip irrigation and is used by Burke et al. (1997) in this volume.

The Hamiltonian for the dynamic problem in current value is

$$H = \hat{H} e^{\beta t} = f[g_t, s_t, p(G_t)] - z(G_t)g_t - w_t s_t + m_t [-g_t + \alpha(g_t + s_t)] \tag{4}$$

where the co-state variable m_t is the imputed shadow price of the level of the aquifer at period t, i.e., it measures the contribution of one cubic meter of G_t to the regional welfare function from period t to the end of the planning horizon. Assuming an internal solution, the first order conditions are

$$\frac{\partial H}{\partial g_t} = f_{g_t} - z(G_t) - m_t(1-\alpha) = 0, \tag{5}$$

$$\frac{\partial H}{\partial s_t} = f_{s_t} - w_t + m_t\alpha = 0. \tag{6}$$

Sufficient conditions for internal maximization are

$$\frac{\partial^2 H}{\partial g_t \partial g_t} < 0, \quad \frac{\partial^2 H}{\partial s_t \partial s_t} < 0.$$

The Maximum Theorem implies

$$\dot{m}_t = \beta m_t - (f_{p_t} pG_t - zG_t g_t). \tag{7}$$

Equation (5) implies that at the optimum the marginal contribution of a cubic meter of g_t is equal to the average cost of pumping water from the local aquifer, plus the cost of reduction in the aquifer level, $f_{g_t} = z(G_t) + m_t(1 - \alpha)$, while equation (6) implies that the marginal value of surface water at the optimum is equal to the price of surface water at the region border minus the benefit of additional surface water to the local aquifer, $f_{s_t} = w_t - m_t\alpha$. The derivative of the Hamiltonian is strictly monotone along the optimum path, from the second order condition. The optimum equations, (5) and (6), implicitly give the control variables,

$$g_t = \phi(G_t, m_t), \tag{8}$$

$$s_t = \varphi(G_t, m_t). \tag{9}$$

The properties of equations (8) and (9) are assumed to satisfy

Assertion 1.

(A) The higher the level of the aquifer, G_t, the larger the amount of ground water used, g_t, i.e., $\partial g_t/\partial G_t = \phi_G \geq 0$. The higher the co-state variable, m_t, the lower the amount of groundwater used, $\partial g_t/\partial m_t = \phi_m \leq 0$.
(B) The higher the level of the aquifer, G_t, the lower the amount of surface water used, s_t, i.e., $\partial s_t/\partial G_t = \varphi_G \leq 0$. The higher the co-state variable, m_t, the larger the amount of surface water used, $\partial s_t/\partial m_t = \phi_m \geq 0$.

From Assertion 1, it is assumed that the higher the level of the aquifer, the lower is the co-state. The economic reasoning for this Assertion is that depleting the aquifer makes one more unit of pumped groundwater more valuable due to the pollution it causes.

From Equations (8) and (9), the equation of motion can be rewritten as

$$\dot{G}_t = (\alpha - 1)\phi_t(G_t, m_t) + \alpha\varphi_t(G_t, m_t). \tag{10}$$

Thus, at the steady state (denoted by an 'ss' subscript), the two singular equations in the (G, m) plane are[2]

$$0 = (\alpha - 1)\phi_{ss} + \alpha\varphi_{ss}, \tag{11}$$

$$0 = \beta m_{ss} - \gamma_{ss}, \tag{12}$$

where $f_p p_{G_{SS}} - z_{G_{SS}} g_{ss} = \gamma_{ss.}$ To determine the nature of the equilibrium, a first order approximation of the linear equations (11) and (12) at the steady state yields

$$\dot{G} = a(G - G_{ss}) + b(m - m_{ss}), \tag{13}$$

$$\dot{m} = c(G - G_{ss}) + d(m - m_{ss}), \tag{14}$$

where

$$a = (\alpha - 1)\phi_{G_{ss}} + \alpha\varphi_{G_{ss}},$$
$$b = (\alpha - 1)\phi_{m_{ss}} + \alpha\varphi_{m_{ss}},$$

---

[2] Note that the singular equations (11) and (12) are autonomous with respect to the time variable t.

$$c = -[\gamma_{G_{ss}} + \gamma_{s_{ss}} \varphi_{G_{ss}} + \gamma_{g_{ss}} \phi_{G_{ss}}],$$
$$d = \beta - \gamma_{s_{ss}} \varphi_{m_{ss}} - \gamma_{g_{ss}} \phi_{m_{ss}} .$$

In order to prove that the dynamic system is of a saddle point nature, the following Assertion is made.

Assertion 2.

The benefit function $f[g_t, s_t, p(G_t)]$ has the following properties:

(A) first,

$$f_{sp} = \frac{\partial}{\partial s}\left[\frac{\partial f}{\partial p}\right] > 0, \; f_{gp} = \frac{\partial}{\partial g}\left[\frac{\partial f}{\partial p}\right] < 0$$

because pollution in the model is not affected by the amount of water used, whereas it is affected by the level of the aquifer;

(B) second,

$$z_{GG} > 0, \;\; f_{pp} < 0 \;\text{ and }\; p_{GG} > 0$$

which ensure

$$\gamma_G = f_p\, p_{GG} + f_{pp}\, (p_G)^2 - z_{GG}\, g < 0;$$

(C) third,

$$\gamma_{S_{SS}} = f_{ps}\, p_G \leq 0,$$

i.e., the more water that comes from an outside source, the smaller the marginal negative effect of pollution on the benefit function (this Assertion is based on part A of Assertion 1, and the Assertion $p_G \leq 0$);

(D) fourth,

$$\gamma_{g_{ss}} = f_{gp}\, p_G - z_G \geq 0, \;\; z_G \leq 0,$$

i.e., the higher the level of the aquifer the cheaper to pump water from it. The pollution effect, $f_{gp}\, p_G \geq 0$, is positive by part A of the Assertion 1 and $p_G \leq 0$.

Using the roots of the characteristic equation yields

$$\lambda_1, \lambda_2 = \frac{a + d \pm \sqrt{(a+d)^2 - 4(ad - bc)}}{2} \; .$$

Using Assertions 1 and 2, the following proposition proves that the dynamic system is of a saddle-point nature.

Proposition 1. If c > 0, the dynamic system described by (13) and (14) results in a steady state unique solution (a saddle-point).

Proof. For a saddle-point solution, the two characteristic roots should be real numbers with opposite signs, i.e., $\lambda_1 > 0, \lambda_2 < 0$, which is equivalent to the condition (ad - bc) < 0. From Assertion 2, $\gamma_{s_{ss}} \leq 0$, $\gamma_{g_{ss}} \geq 0$, $\gamma_{G_{ss}} \leq 0$, and using Assertion 1 results in a < 0, b > 0, and d > 0. Thus, c > 0 is a sufficient condition for (ad - bc) < 0, which obtains a saddle-point. Q.E.D.

The phase diagram in Figure 2 illustrates these results. From the phase diagram, it can be shown that if $G_0 < G_{ss}$ then $m_0 > m_{ss}$, and as the level of the aquifer increases the shadow price decreases. Alternatively, if $G_0 > G_{ss}$ then $m_0 < m_{ss}$, and the level of the aquifer decreases as the value of its co-state variable increases. Thus, while approaching the steady state equilibrium point (G_{ss}, m_{ss}) from either direction along the optimal trajectory, $\partial m/\partial G < 0$.

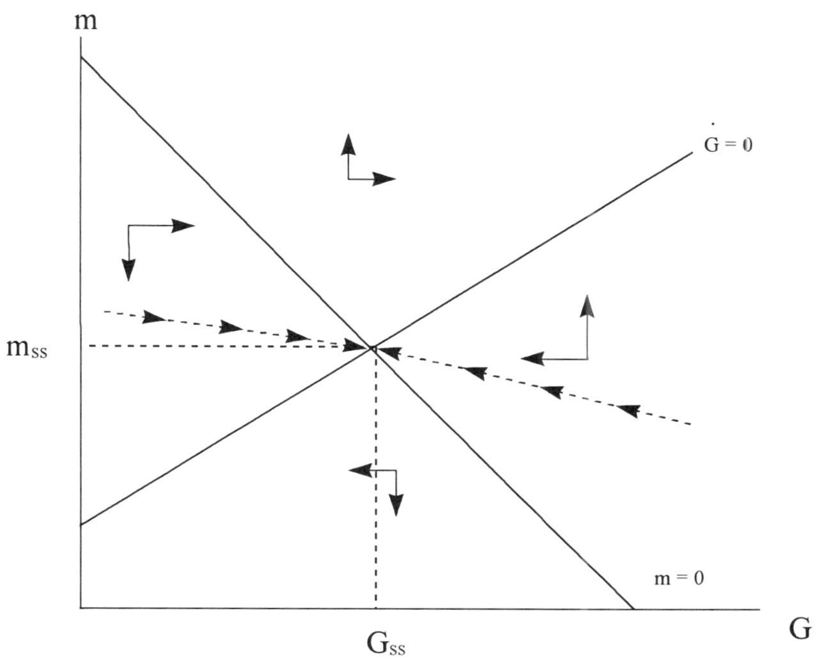

Figure 2. Phase Diagram: Optimal Trajectory and Steady State

Dividing equation (7) by the shadow price and rearranging terms results in

$$\frac{f_{p_t} p_{G_t} - z_{G_t} g_t}{\beta - \rho_t} = m_t . \tag{15}$$

The numerator of the ratio in the left-hand side of (15) measures the current marginal costs of two components: a qualitative one, which measures the damage caused by diminishing quality, $f_p p_G$; and a quantitative one, which measures the increase in the costs of generating water from the aquifer, zG_g. The marginal externality console value of the left-hand side of (15) is obtained by discounting it by the current discount factor β "corrected" by the change in the shadow price, $\rho = \dot{m} / m$.

The economic relation between G and m along the optimal trajectory can be determined from the ratio in (15). The numerator measures the cost of changes in the level of G. It depends on the convexity of the cost of pollution and the cost of pumping the aquifer water with respect to reducing its level G, i.e., $Z_{GG} > 0$ and $P_{GG} > 0$. It also depends on the concavity of the benefit function with regard to pollution, $f_{pp} < 0$. This is according to conventional economic wisdom and also coincides with the Assertions. The denominator includes the opportunity cost of time as measured by the market interest rate, β, minus the effect of the depletion of the aquifer on future generations, measured by the change in its dynamic shadow price, ρ_t. Thus, the "corrected" discount factor, $\beta - \rho_t$, increases (decreases) as G increases (decreases).

2.1 Analysis of the Steady State Results

In the steady state, it can be shown that the shadow price has the property

$$m_{ss} = \frac{f_{p_{ss}} p_{G_{ss}} - z_{G_{ss}} g_{ss}}{\beta} . \tag{16}$$

The marginal externality console value of the right-hand side of (16) is obtained by discounting it by the current discount factor β. It should be noted that when the system is not at the steady state this discount factor is "corrected" by the change in the dynamic shadow price. The marginal product of the two intermediate inputs is

$$f_{g_{ss}} = z(G_{ss}) + \frac{f_{p_{ss}} p_{G_{ss}} - z_{G_{ss}} g_{ss}}{\beta} (1-\alpha) , \tag{17}$$

$$f_{s_{ss}} = w_{ss} - \frac{f_{p_{ss}} p_{Gss} - z_{G_{ss}} g_{ss}}{\beta} \alpha . \tag{18}$$

The marginal product of the two intermediary inputs is the cost of purchasing (or pumping) one more unit of water plus the externality caused by that unit (positive externality for surface water or negative externality for groundwater). This externality is affected by the coefficient of return.

Figure 3 illustrates the existence of a unique internal equilibrium of conjunctive use of g and s by depicting the marginal productivity of surface water and groundwater as a function of ground water at the steady state.

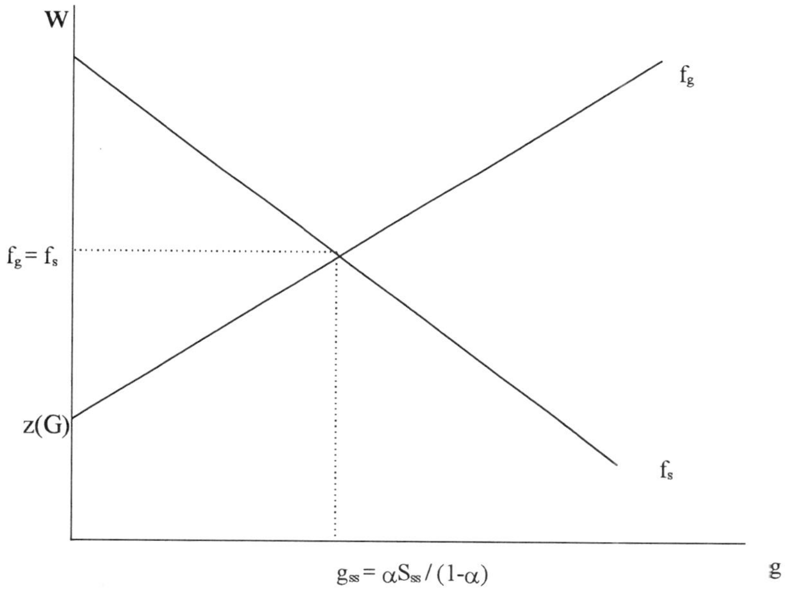

Figure 3. Marginal Benefits of Abating and Polluting Inputs at the Steady State

From Figure 3

1. The economic cost of surface water decreases with the use of ground water,
2. The economic cost of ground water increases with the use of ground water.

These ensure a unique intersection at s_{ss} and $g_{ss} = \alpha s_{ss}/(1 - \alpha)$. The equality of the marginal benefit from groundwater and surface water is also present along the trajectory path when both types of intermediary inputs are used in positive amounts.

3. WATER AGENCY PRIVATIZATION: MONITORING AND REGULATION

In this section, it is assumed that the public agency responsible for the allocation of water at the regional level is undergoing a privatization process. The reason for privatization is not addressed in this chapter (for a discussion on this topic, see Vickers and Yarrow, 1988). Vickers and Yarrow examine privatization of public utilities and argue that a necessary condition for successful privatization is an efficient regulation of the firm after its privatization.

In the case of a regional private water firm, regulation involves two elements. The first one is its monopoly power, which may lead to pricing water above its marginal costs. The second one is the externality costs of conjunctive use of an abating/polluting-input mix. A myopic firm may overuse the "polluting" input to achieve short-term profits, ignoring the long-run externality effects on the local aquifer.

It is argued that efficient regulation of the privatized firm can be accomplished by a double-tool policy. The first one is distribution of equity for regional water users, while the second one is implementation of a tax scheme based on the results obtained in the previous section. In what follows, the efficiency of such a policy is discussed.

3.1 Monitoring the Monopolistic Power in Water Pricing

Assume a local water enterprise owned by shareholders who are the water users in the region. The optimal aggregate quantity Q_t is determined by the intersection of the marginal cost curve with the aggregate demand curve at price w_t, which consists of g_t and s_t. If the private firm sets the local water price equal to w_t, and no tax is imposed on local water, the rent of the private firm will be $R_t = w_tQ_t - C_t(Q_t)$, where $C_t(Q_t)$ is the cost of generating water by the private firm based on the cost of surface water s_t and groundwater g_t (the cost of generating water changes over time due to the change over time in the optimal ratio of groundwater to surface water).

Now consider a model in which the price of water in period t is determined by using a voting system by the shareholders of the private firm. In such a model, subject to the framework explained henceforth, the price of water will be equal to the marginal cost. Shareholders are the water consumers where as their equity claim is determined by their historical water rights (their water rights before privatization). Hence, given N water consumers, $\sum_{n=1}^{N} \alpha_n = 1$, where α_n denotes the nth water user share. Such a user may also be a municipality, which supplies water for households.

The nth water user/shareholder votes for a water price, w_t, to maximize individual benefits from water,

$$\underset{w_t}{\text{Max}}\, \Pi_t^n = u^n(q_t^n) - w_t q_t^n + \alpha_n \{w_t Q_t(w_t) - C_t[Q_t(w_t)]\} \tag{19}$$

where q_t^n is the individual water demand function at time t, $Q_t = \sum_{n=1}^{N} q_t^n$ is the total quantity of water used at time t, $C_t[Q_t(w_t)]$ is the total cost of generating Q_t,

$w_tQ_t(w_t)$ is the revenue from selling this quantity at a price w_t, and Π_t^n is the benefits the nth user has from water based on his direct utility from using the water, $u^n(q_t^n)$, his payment for water, $w_tq_t^n$, and his share of the aggregate rent R_t according to his share α_n of the equity of the water supply firm.

The water user is in a competitive market, so

$$\frac{\partial u^n(q_t^n)}{\partial q_t^n} = w_t .$$

From the definition of $Q_t = \Sigma_{n=1}^N q_t^n$,

$$\frac{\partial Q(q_t^n, q_t^{-n})}{\partial q_t^n} = 1$$

where q_t^{-n} is the amount of water used by all water users except user n at time t.

The first order condition, after rearranging arguments, is

$$0 = -q_t^n + \alpha_n \left[Q_t + (w_t - MC_t) \frac{\partial q_t^n}{\partial w_t} \right]. \tag{20}$$

Suppose user/shareholder n's demand function is downward sloping, i.e., $\partial q_t^n / \partial w_t \le 0$. Using (20), three types of shareholders (Π_1, Π_2 and Π_3) are depicted in Figure 4. The benefit function of the Π_1 type is characterized by a maximum point at $w_t = MC_t$. Type 1 demand for water, q_t^n, is equal to his relative share in the water supply firm, α_nQ_t. The benefit function of the Π_2 type and the Π_3 type are characterized by maxima at $w_t < MC_t$ ($q_t^n > \alpha_nQ_t$) and $w_t > MC_t$ ($q_t^n < \alpha_nQ_t$), respectively. If the equity shares are normally distributed, the median voter is a Π_1 user type. Hence, according to the theory of the median voter, the voting process will result in a marginal pricing policy and therefore an optimal allocation of aggregate quantity, Q_t.

The ratio of groundwater to surface water, g/s, on the optimal trajectory path changes over time. This causes the cost of generating Q_t, represented by $C_t[Q_t(w_t)]$, to change over time as mentioned above. Due to this fact the user/shareholder votes every period for the price of water. It is assumed that the changing ratio's effect on the cost function in a given period, as opposed to between periods, is negligible.

When ownership is separated from management, the monopolistic power, under certain Assertions, is regulated optimally. But because of the nature of the intermediary inputs, a conjunctive monitoring is also required when the firm is myopic. It should be noted that, if the firm is not myopic, conjunctive monitoring may not be needed. Section 3.2 addresses conjunctive monitoring.

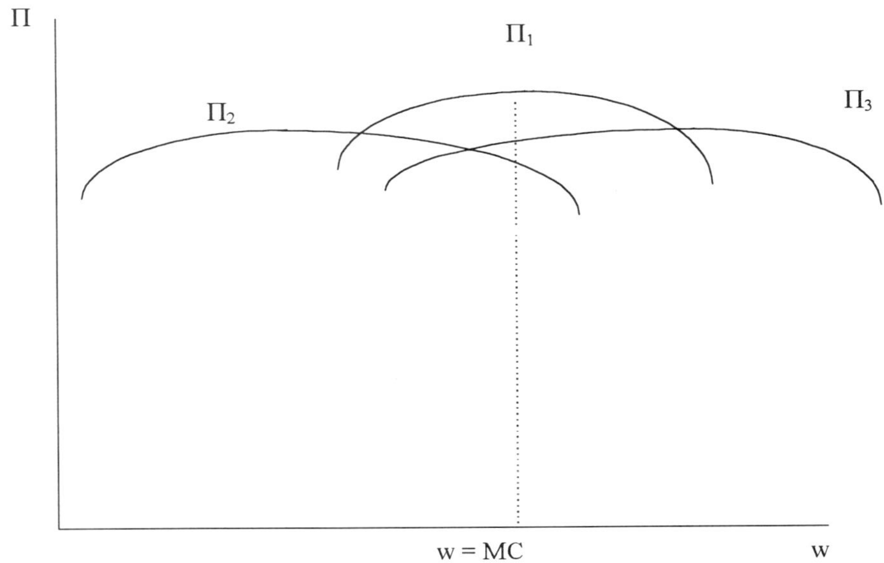

Figure 4. Three Water Users/Shareholders

3.2. Conjunctive Monitoring

It should be noted that the driving force behind any privatization process is the existence of rent, R, which private concerns would like to capture. However, increases in rent are often achieved by over-utilizing water. Hence, a complete monitoring system is needed to resolve the conflict between short-term maximization and the long-term objective of preserving the environment from suboptimal conjunctive use of an abating/polluting input mix. The management of conjunctive use can be done by a Pigouvian tax, which is computed from the dynamic system described in Section 2. Using (15), (5), and (6), it can be shown that

$$f_{g_t} = z(G_t) + \frac{f_{p_t} p_{G_t} - z_{G_t} g_t}{\beta - \rho_t}(1-\alpha), \tag{21}$$

$$f_{s_t} = w_t - \frac{f_{p_t} p_{G_t} - z_{G_t} g_t}{\beta - \rho_t}\alpha \quad , \tag{22}$$

$$w_t = Z(G_t) + \frac{f_{p_t} p_{G_t} - z_{G_t} g_t}{\beta - \rho_t}. \tag{23}$$

Thus, the optimal tax through the trajectory path is the shadow price in equation (15). Tsur (1991) obtains the same conclusion.

Note that unlike a static Pigouvian tax (PT), in the case of a dynamic system the optimal tax changes along the optimal trajectory according to equation (15). Moreover, as can be seen from the phase diagram in Figure 2, imposing a static Pigouvian tax in a dynamic environment does not ensure convergence of the system to the steady state. The Pigouvian tax is larger the lower the level of the aquifer; this is because the lower the level of the aquifer, the higher the marginal externality.

In the following section, two policies will be compared around the steady state when there is a precision problem caused by an error in the coefficient of return.

4. MONITORING CONJUNCTIVE USE AROUND THE STEADY STATE: RELATIVE STANDARDS VERSUS TAXES

The previous section implicitly assumed implementation of a tax/subsidy as a policy tool for achieving optimal allocation of the abating/polluting input mix. Under perfect information, tax results may be identical to the results obtained by using a standard as the policy tool, i.e., determining g_t and s_t by a central planner. However, under imperfect information, the tax and standard may differ (see, for example, Baumol and Oates, 1975). This section discusses the difference between the two policy tools for monitoring conjunctive use of the abating/polluting input mix under privatization when there is a precision problem in the coefficient of return.

Two policy tools are available in order to obtain optimal use of g_t along the time path. The first one is an optimal tax scheme along the time trajectory as in the previous section [see equations (15) and (21)-(23)], i.e., measuring m_t along the optimal trajectory. The second policy tool is setting the optimal input ratio at the steady state as a standard. This relative standard (RS),

$$\frac{\hat{g}}{\hat{s}} = \frac{\alpha}{1-\alpha},$$

is obtained by equating the right-hand side of the equation of motion (2) to zero. The system is at a steady state, so the level of the aquifer does not change.

It is important to note that taxing local water according to the optimal tax scheme eliminates a substantial portion of this rent and may discourage private ownership. Therefore, it is worthwhile to examine an alternative policy which implements a standard. Assuming the regulating authority regulates the private firm by imposing the required RS ratio, a profit-maximizing firm will utilize an aggregate quantity $Q_{ss} = g_{ss} + s_{ss}$. Such a method has the following properties:

1. Optimal allocation of conjunctive inputs coincides with maximization of R, i.e., maximizes shareholders' motivation to vote for marginal pricing.
2. The only information required for its implementation is the estimation of α.[3]

Since both policy tools (taxes and standards) depend on the precision of information on α, it is important to compare the deviation from the optimal allocation around the steady state—resulting from errors in estimation of. Such a comparison within a small neighborhood of the optimal point is illustrated in Figure 5. The diagram illustrates schematically changes in g under the RS and PT schemes resulting from changes in m. These changes result from errors in the estimation of α. Let $\Delta\alpha$ be the deviation of α from its optimal value, i.e., $\Delta\alpha$ is an error in the estimation of α by the regulator as a result of imperfect information. Using equation (21),

$$\frac{\Delta m}{\Delta \alpha} \cong \frac{-[f_g - z(G_t)]}{(1-\alpha)^2},$$

while using the standard

$$\frac{\Delta g}{\Delta \alpha} \cong \frac{s}{(1-\alpha)^2},$$

i.e., the value of g under the RS scheme is upward-sloping as α increases, while the value of g under the PT scheme is downward-sloping in the same direction. Assuming that $\alpha_1 < \alpha_{ss}$, i.e., the regulator underestimates α, the corresponding tax is $m < m_{ss}$ which results in $g_{PT} > g_{ss}$, while the corresponding standard is $g_{RS} < g_{ss}$. The relative deviation of g_{RS} and m from their optimal values (g_{ss} and m_{ss}) are, correspondingly,

$$\frac{\Delta g}{\Delta \alpha}\frac{1}{g} = \xi_1 \cong \frac{1}{\alpha(1-\alpha)}$$

under the RS method and

$$\frac{\Delta m}{\Delta \alpha}\frac{1}{m} = \xi_2 \cong \frac{-1}{1-\alpha}$$

under the PT method, which results in $\xi_2 < \xi_1$. Proposition 2 follows.

Proposition 2. Let $\eta \equiv (\Delta g/\Delta m)(m/g)$ be the elasticity of demand for underground water. Let Δg_{RS} be the deviation of the RS scheme from the optimal point resulting from an error in the estimation of α. Let Δg_{PT} be the

---

[3] A Pigouvian tax requires more information about z(G), p(G) and f[g, s, p(G)].

deviation of the PT scheme from the optimal point under a similar error in α. Then $\Delta g_{RS} \leq (>) \Delta g_{PT}$ if and only if $|\eta| \geq (<) \xi_1/\xi_2$.

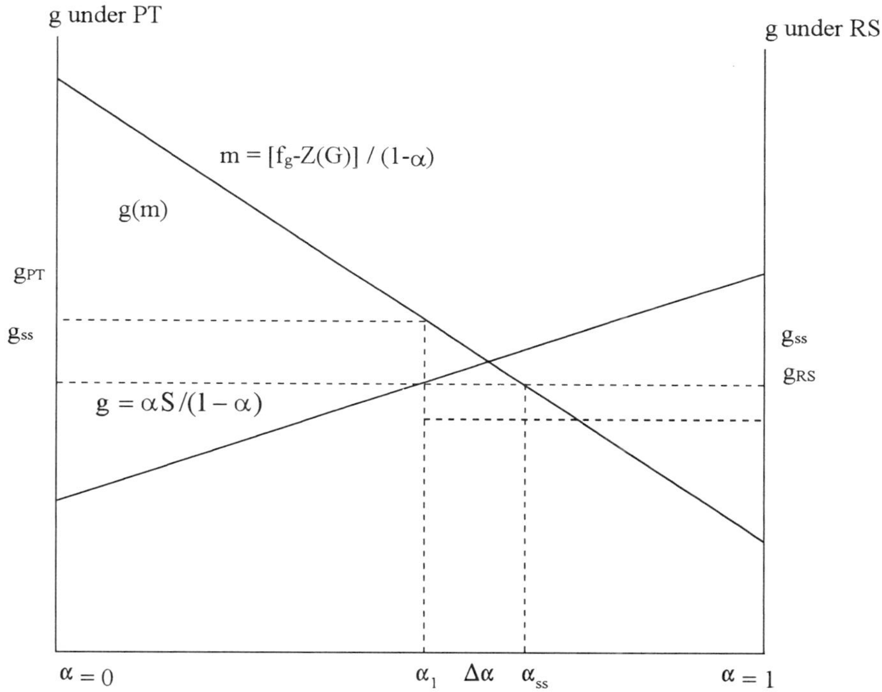

Figure 5. Comparison of the Relative Standard and Pigouvian Tax

Proof. Note that $\Delta g_{RS} = \Delta\alpha g \xi_1$ and $\Delta g_{PT} = \Delta\alpha g \eta \xi_2$. Thus, $|\eta| = |\xi_1/\xi_2|$ results in $\Delta g_{RS} = \Delta g_{PT}$, and $|\eta| \geq (<) |\xi_1/\xi_2|$ results in $\Delta g_{RS} \leq (>) \Delta g_{PT}$. Q.E.D.

Weitzman (1974) and Zusman (1997) obtain a similar result. Weitzman's paper deals with taxes versus standards under uncertainty. He concludes that taxes (standards) are preferred whenever the demand is inelastic. Zusman's paper deals with taxes versus standards under political lobbying. He concludes that taxes (standards) are preferred whenever the demand of the lobbying group is inelastic. Appendix A contains critical values of α and their corresponding values of demand elasticity for which the RS method and the PT method result in identical error.

5. SUMMARY

This chapter deals with two types of conflicts. The first type of conflict is between man and the environment, or more specifically, damage to the environment by water caused by water use. The second type of conflict is the danger of market exploitation created by the privatization of a water public utility.

The resolution to these conflicts includes the following remedies. In the case of centralized management of the public utility, optimal control is used to obtain the optimal mix of abating and polluting inputs over time. In the case of privatization of the public utility, the regulation for efficient allocation is done either by a dynamic Pigouvian tax derived along the optimal path or by a relative standard derived at the steady state. It is shown that in the vicinity of the steady state equilibrium, a deviation caused by incomplete information (represented by an error in the estimation of the coefficient of return) results in the preferred choice between taxes versus standards to depend on the elasticity of water demand. This, in turn, depends on the heterogeneity of the distribution of the derived individual water demands.

This chapter deals with the conjunctive use *over time* of recycled and fresh water and its intertemporal externality effects on the environment. Results suggest that exploitation of a natural resource such as water *over space and time* requires collective action and involves market failure of decentralized mechanisms. Optimal control theory with the aid of simple welfare economics is shown to be helpful in identifying some of the resolutions to the conflicts.

APPENDIX A: CRITICAL VALUES OF α AND η.

ξ_1	ξ_2	α	η
21.05	1.05	0.05	20.00
11.11	1.11	0.10	10.00
7.84	1.18	0.15	6.67
6.25	1.25	0.20	5.00
5.33	1.33	0.25	4.00
4.76	1.43	0.30	3.33
4.40	1.54	0.35	2.86
4.17	1.67	0.40	2.50
4.04	1.82	0.45	2.22
4.00	2.00	0.50	2.00
4.04	2.22	0.55	1.82
4.17	2.50	0.60	1.67
4.40	2.86	0.65	1.54
4.76	3.33	0.70	1.43
5.33	4.00	0.75	1.33
6.25	5.00	0.80	1.25
7.84	6.67	0.85	1.18
11.11	10.00	0.90	1.11
21.05	20.00	0.95	1.05

REFERENCES

Baumol, W.J., and W.E. Oates. 1975. *The Theory of Environmental Policy*. Englewood-Cliffs, NJ: Prentice-Hall, Inc.

Caswell, M., and D. Zilberman. 1985. The Choice of Irrigation Technologies in California, *American Journal of Agricultural Economics* 67: 224-234.

Gisser, M. 1983. Groundwater: Focusing on the Real Issue, *Journal of Political Economy* 91: 1001-27.

Hochman, O., and E. Hochman. 1980. Regeneration, Public Goods and Economic Growth, *Econometrica* 48: 1233-50.

Tsur, Y. 1991. "Managing Drainage Problems in a Conjunctive Ground and Surface Water System." In A. Dinar and D. Zilberman, eds., *The Economics and Management of Water and Drainage in Agriculture*. Norwell, MA: Kluwer Academic Publishers.

Vickers, J., and G. Yarrow. 1988. *Privatization: An Economic Analysis*. Cambridge, MA: The MIT Press.

Weitzman, M.L. 1974. Price vs. Quantities, *Review of Economic Studies* 41: 477-491.

Zusman, P. 1997. "Informational Imperfections in Water Resources Systems and the Political Economy of Water Supply and Pricing in Israel." In D.D. Parker and Y. Tsur, eds., *Decentralization and Coordination of Water Resource Management*. Norwell, MA: Kluwer Academic Publishers.

15

NONCOOPERATIVE AND COOPERATIVE MANAGEMENT OF AN ACCUMULATIVE WATER POLLUTANT

Erik Lichtenberg
Lars J. Olson
University of Maryland, College Park, USA

There are numerous situations where the production decisions of many agents acting in a noncooperative manner have serious and lasting trans-boundary impacts on environmental quality. These include trans-boundary contamination of surface and groundwater, ozone depletion, and global warming. This chapter examines equilibrium outcomes in two-player dynamic games with production and a trans-boundary stock pollutant. The gains from cooperation are evaluated by comparing the noncooperative equilibrium to the cooperative solution. An illustrative example is developed for the problem of regional water quality management and the Chesapeake Bay.

1. INTRODUCTION

There are numerous situations where the production decisions of many agents acting in a noncooperative manner have serious and lasting trans-boundary impacts on environmental quality. These include trans-boundary contamination of surface and groundwater, ozone depletion, and global warming. For trans-boundary environmental quality problems such as these, there are frequently obstacles to achieving a cooperative solution. In addition to strategic and stock externalities, economic agents (countries, states, or individuals) may differ in their incentives to improve environmental quality. Their incentives for production and emissions of pollution may be different, and they may suffer differing degrees of damage from trans-boundary pollutants.

This chapter develops a noncooperative dynamic model of a trans-boundary stock pollutant. The setting is an asymmetric version of the two-player differential game developed in Dockner and Long (1993). The stock pollutant is the state variable. Each agent chooses a time-path for output from production. Production involves emissions that contribute to the accumulation of the stock pollutant. The pollutant is persistent in the environment but degrades over time at some fixed rate, which reflects both the physical and chemical properties of the pollutant as well as the environment's capacity for assimilation. This natural rate of degradation determines the degree of irreversibility for the pollutant. Each agent's utility depends on the agent's own production decisions and the level of the stock pollutant at each point in time. The objective of each agent is to maximize the discounted stream of utility subject to the transition equation for the stock pollutant and the production decisions of the other agent.

Dockner and Long (1993) consider the symmetric version of this two-player game with a stock pollutant.[1] Their analysis focuses on trans-boundary pollution control in the case where agents' utility functions are identical and where the unit contribution to pollution of each player's output is the same. For most trans-boundary pollutants, these symmetry properties are unlikely to hold. Indeed, asymmetries in benefits from and costs of pollution control are a major obstacle to negotiated solutions to trans-boundary pollution problems. It is thus necessary to consider non-symmetric games. Three kinds of asymmetry may be important. First, agents typically face different production incentives. Second, the costs of pollution are often distributed unequally across agents. Third, each agent's production decisions may influence the accumulation of pollution in different ways. A good illustration of these effects can be seen in the case of CO_2 emissions where individual countries burn different amounts of fossil fuels, the estimated costs of global warming vary according to geographic location, and the CO_2 emissions per unit of output from each country depends on the mix and CO_2 intensity of the fossil fuels being burned.

The chapter is organized as follows. Section 2 develops a simple two-agent model of production and pollution accumulation. Section 3 focuses on the outcome from cooperative decision-making, while Section 4 examines the Markov-perfect equilibrium to the dynamic game. Section 5 develops an illustrative example for the problem of regional water quality management and the Chesapeake Bay.

2. THE MODEL

Many environmental problems involve the accumulation of pollutants associated with the production decisions of agents acting in a noncooperative manner. This section describes a simple model that captures many of the essential elements of such problems. For simplicity we assume that there are two agents. Each agent receives utility from production and disutility from the stock pollutant. Let $q_i(t)$ denote the output of agent i, and let $z(t)$ be the accumulated stock of the pollutant at time t. The

[1] Other literature on pollution control in noncooperative dynamic games includes Clemhout and Wan (1991), Kaitala et al. (1991), Long (1992), Martin et al. (1993), Tahvonen et al. (1993), and Wirl (1994). The optimal control of stock pollutants is examined in Plourde (1972), Smith (1972), Keeler et al. (1972), Forster (1975), Kitabatake (1989), and Conrad and Olson (1992), among others.

utility of agent i is assumed to be a quadratic function of that agent's output and the pollution stock, and is denoted by $U_i(q_i(t),z(t)) = a_iq_i(t)-b_iq_i(t)^2/2-c_iz(t)^2/2$.

The rate of change in the stock pollutant is given by

$$z' = \theta - \gamma z(t) + \eta_1 q_1(t) + \eta_2 q_2(t), \tag{1}$$

where $z' = dz/dt$, θ denotes the contribution of natural deposition or exogenous pollution sources, η_1 and η_2 are the emission transformation coefficients associated with the production of each agent, and γ represents the constant natural rate of degradation. In a steady state, the pollution stock is $z = (\theta + \eta_1 q_1 + \eta_2 q_2)/\gamma$. The degree of reversibility of pollution is determined by the degradation parameter, γ. As $\gamma \to 0$, pollution becomes completely irreversible. As $\gamma \to \infty$, the pollutant becomes a flow rather than a stock pollutant and can be handled in a static context.

3. THE COOPERATIVE SOLUTION

The cooperative or first-best solution maximizes the joint utility of both agents, or the sum of their individual utilities. The first-best output policies for each country can be found by solving a standard optimal control problem. The current value Hamiltonian is

$$H(q_1,q_2,z,\mu) = \Sigma_{i=1}^{2} [a_iq_i(t)-b_iq_i(t)^2/2-c_iz(t)^2/2] + \mu(t)[\theta-\gamma z(t)+\eta_1 q_1(t)+\eta_2 q_2(t)].$$

Assuming interiority, the necessary and sufficient conditions for an optimal solution are given by

$$a_1 - b_1q_1(t) + \mu(t)\eta_1 = 0 \tag{2a}$$

$$a_2 - b_2q_2(t) + \mu(t)\eta_2 = 0. \tag{2b}$$

$$\mu' - \delta\mu = -H_z = (c_1+c_2)z + \gamma\mu \tag{2c}$$

$$z' = \theta - \gamma z(t) + \eta_1 q_1(t) + \eta_2 q_2(t) \tag{2d}$$

and

$$\lim_{t\to\infty} e^{-\delta t}\mu(t)z(t) = 0, \tag{2e}$$

where (2e) is the transversality condition. There is a unique optimal steady state pollution stock given by

$$z^* = (\theta+A_1+A_2)/(\gamma+B_1+B_2), \tag{3}$$

where $A_i = \eta_i a_i/b_i$ and $B_i = \eta_i^2(c_1+c_2)/(b_i(\delta+\gamma))$, i=1,2.

Comparative statics of the optimal steady state pollutant stock are given in Table 1.

Table 1. Comparative States of the Optimal Steady State Pollutant Stock

Parameter	a_i	b_i	c_i	δ	θ	γ
Effect on z*	+	?	?	?	+	?

4. MARKOV-PERFECT EQUILIBRIA

Markov-perfect strategies are a set of decision rules, $Q_i(z)$, for each agent that specify the agent's production decision as a function of the current pollution stock. Formally, the set of admissible Markov-perfect strategies is given by

$$S_i = \{Q_i(z(t),t)|Q \text{ is Lipshitz-continuous in z and continuous in t}\}.$$

Such strategies are sometimes referred to as closed-loop no-memory strategies. Under Markov-perfect strategies, each agent's action depends on the previous history only through the value of the current state, z(t). Markov-perfection implies that each player can observe the current value of the state (the pollution stock) and that players make their decisions accordingly.

The optimization problem for each agent can be expressed as

$$V_i(z(t)) = \max_{Q_i(z)} \int e^{-\delta t} U(q_i(t),z(t))dt$$

subject to

$$\text{s.t. } z' = \theta - \gamma z(t) + \eta_1 q_1(t) + \eta_2 q_2(t) \text{ and } q_j \text{ given by } Q_j.$$

Define $V_i(z) = \alpha_{0i} + \alpha_{1i}z - \alpha_{2i}z^2/2$ to be the Markov-perfect equilibrium payoff to agent i, and let $Q(z) = \{Q_1(z),Q_2(z)\}$. For linear-quadratic games, Lockwood (1996, Proposition, p. 754) shows that Q[z(t)] is a Markov-perfect equilibrium if and only if Q satisfies the Hamilton-Jacobi-Bellman equations,

$$\delta V_i(z) = \max_{q_i} U_i + (\partial V_i/\partial z)z' \quad \text{s.t. } Q_j(z) \text{ fixed, } i,j = 1,2, \tag{4}$$

and the transversality condition,

$$\lim_{t\to\infty} -\gamma + \eta_1(\partial q_1/\partial z) + \eta_2(\partial q_2/\partial z) < \delta/2.$$

The current value Hamiltonians for agents 1 and 2 are

$$H^1(q_1,z,\mu_1) = [a_1q_1(t) - b_1q_1(t)^2/2 - c_1z(t)^2/2] + \mu_1(t)[\theta - \gamma z(t) + \eta_1q_1(t) + \eta_2q_2(t)]$$

and

$$H^2(q_2,z,\mu_2) = [a_2q_2(t) - b_2q_2(t)^2/2 - c_2z(t)^2/2] + \mu_2(t)[\theta - \gamma z(t) + \eta_1q_1(t) + \eta_2q_2(t)].$$

Maximizing H^i with respect to q_i is equivalent to solving the maximization on the right-hand side of equation (4). The first order conditions are

$$a_1 - b_1q_1(t) + \mu_1(t)\eta_1 = 0 \tag{5a}$$

and

$$a_2 - b_2q_2(t) + \mu_2(t)\eta_2 = 0. \tag{5b}$$

The marginal benefits of production are equated to the marginal costs of contributions to the stock pollutant. The shadow prices, μ_i, evolve according to

$$\mu_1' - \delta\mu_1 = -H^1_z = c_1z + [\gamma-\eta_2(\partial q_2/\partial z)]\mu_1 \tag{6a}$$

and

$$\mu_2' - \delta\mu_2 = -H^2_z = c_2z + [\gamma-\eta_1(\partial q_1/\partial z)]\mu_2. \tag{6b}$$

The difference between Markov-perfect equilibria and the cooperative solution can be seen by comparing equations (6) with (2c).

Equations (5) can be rewritten by substituting for $\mu_i(t) = \partial V_i/\partial z = \alpha_{1i} - \alpha_{2i}z(t)$ to obtain

$$a_i - b_iq_i(t) + [\alpha_{1i}-\alpha_{2i}z(t)]\eta_i = 0.$$

This can be solved for

$$q_i(t) = [a_i+\eta_i\alpha_{1i} - \eta_i\alpha_{2i}z(t)]/b_i, \quad i=1,2. \tag{7}$$

Let α be the matrix of coefficients of the value functions. Substituting the value functions and the above solutions for q_1 and q_2 into (4) for each agent gives two equations in the parameters, the undetermined coefficients α, and the state variable z. In each equation the constant terms, terms in z, and terms in z^2 must be the same on the left-hand and right-hand side of the equation. This yields three pairs of Ricatti equations in α that can be represented as $g^i_{const}(\alpha) = 0$, $g^i_z(\alpha) = 0$, and $g^i_{z2}(\alpha) = 0$, i=1,2. In theory, these equations can be solved recursively following the approach discussed in Lockwood (1996, p. 756). Once the coefficients of the value functions are determined, the Markov-perfect equilibrium output policies are given by (7). Unfortunately, the closed-form algebraic solution is a complicated function of the model parameters. For example, the solution for α_{2i} involves finding the appropriate root of a fourth-order polynomial. In principle this is possible but, unless symmetry is imposed, the solution is typically so complicated that it yields little economic insight. For this reason the full system of Ricatti equations is solved numerically for the illustrative application considered in the next section of the chapter.

5. PHOSPHATE POLLUTION IN THE CHESAPEAKE BAY

This section presents an empirical example dealing with the problem of phosphorus pollution of the Chesapeake Bay. We do not attempt a complete treatment of this problem. Instead, we draw on data and the present state of scientific knowledge about this problem to create a stylized version of the problem to illustrate the theory presented in the preceding section.

The Chesapeake Bay is the estuary of the Susquehanna, Potomac, and numerous smaller rivers. It drains portions of six states—New York, Pennsylvania, Maryland, Delaware, Virginia, and West Virginia—plus the District of Columbia. It supports major commercial fisheries, the best-known of which is for Maryland blue crab. Its proximity to major urban centers like Philadelphia, Baltimore, and Washington make it a heavily used recreation site, supporting activities such as sailing and boating, swimming, and recreational fishing. It is a major shipping route as well, serving the port of Baltimore.

The Bay's capacity to provide these services has deteriorated over the last century due to a combination of overexploitation, habitat destruction, and pollution. Harvests in major fisheries such as striped bass, American shad, and oysters have declined to small fractions of levels fifty years ago. Causes include overharvesting, disease, and habitat loss (e.g., from dam construction), but pollution has played an important role as well. Emissions of nutrients (nitrogen, phosphorus) and sediment from farms, construction sites, and urban waste disposal have reduced submerged aquatic vegetation, which provides essential habitat for the young of these and other species. Eutrophication due to nutrient pollution creates anoxic conditions on the Bay bottom as well, harming aquatic life harvested directly and indirectly by reducing food sources for harvested species. Nutrients and sediment have impaired water quality for swimming and boating as well, producing algal blooms and turbidity.

Nutrients are the major pollution problem in the Bay as a whole, although heavy metals and other toxic substances cause severe problems in certain localized areas. Phosphorus is thought to be the limiting nutrient governing algal growth in the upper third of the Bay, from the mouth of the Susquehanna to near Annapolis, Maryland. Nitrogen appears to be limiting in the lowest third of the Bay, while the remaining third is a transition zone where the limiting nutrient varies according to season, freshwater flows, and other factors (Chesapeake Bay Program 1994).

The major jurisdictions of the Bay watershed have attempted to address nutrient and other types of pollution problems through a series of Chesapeake Bay Agreements. These agreements commit the states to reduce pollutant loadings by 40 percent from a 1985 baseline by the year 2000. Some progress has been made in meeting those goals, due mainly to upgrades of sewage treatment facilities and a ban on phosphate-containing detergents in Maryland. By 1990, total phosphorus in the mainstem Bay appears to have fallen by 19 percent, although total nitrogen was estimated to have increased by 2 percent. Subsequent statistical modeling conducted by the Maryland

Department of Environment (1995) indicates downward trends in total phosphorus and nitrogen concentrations in the Susquehanna, Potomac, and Choptank Rivers, although decreases in nitrogen in particular appeared to be leveling off by 1993.

Agriculture is considered the principal source of both nutrients, accounting for an estimated 49 percent of phosphorus and 39 percent of nitrogen. It accounts for larger shares of controllable nutrients: about 10 percent of phosphorus emissions and almost 30 percent of nitrogen emissions are considered uncontrollable and therefore not subject to policy influence (Chesapeake Bay Program 1992). The principal sources of nutrient emissions are cropland and animal waste, mainly from cattle. Cropland is the largest source in Maryland, accounting for about 60 percent of controllable nonpoint nitrogen emissions and 55 percent of controllable nonpoint source phosphorus emissions, while animal wastes account for about 1 percent of controllable nonpoint source nitrogen emissions and 2 percent of controllable nonpoint source phosphorus emissions (Maryland Department of Natural Resources 1996). Animal waste, mainly from dairy cattle, is the main nonpoint source from Pennsylvania.

This chapter focuses on phosphorus, mainly because it has a significant cumulative effect while nitrogen does not. Phosphate binds tightly to soil and is thus deposited on the Bay bottom with sediment. Some of this deposited phosphorus redissolves in the Bay, while the remainder makes up a pollutant stock. The forms of nitrogen entering the Bay, by contrast, are highly soluble and are subject to conversion to nitrogen gas (denitrification), making nitrogen a flow rather than a stock pollutant.

Our stylized version of the phosphorus pollution problem features two jurisdictions. Pennsylvania represents the upstream jurisdictions that rely more heavily on phosphorus to generate income and suffer less damage from phosphorus pollution because the Bay lies outside of its borders. Maryland represents the downstream jurisdictions that earn income from phosphorus pollution but suffer greater damage because of greater reliance on the Bay for recreational and commercial activities.

The empirical analysis is restricted to agriculture, because agriculture is the largest single source of phosphorus emissions from both states. Waste from dairy cattle is assumed to be the sole controllable source of phosphorus emissions from Pennsylvania. Cropland was assumed to be the sole controllable source of phosphorus emissions from Maryland. Emissions from urban sewage, construction sites, cropland in Pennsylvania, and animal wastes in Maryland were all assumed to be constant.

The control variables in the model were the number of dairy cows in Pennsylvania and cropland acreage in Maryland. Phosphorus emissions were assumed to be proportional to each. In Pennsylvania, all controllable nonpoint source phosphorus emissions were attributed to dairy cattle. The Chesapeake Bay Program (1992) estimated that 1985 baseline controllable nonpoint phosphorus emissions from Pennsylvania amounted to 3.92 million pounds annually. The U.S. Department of Agriculture reported that there were 740,000 dairy cows in Pennsylvania that year, suggesting that emissions per 1,000 cows (η_2) amounted to 0.00529 pounds annually. The Maryland Department of Natural Resources (1996) estimated that 1985 baseline emissions of phosphorus from cropland amounted to 2.168 million pounds annually. The Maryland Department of Agriculture reported 1.53 million acres of cropland in the state that year, giving 0.001417 million pounds of phosphorus per 1,000 acres of cropland (η_1). Background phosphorus loadings (θ) were assumed to equal to 7.002 million pounds. This was calculated by taking total loadings from Maryland and Pennsylvania, and subtracting out the controllable loadings attributed to Pennsylvania dairy and Maryland cropland.

Regression analysis was used to estimate quadratic relationships (i.e., the parameters a_i and b_i) between dairy cattle and farm income in Pennsylvania and cropland and farm income in Maryland. Income from farming was taken to equal to state-level net farm income less non-cash income as reported by the U.S. Department of Agriculture, adjusted to 1992 dollars using the implicit GNP deflator reported by the Council of Economic Advisers (1997). The number of dairy cattle in the state of Pennsylvania as reported by the U.S. Department of Agriculture and acreage of major crops in Maryland reported by the Maryland Department of Agriculture were the independent variables in the regressions. The estimated parameters are reported in Table 2.

Table 2. Estimated Parameters of the Chesapeake Bay Phosphorus Model

Model Parameter	Estimate
η_1	0.001417 million pounds per year per 1,000 acres
η_2	0.005297 million pounds per year per 1,000 cows
θ	7.002 million pounds per year
c_1	\$1.4306 million per million pounds^2 phosphorus (1992 dollars)
c_2	\$0.3576 million per million pounds^2 phosphorus (1992 dollars)
a_1	\$0.1213 million per 1,000 acres (1992 dollars)
b_1	\$0.000043 million per 1,000 acres squared (1992 dollars)
R^2	0.09
a_2	\$0.624 million per 1,000 cows (1992 dollars)
b_2	\$0.0087 million per 1,000 acres squared (1992 dollars)
R^2	0.16

The costs of phosphorus pollution include impaired recreational opportunities and reductions in fishery harvests. For this analysis, we analyzed only costs of reduced recreational uses of the Bay. This may not be too restrictive. For example, McConnell and Strand (1989) have argued that, under conditions of open access like those characterizing commercial fishing in the Bay, fishing effort will adjust to changes in harvests induced by changes in environmental conditions so that the economic value of reductions in commercial harvests may be negligible. Bockstael, McConnell, and Strand (1988) estimated that a 20 percent decrease in nitrogen and phosphorus loadings in the Bay would increase Maryland residents' recreational benefits from beach use, boating, and fishing by \$40.74 million, or \$40.03 million in 1992 dollars. We assumed that Maryland residents accounted for 80 percent of recreational uses of the Bay and Pennsylvania residents for the remaining 20 percent. Adjusting the Bockstael, McConnell, and Strand (1988) estimates proportionally gives an estimate of \$61.28 million (in 1992 dollars) for a 20 percent decrease in nutrient pollution. For simplicity, we attributed all benefits of pollution reduction to phosphorus. In 1985, phosphorus emissions were estimated to be 13.09 million pounds. Using the simple quadratic form presented in the preceding section, these figures imply that the sum of the Maryland and Pennsylvania environmental damage coefficients ($c_1 + c_2$) is equal to 1.78822 million dollars per million pounds squared of phosphorus. The Maryland coefficient (c_1) was assumed to be 80 percent of this total, and the Pennsylvania damage coefficient (c_2) was assumed to be the remaining 20 percent.

Reductions in phosphorus emissions were assumed to be achieved by reducing the number of dairy cows in Pennsylvania and cropland acreage in Maryland, with corresponding losses in farm income. Reductions in either dairy herd size or cropland can be interpreted either literally or as adoption of best management practices that reduce phosphorus emissions but are more costly than standard farming or dairying practices.

Estimates of the degradation rate of the phosphorus stock (γ) in the Chesapeake Bay were not available. The focus of the analysis, however, is on how the cooperative and noncooperative solutions change as the pollution problem becomes more or less irreversible, that is, as γ varies from zero to infinity, respectively. We thus analyzed steady state solutions for the cooperative and noncooperative cases under values of γ ranging upward from zero.

When pollution is highly irreversible, a unit of emissions imposes larger environmental costs on both agents. This means that emissions in both the cooperative and noncooperative steady state eventually decline as γ increases. When the pollutant is very persistent, the environmental damages may be so high that the equilibrium outcome is for one or both states to completely curtail their emissions. This is the case for the example considered here when γ is less than 0.4, which is equivalent to a half-life for phosphorus of greater than one year and nine months.

Figure 1 shows the steady-state phosphorus stock under the cooperative and noncooperative outcomes, plotted as a function of the degradation rate, γ. As the irreversibility of pollution decreases (γ increases), the cooperative and noncooperative outcomes converge. For values of γ greater than 4, which corresponds to a pollutant half-life of slightly over two months, the pollution stock and nutrient loadings from each state differ by less than one percent under the cooperative and noncooperative solutions. As expected, the pollution stock is higher in the noncooperative equilibrium and the difference from the cooperative outcome is greater when the pollutant persists longer in the environment. The difference between the two increases as the pollutant becomes more irreversible (γ approaches zero).

The contributions of each state to steady state nutrient loadings under the cooperative and noncooperative equilibria are shown in Figure 2. Several observations are worth noting. First, Pennsylvania contributes a larger amount to nutrient loadings in both outcomes. This result is as expected, since Pennsylvania's benefits from dairy production are greater and its losses from nutrient pollution are smaller than Maryland's, indicating weaker incentives for cutting back on pollution. Second, Pennsylvania is relatively insensitive to changes in the degradation rate under the noncooperative solution, while the difference in Maryland cropland between the cooperative and noncooperative case is relatively small. While cropland declines as the degradation rate decreases, it does so at about the same rate under both cooperation and noncooperation. Evidently, the weakness of Maryland's strategic position relative to Pennsylvania forces it to shoulder a larger share of pollution control effort in the noncooperative situation Third, at low degradation rates, there is a marked difference in the phosphorus loadings from both Pennsylvania and Maryland between the two cases. When γ=0.4, the cooperative equilibrium involves phosphorus loadings from Pennsylvania dairies that are approximately one-third those in the noncooperative equilibrium, while cooperative emissions from Maryland cropland are about 10 percent of their noncooperative levels. A final note about welfare comparisons is that at high degradation rates, the noncooperative equilibrium involves relatively small losses in welfare. When pollution is more irreversible, this is no longer true.

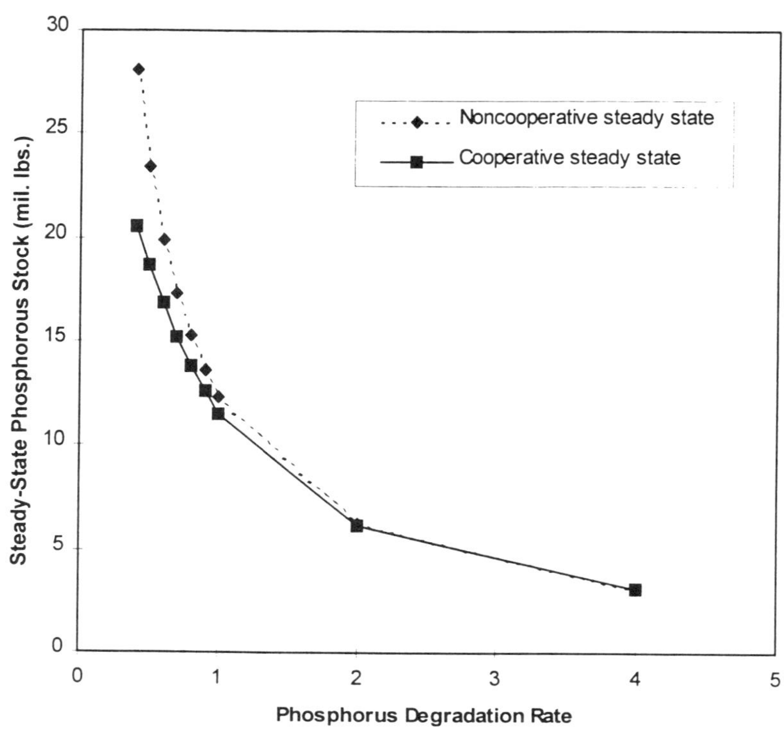

Figure 1. Steady State Cooperative and Noncooperative Phosphorus Stocks as a Function of the Phosphorus Degradation Rate

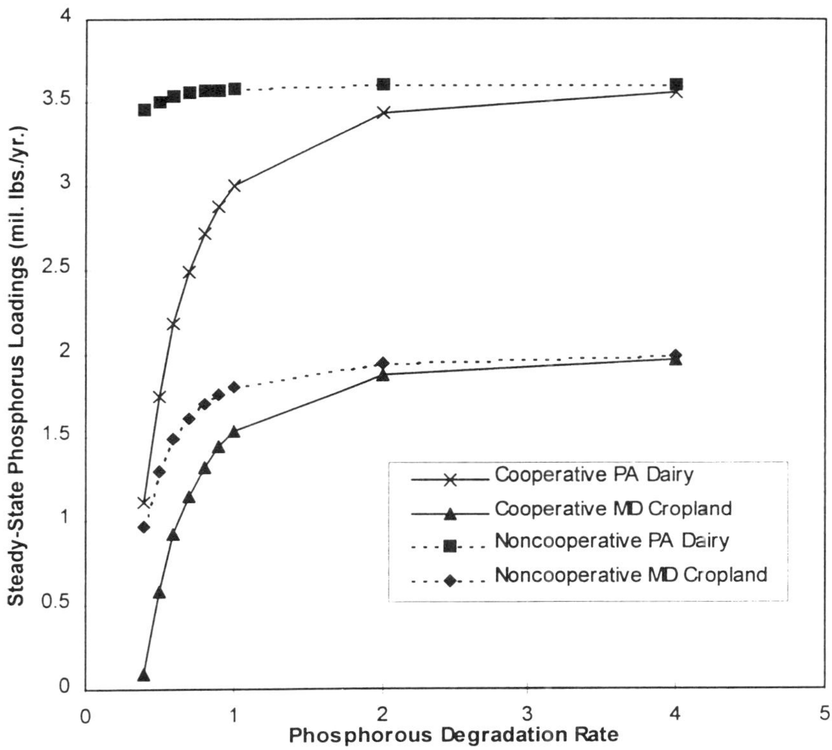

Figure 2. Contribution to Steady State Phosphorus Loadings by Pennsylvania (PA) Dairy and Maryland (MD) Cropland under Cooperation and Noncooperation as Functions of the Phosphorus Degradation Rate

6. CONCLUSIONS

This chapter develops a noncooperative dynamic model of a trans-boundary stock pollutant with asymmetric agents. An application is developed for the problem of phosphate pollution in the Chesapeake Bay. Numerical methods are used to solve for the linear Markov equilibrium of a linear-quadratic differential game. The chapter examines how the degradation rate influences the steady state phosphorus stock and the relative contribution to the steady state by Maryland cropland and Pennsylvania dairy under cooperative and noncooperative solutions. It would be interesting to extend the analysis to encompass additional welfare issues and to develop an improved model of the processes that lead to phosphate contamination in the Chesapeake Bay.

REFERENCES

Bockstael, N.E., K.E. McConnell, and I.E. Strand. 1988. Benefits from Improvements in Chesapeake Bay Water Quality. Department of Agricultural and Resource Economics, University of Maryland, College Park.

Chesapeake Bay Program. 1992. Progress Report of the Baywide Nutrient Evaluation Reduction Reevaluation. Annapolis, MD.

Chesapeake Bay Program. 1994. Response of the Chesapeake Bay Water Quality Model to Loading Scenarios. Annapolis, MD.

Clemhout, S., and H.Y. Wan, Jr. 1991. Environmental Problem as a Common Property Resource Game. In R.P. Hamalainen and H.K. Ehtamo, eds., *Dynamic Games in Economic Analysis*. Berlin: Springer-Verlag.

Council of Economic Advisers. 1997. Economic Report of the President. Washington, D.C.

Conrad, J.M, and L.J. Olson. 1992. The Economics of a Stock Pollutant: Aldicarb on Long Island, *Environmental Resource Economics* 2: 245-258.

Dockner, E.J., and N.V. Long. 1993. International Pollution Control: Cooperative Versus Noncooperative Strategies, *Journal of Environmental Economics Management* 25: 13-29.

Forster, B.A. 1975. Optimal Pollution Control with a Nonconstant Exponential Rate of Decay, *Journal of Environmental Economics Management* 2: 1-6.

Kaitala, V., M. Pohjola, and O. Tahvonen. 1991. Transboundary Air Pollution between Finland and the USSR: A Dynamic Acid Rain Game. In R.P. Hamalainen and H.K. Ehtamo, eds., *Dynamic Games in Economic Analysis*. Berlin: Springer-Verlag.

Keeler, E., A.M. Spence, and R. Zeckhauser. 1972. The Optimal Control of Pollution, *Journal of Economic Theory* 4: 19-34.

Kitabatake, Y. 1989. Optimal Exploitation and Enhancement of Environmental Resources, *Journal of Environmental Economics Management* 16: 224-241.

Lockwood, B. 1996. Uniqueness of Markov-perfect Equilibrium in Infinite-time Affine Quadratic Differential Games, *Journal of Economic Dynamics Control* 20: 751-765.

Long, N.V. 1992. Pollution Control: A Differential Game Approach, *Annals of Operations Research* 37: 283-296.

Martin, W.E., R.H. Patrick, and B. Tolwinski. 1993. A Dynamic Game of a Transboundary Pollutant with Asymmetric Players, *Journal of Environmental Economics Management* 25: 1-12.

Maryland Department of Agriculture. Various years. Maryland Agricultural Statistics. Annapolis, MD.

Maryland Department of Environment. 1995. Chesapeake Bay Water Quality Monitoring River Input Monitoring Component: Level 1 Data Summary Report. Annapolis, MD.

Maryland Department of Natural Resources. 1996. Technical Appendix for Maryland's Tributary Strategies. Annapolis, MD.

McConnell, K.E., and I.E. Strand. 1989. Benefits from Commercial Fisheries When Demand and Supply Depend on Water Quality, *Journal of Environmental Economics Management* 17: 284-292.

Plourde, C.G. 1972. A Model of Waste Accumulation and Disposal, *Canadian Journal of Economics* 5: 119-125.

Smith, V.L. 1972. Dynamics of Waste Accumulation: Disposal Versus Recycling, *Quarterly Journal of Economics* 86: 600-616.

Tahvonen, O., V. Kaitala, and M. Pohjola. 1993. A Finnish-Soviet Acid Rain Game: Noncooperative Equilibria, Cost Efficiency, and Sulfur Agreements, *Journal of Environmental Economics Management* 24: 87-99.

U.S. Department of Agriculture. Various years. Agricultural Statistics. Washington, D.C.

Wirl, F. 1994. Pigouvian Taxation of Energy for Flow and Stock Externalities and Strategic, Noncooperative Energy Pricing, *Journal of Environmental Economics Management* 261: 1-18.

16 TRANS-BOUNDARY WATER PROJECTS AND POLITICAL UNCERTAINTY

Yacov Tsur
Hebrew University of Jerusalem, Israel
and University of Minnesota, USA
Amos Zemel
Ben Gurion University of the Negev, Israel

We investigate effects of political uncertainty on economic viability and time profile of international water development projects. The political uncertainty considered is manifested in the form of sudden, discrete events that upon occurrence may irreversibly damage or terminate the project. Occurrence risk is treated as an exogenous hazard, increasing the effective discount rate. One class of events may occur before or after the project completion date. Other projects are exposed to risk only until completion, following which the benefits are guaranteed. This distinction bears important consequences regarding the project's time profile (duration) and whether or not it should be undertaken.

1. INTRODUCTION

Joint exploitation of a stock of natural resource, such as a water basin, by several nations (states or provinces) typically involves considerations that are not restricted to the economic and technical aspects of the project. If all the parties involved are rational, i.e., base their decisions on economic cost-benefit criteria, and when no uncertainty prevails, an optimal development strategy meeting the needs and interests of the parties can be worked out in a rather straightforward (though far from simple) fashion. The situation is different if normal development of the project may be interrupted at any given time as sudden changes in the political atmosphere bring non-economic considerations to the focus of attention. Often, the interruption decision of one party involves uncertainty, e.g., when it is instigated by political instability internal to the interrupting state.

If development of a resource requires little or no initial investment, the presence of interruption risk may not be of much importance as not much is at stake. However, a development program typically involves considerable initial investment, and the interruption risk can easily tilt the decision against the project. It is this latter situation that we investigate in this work.

Trans-boundary water projects appear to be the natural approach when political and natural resource boundaries cross each other. For a large number of arid and semi-arid countries, whose in-house resources are nearly exhausted, international water bodies are the only major new source of water which can yet be economically developed (Biswas 1994).

The Middle East is abundant with examples of potential trans-boundary development projects, including the integrated development of the Nile basin (Mageed 1994) and the resolution of the long standing Iraqi-Syrian-Turkish water dispute (Kolars 1994; Bilen 1994). The development of joint Jordanian-Israeli water projects currently appears promising, since these issues are an important part of the peace treaty recently signed between the two nations. At different times, the construction of large-scale desalination plants along the Arava valley as well as the development of a desalination-hydroelectricity co-generation plant in the Dead Sea basin have been proposed (Murakami and Musiake 1994). Clearly, an acceptable sharing of the scarce water supplies will become a major issue on the Israeli-Palestinian agenda. Indeed, the various plans addressing exploitation of the Jordan River water (including Lebanon and Syria, as well as Jordan, Israel and the West Bank) have marked the development of Israeli-Arab relations from their early stages (Wolf 1994; Murakami and Musiake 1994).

Many examples from other regions around the globe are listed in the works compiled by Dinar and Loehman (1995), while the Center for Natural Resources, Energy and Transportation (CNRET) Register lists as many as 214 international river and lake basins (CNRET 1978). In most cases, joint projects are not favorably viewed by some sectors of the population and of the political establishments. Rather, water resources are seen as national assets, not to be shared with neighbors/rivals. Just as the mutual benefit derived from joint projects is believed to push forward the political processes of peace and reconciliation, hampering these projects can play into the hands of those who disapprove of these processes. It is this tension that underlies the uncertainty we wish to analyze.

These potential projects share common features in that they (i) involve two or more states (i.e., are trans-boundary), (ii) require considerable initial investment, (iii) require construction over a prolonged period of time (before exploitation begins), (iv) can be interrupted due to conditions which are exogenous to the project's costs and benefits, and (v) can be interrupted by events that are largely uncertain at the time the decision to undertake the project is made.

Not surprisingly, interruption uncertainty often casts doubt on prospects for trans-boundary development of natural resources which otherwise pass the cost-benefit criterion. While the reduction in a project's valuation appears intuitively obvious, we wish to draw attention also to the temporal aspects of the problem. How should the investment scheduling respond to the interruption risk? On the one hand, delayed investments decrease the damage in case the project is abandoned. On the other hand, the benefits are also delayed and, under some scenarios, early completion signals the end of the risky period. It is our purpose here to describe precisely the manifestation of these constraints. Understanding them is useful for the design of arrangements that mitigate the impeding effects of political uncertainty.

2. THE INVESTMENT PROBLEM

In this section, we consider the investment policy project owners should adopt in order to best derive the advantages offered by the resource available to them. Rather than analyze a specific project in detail, we construct the simplest model that incorporates the uncertainty and temporal aspects we wish to highlight. The model thus admits analytical solutions, yet is rich enough to illustrate the delicate effects of uncertainty on investment scheduling—in particular on project duration.

The status of the project is defined in terms of the state variable N_t. The project is complete and begins to bear fruits when its state reaches the level $\overline{N}$, at which time no further investment is required and the project owners enjoy a stream of revenues at constant rate B. The project construction progress depends on the investment rate R_t via the production function y(R) according to

$$dN/dt = y(R). \tag{1}$$

The function y is increasing, concave and bounded with finite initial slope: $y(0) = 0$, $y'(R) > 0$, $y''(R) < 0$, $y(\infty) = \bar{y} < \infty$ and $y'(0) < \infty$. Thus, the inverse function $Q(z) \equiv y'^{-1}(z)$ is well defined and decreasing over $[0,y'(0)]$, with $Q[y'(0)] = 0$ and $Q(0) = \infty$.

When it is certain that the project can proceed uninterrupted until fruition, project planning involves determining the investment profile R_t, $t \in [0, T]$ and the completion date T according to

$$V^c(\psi,\overline{N}) = \operatorname{Max}_{\{R_t\},T}\left\{\int_0^T -R_t e^{-\rho t}dt + e^{-\rho T}\psi\right\} \tag{2}$$

subject to (1), $R_t \geq 0$, $N_0 = 0$, $N_T = \overline{N}$ (or $R_t \equiv 0$ and $T = \infty$). In (2), ρ is the discount rate and $\psi = B/\rho$ is the project's value at completion time The choice $R_t = 0$ for all t and $T = \infty$ corresponds to the decision not to undertake the project which is adopted when ψ is too low to justify the investment required to complete the project.

This simple structure permits solving the problem in a closed form, yielding an explicit expression for the optimal investment rate. This task is carried out in Appendix A. We list the salient properties required for our discussion:

(P1) Let $W(z) \equiv \{y[Q(z)]/z - Q(z)\}/\rho$ be defined over $[0, y'(0)]$ and let z_ψ be the unique root of the equation $W(z_\psi) = \psi$ [or $\rho W(z_\psi) = B$]. Then, the benefit rate B affects the solution through the parameter z_ψ, which is independent of $\overline{N}$.

(P2) The completion level $\overline{N}$ determines the project duration T according to the terminal condition,

$$\int_0^T y[Q(z_\psi e^{\rho t})]dt = \overline{N}.$$

(P3) The project value is given by $V^c(\psi, \overline{N}) = W(z_\psi e^{\rho T})$.

(P4) The value $V^c(\psi, \overline{N})$ increases with ψ and decreases with $\overline{N}$.

(P5) The project duration T increases with $\overline{N}$ and decreases with ψ.

(P6) With any completion level $\overline{N}$ there is associated a critical value $\underline{\psi}$ such that the project should not be undertaken for $\psi < \underline{\psi}$. The corresponding critical duration is given by $T_{max} = \log[y'(0)/z_{\underline{\psi}}]/\rho$. For fixed values of $\underline{\psi}$ (and varying values of $\overline{N}$), T_{max} is the maximum possible duration for a profitable project.

3. INTERRUPTION RISK

Suppose now that the region is under a constant threat of a sudden breakup of some hostilities which force project abandonment and effectively render all previous investments worthless. We refer to this kind of interruption as the *event*. The event occurrence conditions are exogenous and can be viewed, from the planner's point of view, as random. In general, the occurrence hazard can depend on the status of the project. It is expedient, however, to restrict this dependence to two extreme scenarios. In the *permanent-risk* scenario, the hazard is completely blind to the project's status and the event can occur both before and after the project's completion—with the same devastating result of terminating the project construction (if it has not been completed yet) and eliminating all future benefits.

At the other extreme, occurrence hazard holds only during the construction period and, once completed, the project's benefit ψ is guaranteed. In this case, the event is influential only as long as $N < \overline{N}$. This situation is denoted as *temporary-risk*. Under both scenarios, the planners must take into account the hazard of an untimely irreversible termination of the investment problem. Yet, the distinction between durations of the risk periods turns out to be important to the effect of occurrence uncertainty on optimal investment policies.

Let us denote by λ_t the occurrence hazard rate, representing the probability that the event will immediately occur following time t given that it has not occurred at or before time t. When λ is constant (independent of t), the event occurrence time T^e is distributed according to

$$\Pr\{T^e > s\} = \exp\left\{-\int_0^s \lambda dt\right\} = e^{-\lambda s}$$

with the associated density $\lambda e^{-\lambda s}$. For a project planned to be completed by time T, the expected investment cost is

$$\int_0^T \lambda e^{-\lambda s}\left\{\int_0^s R_t e^{-\rho t} dt\right\} ds + e^{-\lambda T}\int_0^T R_t e^{-\rho t} dt = \int_0^T R_t e^{-(\rho+\lambda)t} dt.$$

Similarly, the expected benefit is decreased by the probability $e^{-\lambda T}$ that the event will not occur until T, yielding $e^{-\lambda T}e^{-\rho T}\psi = e^{-(\rho+\lambda)T}\psi$. However, the relevant value of ψ differs depending on the specification of the post-completion hazard. In the permanent-risk scenario $\psi = B/(\rho + \lambda)$, whereas in the temporary-risk case the benefit is not affected by the risk and $\psi = B/\rho$. For both scenarios, we see that the optimization problem under uncertainty reduces to that under certainty, with ρ replaced by $(\rho + \lambda)$ as the effective discount rate. Next, we investigate the implications of this difference on the optimal project duration as well as on its value.

4. PROJECT VALUE AND DURATION UNDER UNCERTAINTY

Let the superscripts "c" and "uc" relate to the various quantities associated with the certainty and uncertainty problems, respectively. We formulate the uncertainty problem with the parameters $\rho^{uc} = \rho + \lambda$, ψ^{uc} and $\overline{N}^{uc}$ as a certainty equivalent problem (with $\rho^c = \rho$ and modified values for ψ^c and $\overline{N}^c$). This can be done because increasing the discount rate is equivalent to rescaling the time variable $\tau = t(\rho + \lambda)/\rho$,

$$V^{uc}(\psi^{uc}, \overline{N}^{uc}) = \text{Max}_{\{R_t\},T}\left\{\int_0^T -R_t e^{-(\rho+\lambda)t} dt + e^{-(\rho+\lambda)T}\psi^{uc}\right\}$$

$$= \text{Max}_{\{R_\tau^c\},T^c}\left\{\frac{\rho}{\rho+\lambda}\int_0^{T^c} -R_\tau^c e^{-\rho\tau} d\tau + e^{-\rho T^c}\psi^{uc}\right\},$$

subject to (1), $R_t \geq 0$, $N_0 = 0$, $N_T = \overline{N}^{uc}$ (or $R_t \equiv 0$ and $T = \infty$), where

$$R_\tau^c = R_t \text{ and } T^c = T(\rho + \lambda)/\rho. \tag{3}$$

Observe that (1) is written in terms of the original time variable t. Transforming the time variable to τ entails a similar transformation on the state variable, $N^c = N(\rho + \lambda)/\rho$, to ensure that $dN^c/d\tau = dN/dt = y(R_\tau^c)$. Expressed in terms of the scaled variables, the other constraints are $R_\tau^c \geq 0$, $N_0^c = 0$, and $N_{T^c}^c = \overline{N}^c$. Setting $\psi^c = \psi^{uc}(\rho + \lambda)/\rho$, we find

$$V^{uc}(\psi^{uc}, \overline{N}^{uc}) = \frac{\rho}{\rho+\lambda} \operatorname{Max}_{\{R_\tau^c\},T^c} \left\{ \int_0^{T^c} -R_\tau^c e^{-\rho\tau} d\tau + e^{-\rho T^c} \psi^c \right\} \tag{4}$$

$$= \frac{\rho}{\rho+\lambda} V^c(\psi^c, \overline{N}^c).$$

The last equality follows because the latter optimization problem is formulated in terms of the "certainty" rate ρ and is therefore equivalent to the certainty problem (with the "scaled" benefit and completion level). Thus,

$$\rho^{uc} V^{uc}(\psi^{uc}, \overline{N}^{uc}) = \rho^c V^c(\psi^c, \overline{N}^c). \tag{5}$$

This scaling relation between the uncertainty and certainty value functions induces a similar relation between the corresponding optimal completion times. Denoting the optimal value of T by $T^{uc}(\psi^{uc}\overline{N}^{uc})$, we find from (3)

$$\rho^{uc} T^{uc}(\psi^{uc}, \overline{N}^{uc}) = \rho^c T^c(\psi^c, \overline{N}^c). \tag{6}$$

To investigate how uncertainty affects the value V and the completion date T requires evaluating the sensitivity of these functions to the discount rate (because the effects of uncertainty are manifested through the discount rate). Equations (5) and (6) do not explicitly provide the desired changes of V and T with ρ because they relate quantities corresponding to different values of ψ and $\overline{N}$. We seek the effect of increasing ρ, holding the latter parameters constant. Recall from (P4) and (P5) that increasing these parameters entails conflicting trends, $\partial V/\partial\psi > 0$ but $\partial V/\partial \overline{N} < 0$. Similarly, $\partial T/\partial \overline{N} > 0$ but $\partial T/\partial\psi < 0$.

To determine the trend entailed by increasing ρ alone, we consider first the "permanent-risk" scenario, for which the analysis is simpler. In this case, $\psi^{uc} = B/(\rho + \lambda) = B/\rho^{uc}$. Hence, $\psi^c = \psi^{uc}(\rho + \lambda)/\rho = B/\rho^c$, and rescaling does not affect the value of B.

Consider now solutions of the optimization problems with the two different discount rates, but the same values of B and $\overline{N}$. According to (P1), the parameter z_ψ depends on B but is independent of ρ in this scenario. The completion time T is determined by (P2). Since y is increasing and Q is decreasing, increasing ρ to ρ^{uc} in the uncertainty problem and keeping $\overline{N}$ constant implies a *larger* value of T^{uc}. The increase in the effective discount rate encourages delayed investments and extends the project duration.

The project value should decrease when interruption risk is introduced. To verify this, use (P3) noting that the product $\rho^{uc}T^{uc}$ is larger than the certainty value of ρT while z_ψ remains unchanged. Since W(z) is a decreasing function, the uncertainty value must be lower than its certainty counterpart.

These results are illustrated in Figure 1, in which the project duration, T, and value, V, are plotted against the effective discount rate, $\rho + \lambda$, under permanent risk. The form $y(R) = aR/(R + b)$, where a and b are given positive parameters, is assumed for the production function. The project duration increases from about eight years to over eleven years as the project value decreases from \$2 billion at certainty (with ρ = 5 percent per annum) down to zero.

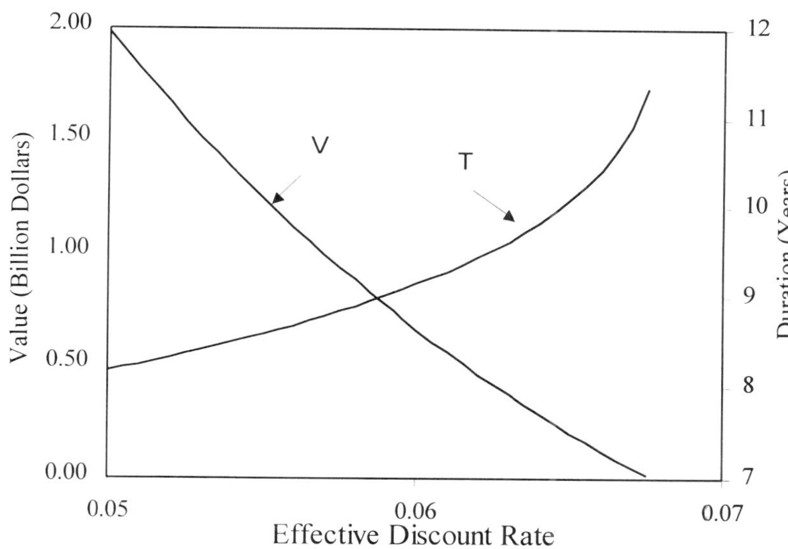

Figure 1. Project Duration T (right scale) and Value V (left scale) versus the Effective Discount Rate $\rho + \lambda$ under Permanent Risk

While it is easy to verify that the latter conclusion concerning the decrease in value associated with uncertainty is robust under scenario variations, the effect on project duration is more subtle. For the temporary-risk scenario we can identify situations in which uncertainty actually implies reduced project durations. Under this scenario, it is the post-project completion value, ψ, which remains invariant when the discount rate is increased to the uncertainty rate of $\rho^{uc} = \rho + \lambda$. We compare, therefore, the values of V^c and V^{uc} holding ψ and $\overline{N}$ constant. The uncertainty duration, T^{uc}, is suboptimal for the certainty problem. Hence,

$$V^c \geq \text{Max}_{\{R_t\}}\left\{\int_0^{T^{uc}} -R_t e^{-\rho t}dt + e^{-\rho T^{uc}}\psi\right\}$$

$$= \text{Max}_{\{R_t\}}\left\{\int_0^{T^{uc}} -R_t e^{-\rho^{uc}t}e^{\lambda t}dt + e^{-\rho^{uc}T^{uc}}e^{\lambda T^{uc}}\psi\right\}$$

$$> \text{Max}_{\{R_t\}} e^{\lambda T^{uc}}\left\{\int_0^{T^{uc}} -R_t e^{-\rho^{uc}t}dt + e^{-\rho^{uc}T^{uc}}\psi\right\} = e^{\lambda T^{uc}}V^{uc} > V^{uc}. \quad (7)$$

Thus, introducing uncertainty or increasing the discount rate decreases the value under temporary risk as well.

In contrast, project duration cannot be a monotonic function of the discount rate. Consider first projects of very short duration, as in the case of infinitesimally small $\overline{N}$. Evidently, if $\overline{N} = 0$ then $T = 0$ and $V = \psi$ regardless of ρ. Alternatively, let $\overline{N} = \delta > 0$ be very small and expand all quantities to $o(\delta)$, where $o(\delta)/\delta \to 0$ when $\delta \to 0$. Then (see Appendix B) the optimal investment rate is given by $R = Q(z_\psi) + o(\delta)$. The corresponding project duration is, up to the same order, $T = \delta/y[Q(z_\psi)]$. Unlike the case of permanent risk, the parameter z_ψ now depends on ρ because $W(z)$ depends on ρ, and this dependence determines the variation of T with ρ,

$$\frac{dT}{d\rho} = (T/\rho)\psi[y'(R)]^3/[y^2(R)y''(R)] < 0. \quad (8)$$

In this limit, T decreases with ρ. For short project duration, uncertainty tends to increase R and decrease T.

At the other extreme, when ψ and T are close to their critical values [see (P6)], the behavior is reversed. Assume the project is barely profitable at the discount rate ρ and that an infinitesimal hazard rate $\lambda = \delta$ is enough to drive the value to zero. Using the scaling laws (5) and (6), we find in Appendix B that

$$T^{uc}(\psi,\overline{N}) - T^c(\psi,\overline{N}) = \frac{z(T^c)V^c(\psi,\overline{N})}{y(R_0)} + O(\delta), \quad (9)$$

where the first right-hand term of (9) is positive and of order $O(\delta^{1/2})$. Thus, project duration increases with uncertainty towards its critical value with an ever increasing slope. Contrasting this result with small $\overline{N}$ behavior we see that, unlike the value V, optimal project duration is not a monotonic function of the discount rate. For temporary risk with fixed values of ψ and $\overline{N}$, it obtains some minimum value. Let the discount rate of ρ_{min} correspond to this minimum duration. Then the effect of uncertainty depends on whether ρ lies below or above ρ_{min}. In the former case, uncertainty entails shorter duration, while in the latter case it implies longer projects. Of course, if the hazard rate is large enough to bring the effective discount rate above the critical value, the project is no longer profitable and T is infinite.

As an example, we illustrate these findings in Figure 2 using again $y(R) = aR/(R + b)$ for the production function and a certainty discount rate of 5 percent per annum. While the project value is seen again to decrease monotonically (albeit at a lower rate relative to permanent risk), the project duration is found to decrease by nearly a year until it begins to rise again at the effective discount rate of 12 percent per annum, at which point the project would have lost some 75 percent of its certainty value.

It is easy to understand the robustness of the results concerning the value function. Occurrence is never profitable. Hence, the hazard associated with it can only decrease the expected value. How should one interpret the sensitivity of the project duration to variations in the parameters? In the temporary-risk scenario, two conflicting considerations call for a trade off. Increased early investments can bring about early completion and ensure a safe stream of benefits so the overall risk is reduced. On the other hand, the expected loss due to interruption prior to completion should encourage prudence and delay investments. For highly profitable short projects, the first consideration dominates while, for long projects of insignificant value, the prudent policy of reduced investment rates is more advantageous. In contrast, under permanent risk even an early completion cannot guarantee the profit. Hence uncertainty is always associated with the more prudent investment policy.

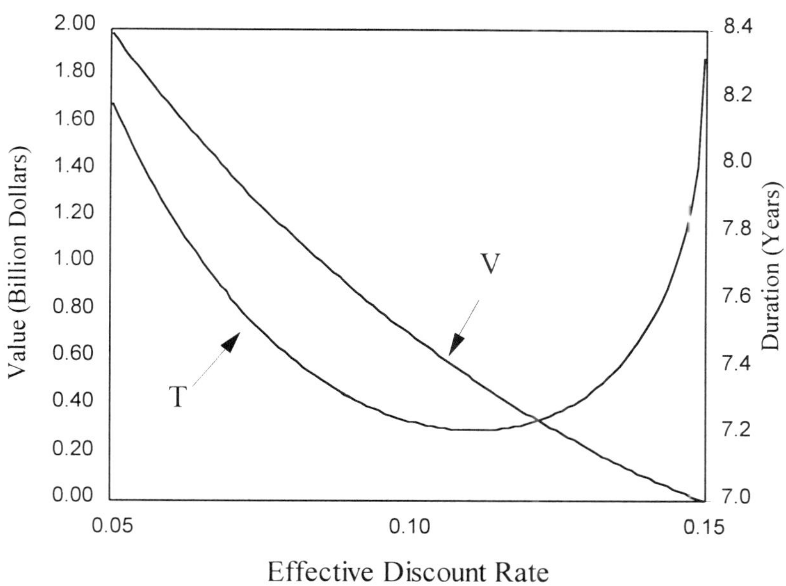

Figure 2. Project Duration T (right scale) and Value V (left scale) versus the Effective Discount Rate $\rho + \lambda$ under Temporary Risk

5. CONCLUDING COMMENTS

The water crisis approaching many countries in the Middle East may increase competition and hostility among nations sharing an international water body. But it may also be an impetus towards cooperation in trans-boundary projects that better exploit the common resource (Naff and Matson 1984). Internal debates on such projects tend to be emotionally loaded and compounded with arguments that do not relate directly to water issues alone. The resolution of such debates often depends on a delicate balance of political powers which introduces uncertainty regarding the will and power of governments to carry on development projects until completion. The political uncertainty considered in this work takes the form of irreversible interrupting events for which the occurrence hazard is exogenous to the planning task. As is well known, event uncertainty of this type is equivalent to an increased effective discount rate. In the present context, this observation gives rise to simple scaling laws for both the project value as well as its duration. Expressing the optimization problem under uncertainty as an equivalent certainty problem (with scaled parameters), one is able to derive the loss of value associated with uncertainty in a straightforward manner.

The loss of value comes as no surprise. The interrupting events represent total investment loss. Hence, the threat of their occurrence can only reduce value. In contrast, the effect of uncertainty on the optimal project duration is more subtle. This observation may be of practical importance. Realizing the losses associated with the interruption hazard, project owners will naturally consider incentive schemes to encourage development firms to expedite their operations. As is shown above, this strategy is not always optimal and certain scenarios call for a more prudent investment policy under political uncertainty.

It is clear, however, that the resolution of problems associated with this type of uncertainty cannot be restricted to economic measures. Perhaps the most important message stemming from studies like the one presented here should be directed at political decision makers. A clear presentation of the prohibitive costs of political uncertainty on development programs and how they dramatically affect social well being might provide the drive for effective measures towards the elimination of its political sources.

APPENDIX A: SOLUTION OF THE INVESTMENT PROBLEM

In this Appendix we evaluate the project duration $T(\psi, \overline{N})$ and the corresponding value function $V^c(\psi, \overline{N})$ for the investment problem under certainty. The investors seek to maximize the benefit from the project,

$$V^c(\psi, \overline{N}) = \text{Max}_{\{R_t\},T}\left\{\int_0^T - R_t e^{-\rho t} dt + e^{-\rho T}\psi\right\}, \tag{A1}$$

subject to

$$dN/dt = y(R_t),\ R_t \geq 0,\ N_o = 0,\ N_T = \overline{N}\ \ (\text{or } R_t \equiv 0 \text{ and } T = \infty). \tag{A2}$$

The Hamiltonian for the optimization problem (A1) is

$$H_t = -R_t e^{-\rho t} + e^{-\rho t}\, p_t y\,(R_t) \tag{A3}$$

where p_t is the current value costate variable. The first order conditions are (Leonard and Long 1992, pp. 334-335)

$$\partial H / \partial R = 0 \Rightarrow y'(R_t) = 1/p_t, \tag{A4}$$

$$\dot{p}_t - \rho p_t = -\frac{\partial H_t}{\partial N_t} = 0 \Rightarrow p_t = p_0 e^{\rho t}. \tag{A5}$$

Combining (A4) and (A5) gives the investment rate

$$R_t = Q(e^{-\rho t}/\rho_0) \tag{A6}$$

with $1/p_0 \leq y'(0)$ to ensure that $R_t \geq 0$ for all t. The transversality condition corresponding to free T is

$$H_T + \frac{\partial(e^{-\rho T}\psi)}{\partial T} = 0,$$

giving $-R_T + p_T y(R_T) = \rho\psi$, or

$$-Q(1/p_T) + p_T y[Q(1/p_T)] = \rho\psi \tag{A7}$$

unless $T = \infty$.

Finally, for a finite T, the terminal condition $N_t = \overline{N}$ implies

$$\int_0^T y(R_t)dt = \int_0^T y[Q(e^{-\rho t}/p_0)]dt = \overline{N}. \tag{A8}$$

Since the solution in (A6) is always available, one needs only to determine the parameters p_0 and T from (A7) and (A8) without violating (A2). Define the function

$$W(z) = \left\{ \frac{y[Q(z)]}{z} - Q(z) \right\} / \rho, \ z \in [0, y'(0)] \tag{A9}$$

with

$$W'(z) = \frac{-y[Q(z)]}{\rho z^2} < 0,$$

$$W''(z) = \frac{-Q'(z)}{\rho z} + \frac{2y[Q(z)]}{\rho z^3} > 0. \tag{A10}$$

It follows that $W[y'(0)] = 0$ and $W(0) = \infty$. Thus, the equation $W(z) = \psi$ must have a single positive root for every ψ, denoted z_ψ, and (A7) implies that $p_T = 1/z_\psi$. Thus, p_T is independent of $\overline{N}$. Evidently, z_ψ decreases continuously with ψ [because $z'_\psi = 1/W'(z_\psi) < 0$] and $z_0 = y'(0)$.

For a feasible solution with $R_t \geq 0$, we must ensure that the argument of Q is always less than y'(0) (see A6). In particular, for t = 0,

$$z_\psi e^{\rho T} = 1/p_0 \leq y'(0)$$

and, unless $T = \infty$ and $N_\infty = 0$, T must satisfy

$$T \leq \log(y'(0)/z_\psi)/\rho \equiv T_{max}(\psi). \tag{A11}$$

Express the final level N_T in the form

$$N_T \equiv K(\psi, T) \equiv \int_0^T y[Q(e^{-\rho t}/p_0)]dt$$

$$= \int_0^T y[Q(z_\psi e^{\rho(T-t)})]dt = \int_0^T y[Q(z_\psi e^{\rho t})]dt. \tag{A12}$$

Evidently, $K(\psi,0) = 0$ and, for all $T < T_{max}(\psi)$,

$$\frac{\partial K}{\partial T} = y[Q(z_{\psi} e^{\rho T})] > 0; \quad \frac{\partial^2 K}{\partial T^2} = \rho z_{\psi}^2 e^{2\rho T} Q'(z_{\psi} e^{\rho T}) < 0. \tag{A13}$$

It follows that for any given $\overline{N}$ and ψ, equation (A8) [recast as $K(\psi,T) = \overline{N}$] must have a unique feasible solution $T(\psi, \overline{N}) \leq T_{max}(\psi)$ if

$$\overline{N} \leq K[\psi, T_{max}(\psi)], \tag{A14}$$

and no solution otherwise. The latter case implies that (A4)-(A8) cannot be solved consistently to provide a feasible plan, and the policy $[R_t \equiv 0, T(\psi, \overline{N}) = \infty]$ forms the optimal plan, yielding $V^c(\psi, \overline{N}) = 0$. Indeed, since $K[\psi, T_{max}(\psi)]$ is a continuous and strictly increasing function of ψ with $K[0, T_{max}(0)] = 0$, equality in (A14) determines the minimal benefit $\underline{\psi}$ for any finite $\overline{N}$. Clearly, the higher the completion level $\overline{N}$, the higher must be the minimum benefit $\underline{\psi}$ that would render the project profitable. Having determined $T(\psi, \overline{N})$, the characterization of the optimization problem (A1) is complete, because the parameter p_0 is readily obtained via

$$\frac{1}{p_0} = z_{\psi} e^{\rho T(\psi, \overline{N})}.$$

We now turn our attention to the value function, $V^c(\psi, \overline{N})$. Consider first the cumulative investment present value,

$$\int_0^T R_t e^{-\rho t} dt = \int_0^T Q(z_{\psi} e^{\rho (T-t)}) e^{-\rho t} dt = e^{-\rho T} \int_0^T Q(z_{\psi} e^{\rho t}) e^{\rho t} dt.$$

Set $z(t) = z_{\psi} e^{\rho t}$ to write the latter integral as

$$(e^{-\rho T} / \rho z_{\psi}) \int_{z(0)}^{z(T)} Q(z) dz = (e^{-\rho T} / z_{\psi}) [zW(z)]_{z(T)}^{z(0)} = \psi e^{-\rho T} - W[z(T)].$$

[The latter equality follows because $z(0) = z_{\psi}$ and $W(z_{\psi}) = \psi$.] Thus,

$$V^c(\psi, \overline{N}) = W[z(T)] = W\left[z_{\psi} e^{\rho T(\psi, \overline{N})}\right]. \tag{A15}$$

In the critical case with equality in (A11) and (A14), $z(T) = y'(0)$ and $W[z(T)]$ vanishes, consistent with the continuity of the value function at $\underline{\psi}$. Otherwise, (A11) implies that $z(T) < y'(0)$, and $W[z(T)] > 0$, justifying identification of $\underline{\psi}$ as the root of (A14).

We now investigate sensitivity of project value and duration to parameters of the problem. Note that T is determined by (A12). Hence,

$$\frac{\partial T(\psi, \overline{N})}{\partial \psi} = -\frac{\partial K / \partial \psi}{\partial K / \partial T} = \frac{z'_{\psi}\left[y[Q(z_{\psi})] - y\{Q[z(T)]\}\right]}{\rho z_{\psi} y\{Q[z(T)]\}} < 0. \tag{A16}$$

Using (A15), we write

$$\frac{\partial V^c}{\partial \psi} = W'[z(T)]\left[\frac{z'_{\psi}\, z(T)}{z_{\psi}} + \rho z(T)\frac{\partial T}{\partial \psi}\right]$$

$$= \frac{W'[z(T)]z'_{\psi}\, z(T)}{z_{\psi}}\left[1 + \frac{y[Q(z_{\psi})] - y\{Q[z(T)]\}}{y\{Q[z(T)]\}}\right]$$

$$= \frac{W'[z(T)]z'_{\psi}\, z(T) y[Q(z_{\psi})]}{z_{\psi}\, y\{Q[z(T)]\}}.$$

However, $z'_{\psi} = 1/W'(z_{\psi})$. Using (A10) to evaluate $W'(z)$, we find

$$\frac{\partial V^c(\psi, \overline{N})}{\partial \psi} = \frac{z_{\psi}}{z(T)} = e^{-\rho T(\psi, \overline{N})} > 0 \tag{A17}$$

and

$$\frac{\partial^2 V^c(\psi, \overline{N})}{\partial \psi^2} = -\rho e^{-\rho T(\psi, \overline{N})}\frac{\partial T}{\partial \psi} > 0. \tag{A18}$$

So far we have been interested in the dependence of the duration T and the value V^c on ψ. It is also illuminating to consider the variation of these parameters with $\overline{N}$. Using (A8) and (A13) we find

$$\frac{\partial T}{\partial \overline{N}} = \left(\frac{\partial K}{\partial T}\right)^{-1} = \frac{1}{y(T)} > 0, \tag{A19}$$

$$\frac{\partial^2 T}{\partial \overline{N}^2} = -\frac{\partial^2 K}{\partial T^2}\left(\frac{\partial K}{\partial T}\right)^{-3} = \frac{-\rho z^2(T) Q'(T)}{y^3(T)} > 0. \tag{A20}$$

Note that $y(T) = y\left[Q\{z[T(\psi, \overline{N})]\}\right], Q(T) = Q\{z[T(\psi, \overline{N})]\}$, etc. Similarly, using (A10), (A13), (A15), and (A19), we obtain

$$\frac{\partial V^c}{\partial \overline{N}} = \rho z(T) W'[z(T)] \frac{\partial T}{\partial \overline{N}} = -\frac{1}{z(T)} < 0 \tag{A21}$$

and

$$\frac{\partial^2 V^c}{\partial \overline{N}^2} = \frac{\rho}{z(T)} \frac{\partial T}{\partial \overline{N}} = \frac{\rho}{z(T) y(T)} > 0. \tag{A22}$$

APPENDIX B: PROJECT DURATION UNDER TEMPORARY RISK

In this Appendix, we derive the dependence of project duration on the hazard rate under the temporary-risk scenario. Consider first projects of very short duration, when $\overline{N} = \delta$ is very small, and expand all quantities to $o(\delta)$.

Neglecting the variation of R in time we write

$$\delta \approx \int_0^T y(R)dt = y(R)T.$$

Hence

$$T = \frac{\delta}{y(R)} + o(\delta)$$

and

$$V = \int_0^T -Re^{-\rho t}\, dt + e^{-\rho T}\psi = -RT + (1-\rho T)\psi + o(\delta) = \psi - (\psi\rho + R)T + o(\delta)$$

$$= \psi - \frac{\delta(\psi\rho + R)}{y(R)} + o(\delta).$$

Maximizing V with respect to R, we find

$$(\psi\rho + R)y'(R)/y(R) = 1 \tag{B1}$$

Transforming to the variable $z = y'(R)$ [so that $R = Q(z)$], we write (B1) as

$$W(z) = \psi. \tag{B2}$$

It follows that $z = z_\psi$ and the optimal investment rate is given by $R = Q(z_\psi)$. The corresponding project duration is, to $o(\delta)$, $T = \delta/y[Q(z_\psi)]$. In this scenario, the parameter z_ψ depends on ρ because W(z) depends on ρ, implying a variation of T with ρ,

$$\frac{dT}{d\rho} = -\frac{\delta}{y^2} y'[Q(z_\psi)]Q'(z_\psi)\frac{dz_\psi}{d\rho} = -\frac{\delta z_\psi Q'}{y^2}\frac{-\partial W/\partial\rho}{\partial W/\partial z}$$

$$= -\frac{\delta z_\psi Q'}{y^2}\frac{W(z_\psi)/\rho^2}{-y/(\rho z_\psi^2)} = \delta[z_\psi / y(R)]^3(\psi/\rho)Q'(z_\psi)$$

$$= (T/\rho)\psi[y'(R)]^3/[y^2(R)y''(R)] < 0. \tag{B3}$$

Thus, in this limit T decreases when the hazard rate is added to ρ. For short project duration, uncertainty tends to increase R and decrease T.

Consider now the other extreme, when the project is barely profitable at discount rate ρ, and an infinitesimal hazard rate λ = δ reduces the value to zero. Using (5), this situation is described by

$$\frac{\rho}{\rho+\delta} V^c(\psi + \psi\delta/\rho, \overline{N} + \overline{N}\delta/\rho) = V^{uc}(\psi, \overline{N}) = 0. \tag{B4}$$

Expanding to o(δ), and using (A18) and (A21) for the partial derivatives of V^c, we rewrite (B4) as

$$V^c(\psi, \overline{N}) = \frac{\delta}{\rho}\frac{\overline{N} - \psi z_\psi}{z(T^c)} + o(\delta). \tag{B5}$$

Using (6) and (B5), as well as the explicit expression for the investment rate $R_t = Q[z(T^c - t)]$, we follow the same procedure for T and find

$$T^{uc}(\psi, \overline{N}) - T^c(\psi, \overline{N})$$

$$= \frac{z(T^c)V^c(\psi, \overline{N})}{y(R_0)} + \frac{\delta}{\rho}\left[\frac{\psi z_\psi}{y(R_{T^c})} - T^c(\psi, \overline{N})\right] + o(\delta). \tag{B6}$$

The sign of $T^{uc}(\psi, \overline{N}) - T^c(\psi, \overline{N})$ is seen to depend on the relative magnitudes of the terms forming the right-hand side of (B6). Obviously, the second term is of order O(δ) and V^c in the first term is of the same order. However, $y(R_0)$ is also of order o(1). Hence, the first term dominates implying $T^{uc}(\psi, \overline{N}) > T^c(\psi, \overline{N})$, and T increases with uncertainty. Indeed, because $W[z(T^c)] = V^c = O(\delta)$, and because $W[y'(0)] = W'[y'(0)] = 0$, it follows that

$$z(T^c) = y'(0) + O(\delta^{1/2})$$

and

$$y(R_0) = y\{Q[z(T^c)]\} = O(\delta^{1/2}).$$

Thus, (B5) and (B6) imply

$$T^{uc}(\psi, \overline{N}) - T^c(\psi, \overline{N}) = \frac{(\delta / \rho)(\overline{N} - \psi z_\psi)}{y(R_0)} + O(\delta) = O(\delta^{1/2}). \qquad \text{(B7)}$$

REFERENCES

Bilen, O. 1994. Prospects for Technical Cooperation in the Euphrates-Tigris Basin. In A.K. Biswas, ed., *International Waters of the Middle East*. Oxford: Oxford University Press.

Biswas, A.K. 1994. Management of International Water Resources: Some Recent Developments. In A.K. Biswas, *International Waters of the Middle East*. Oxford: Oxford University Press.

Centre for Natural Resources, Energy and Transport. 1978. Register of International Rivers. *Water Supply & Management* 2: 1-58.

Dinar, A., and E.T. Loehman, eds. 1995. *Water Quantity/Quality Management and Conflict Resolution*. London: Praeger.

Kolars, J. 1994. Problems of International River Management: The Case of the Euphrates. In A.K. Biswas, ed., *International Waters of the Middle East*. Oxford: Oxford University Press.

Leonard, D., and N.V. Long. 1992. *Optimal Control Theory and Static Optimization in Economics*. Cambridge: Cambridge University Press.

Mageed, A.Y. 1994. The Nile Basin: Lessons from the Past. In A.K. Biswas, ed., *International Waters of the Middle East*. Oxford: Oxford University Press.

Murakami, M., and K. Musiake. 1994. The Jordan River and the Litani. In A.K. Biswas, ed., *International Waters of the Middle East*. Oxford: Oxford University Press.

Naff, T., and R. Matson, eds. 1984. *Water in the Middle East: Conflict or Cooperation?* Boulder, CO: Westview Press.

Wolf, A.T. 1994. A Hydropolitical History of the Nile, Jordan and Euphrates River Basins. In A.K. Biswas, ed., *International Waters of the Middle East*. Oxford: Oxford University Press.

PART IV

ECONOMIC ISSUES IN TRANS-BOUNDARY WATER ALLOCATION

17 THE PERFORMANCE OF WATER MARKETS: TRANSACTION COSTS, INTERJURISDICTIONAL BARRIERS AND INSTITUTIONAL OPTIONS

K. William Easter
University of Minnesota, USA
Ariel Dinar
The World Bank, USA
Mark Rosegrant
International Food Policy Research Institute, USA

Water markets may provide a workable alternative to high administrative pricing strategies for water. Markets can be formal or informal, depending on the country, the water resources, and the type of trade anticipated. In selected areas of northern Chile, formal markets have significantly improved the allocation efficiency of water, both within and between sectors. In other countries such as India and Pakistan, informal markets are more common than was once realized, but occur only within a given irrigation system or district. If the major inefficiencies in water allocation are within irrigation districts, then informal markets may be sufficient. But where interjurisdictional trades offer most of the potential for gains from trade, steps should be taken to establish formal water markets.

1. INTRODUCTION

In areas with irrigation and scarce water supply, research suggests there is a large potential gain in economic efficiency from conservation and the reallocation of water among users (Hearne and Easter 1995; Vaux and Howitt 1984). Traditionally, economists have argued that what is needed is a water pricing system that reflects the scarcity value of water. Their approach has been to harangue irrigation departments and water utilities to charge much higher prices for water (Rogers 1992; Repetto 1986). The problem is that users have been able to use their political power to prevent major increases in water prices, especially for irrigation water. Given this dilemma, a number of economists such as Rosegrant and Binswanger (1994), have suggested that there may be a better way to achieve efficient allocation of water by changing the incentives so that users support the efforts to reallocate water. They argue that if water markets are established where water is scarce, then users can buy and sell water and the whole incentive system is changed. When users can benefit from the reallocation, they are willing to sell water or pay a higher price for new supplies. The numerous informal water markets that have evolved suggest that water users will buy and sell water even if it is illegal or discouraged by governments (Shah 1993; Renfro and Sparling 1986).

Problems arise when governments are asked to encourage or help develop formal markets or allow informal markets to procreate, particularly within government constructed irrigation projects. Many government officials feel that since these are subsidized projects, the users should not be able to sell the water at a profit and that "poor" farmers will be disadvantaged by such water markets. Thus, even though water markets can change the incentives for water users and improve water allocation, there are a number of constraints that may prevent the introduction of water markets.

For formal water markets to work, one key component is for users to have some type of water right or use right they can buy or sell. This may create a serious problem, not only because water rights may be difficult to establish, but also because public water agencies feel they will lose a great deal of power if they allocate water rights to users. If users have rights, this means that system operators (the government officials) have the responsibility to deliver water to the users more or less when the users want it. In contrast, if a government agency holds the water rights, they can dictate to farmers the conditions under which they will receive water, including, in some cases, necessary side payments from farmers. If water rights are made tradable, this creates an even greater dilemma for government agencies. To prevent losing control over tradable water rights, Comision Nacional del Agua in Mexico and some of the water districts in the western United States limit trading among water districts. A water user must obtain special government approval to sell water outside the district or jurisdiction, and any profits from the sale must accrue to the district and not the seller. This, of course, discourages interdistrict trading, but it reduces the chance that trades will have third party effects.

Even if the water management unit that allocates water is set up and hired by the users, the unit may have an incentive to discourage water trades with other jurisdictions. The fear is that if too much water is transferred out of the district, it may reduce economic activity and make it difficult to effectively operate the irrigation system or, at least, parts of the system. Once most of the farmers along a canal have sold their water, it may be difficult and expensive to serve the few remaining farmers who own water rights on the canal. This means that although water markets may change water user incentives and encourage water use efficiency, the management of the system may prevent trades or, at least, raise the transaction costs of interdistrict or interjurisdictional trades.

The organizational problem appears to involve two important aspects. First is the resistance to water trading between or among districts or jurisdictions. Second is the problem of establishing water rights and giving the users more control over water.

Other problems which raise the transaction cost of water trading include legal challenges by third parties who feel they might be damaged by a transfer, the lack of sufficient infrastructure to transfer water among potential buyers, and the lack of an effective means for verifying and enforcing water rights. The question is whether or not it is in the best interest of a country to take action to reduce these transaction costs. If the answer to this question is at least a tentative "yes," then the second question is, How can these transaction costs be lowered?

The problem is illustrated in Figure 1, where transaction costs raise the market price of water and reduce the amount of water transferred. The question is whether or not these transaction costs move us closer to the social optimum. Let us start with a quantity, Q_1, as the initial allocation of water to the urban sector, while the rest of the water is allocated to agriculture. When water becomes tradable, farmers can increase returns by selling water to the urban sector. The farmers' supply curve is S_1, which measures the benefits farmers forgo when they sell water. The urban water demand with no transaction costs is shown by D_1. In the market allocation without transaction costs, the urban demand D_1 is equal to S_1 at Q_2. Under this market allocation, the urban sector receives an additional quantity of Q_2 - Q_1 units of water.

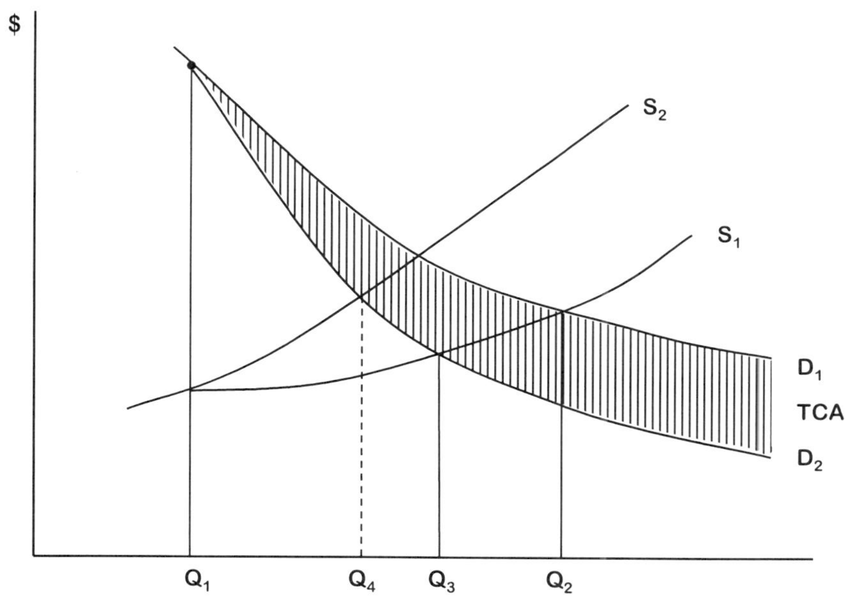

Figure 1. Transaction Costs and the Market Allocation of Water Between Sectors
Source: Adapted from Colby (1990)

If the transaction costs of administering the trades are included, the urban demand drops to D_2, which is the urban demand minus the transaction costs (this assumes the buyer pays the administrative transaction costs, which is the case in a number of countries, such as Chile). The administrative transaction costs are the lined area, TCA, between the two demand curves. The allocation, with transaction costs included, becomes Q_3, which is where S_1 is equal to D_2. In this example, transaction costs increase the water price. Yet this does not include policy-induced transaction costs, which, if included, would further increase water prices and reduce transactions. If these policy-induced transaction costs are paid by sellers, as they are in California, then the supply curve would be shifted up and to the left. Assuming that the shift is to S_2, then the new equilibrium is Q_4, where $S_2 = D_2$. These policy-induced transaction costs result from the political action by third parties who are concerned about instream water use, environmental quality, and the economic impacts on regional sales caused by reduced agricultural production. If, through political action, third parties are able to impose high policy-induced transaction costs on sellers, then trading may be minimal and the equilibrium allocation close to Q_1.

2. ESTABLISHING WATER MARKETS

Before we determine if transaction costs should be reduced and how they can be reduced, it is important to understand the larger social structure of a country in which such economic exchanges might take place. If a country has little experience with private markets for allocating scarce goods and services, then it is unlikely that water will be one of the first goods to be exposed to private markets forces. In contrast, in a country that is exploring new ways to use the private market to improve the allocation of publicly managed resources, scarce water resources might be a good candidate for market trading. Formal water markets developed during the 1990s in California, not only because of the general trend toward the wider use of markets to allocate public resources, but also because of the high economic and environmental costs of developing new sources and the resulting potentially high benefits from trades of existing water supplies. In contrast, one would not expect water markets to develop very soon in the Central Asian republics of the former Soviet Union because of their lack of experience and institutions that would facilitate market development.

There are polar views of how market exchanges occur. One is the neoclassical view that a legal system is required to foster efficient market exchanges. The other is the more pragmatic view that emphasizes the importance of informal contract enforcement to foster market exchanges (Greif 1997, pp. 239-240):

> *This neoclassical view that places the legal system at the center of contract enforcement in market economies has recently been criticized on the basis of evidence indicating that many contemporary exchange relations in the West and elsewhere are informal. The associated contract enforceability is not provided by the legal system but is based on reputation, general morality, and personal trust within social networks. Empirical evidence indicates the importance of two distinct systems of informal contract enforcement: the individualistic system of informal contracts enforcement prevalent in the West, under which the reputation and morality of the individuals are key, and the collectivist system of contract enforcement prevalent in most other societies, under which personal trust within the social network is critical.*

Greif (1997, pp. 240-1) argues that "the neoclassical theory does not imply that a legal system is necessary for anonymous exchange, while the history of the West does not support the view that the legal system is key for the operation of market economies. ... Furthermore, 'asymmetric information' and incomplete contracts limit the extent to which legal systems can facilitate exchanges." He goes on to argue that informal contract enforcement systems are also not costless and can be more expensive than enforcement through a legal framework. In addition, in the collectivist model, exchanges among members of a community can take place, but market exchanges between members of different communities cannot function because the exchanges are based on social ties and personal trust within each small community.

Cooter (1997) comes to a similar conclusion in reviewing the problem of contracting and establishing a rule-of-law that is consistent with a country's social norms. He argues (p. 194),

> *Although enforcement of contracts typically requires coercion by a third party such as a state, the problem can also be solved without recourse to third-party enforcement. Investment in a business network usually occurs among people who maintain enduring relationships... Contracts solve the problem of cooperation through law, and enduring relationships solve the problem of cooperation through repetition. In tentative relationships, the problem of cooperation is solved by social norms.*

At least in the case of markets for irrigation water, it appears that both the formal neoclassic-legalistic system and informal system are at work. The transfer of permanent water rights appears to require the certainty that is provided by a legally based approach where water rights are recorded and can be defended in court. In addition, if water is being transferred among districts and the sale will change return flows, a more formal market may be required with safeguards that prevent losses to those using return flows. In contrast, if the sales are temporary (for one season or less) and do not change return flows, then informal water markets based on "informal" water rights can suffice. These sales will likely be among farmers in the same water district and, in many cases, among farmers served by the same canal. In addition, these sales are not likely to be anonymous, and enforcement of the contracts will not be provided by the legal systems, but will be based on reputation and personal trust within the water user group. This is another reason why trades among districts or jurisdictions may be limited. It also suggests that to obtain more interdistrict or interjurisdictional water trades, a country will have to develop legally based water rights that can be easily verified and defended in court at a reasonable cost. However, if interdistrict trades yield only small efficiency gains for a country, while significant efficiency gains are still possible with intradistrict trades, then informal water markets based on collective enforcement may be all that is needed at this stage of development.

Informal water markets have worked fairly well for groundwater as long as recharge is adequate and there are a sufficient number of sellers in the market (Shah 1993; Palanisami and Easter 1991; Saleth 1997). The "tit for tat" game theory enforcement strategy appears to work. If farmers do not pay, they will not receive water next time, or if a seller does not or will not deliver, then the buyer can go to another supplier. In one area of Gujarat, India, farmers have pipelines from three or four different suppliers coming to their fields (Shah 1993). They can buy from the supplier who offers the best price and service. Shah (1993, pp. 61-62) found that, "while the main beneficiaries of the private investments in pipelines have been the buyers of water, early operators in the water business were motivated mainly by the desire to establish monopoly positions and to overcome topographical constraints in supplying water to a large command."

In many cases, a buyer's choices may be quite limited either because there is not an abundance of suppliers or because the conveyance system is limited by capacity or location (Saleth 1997). In some groundwater-served areas, there are few canals between farmers. Even in an area with canals connecting most farmers, these canals may have limited capacity or be limited by the irrigation authorities to delivering only water from within the irrigation system. The control of the canal system has been used as a way to reduce competition and trading among districts. An inadequate canal system can also limit trades within a district. However, where water is being traded, users have a strong incentive to see that a canal is well maintained since it is transporting "their" purchased water (Palanisami and Easter 1991).

If a country decides to establish a more formal or legalistic approach to water trading, care must be taken in establishing and allocating the water rights to users. For water markets to work, especially in transferring water among sectors or districts, the community of users needs to support the concept as beneficial and fair. The law must be written so that the resulting allocation of rights is considered fair. If the economic rents are concentrated in the hands of a few individuals, or negative third-party effects are large and not mitigated, then the community is not likely to obey the law. According to Cooter (1997, p. 191),

> *A modern economy needs effective laws to promote cooperation among people. Yet states enact many laws that few people obey. People tend to disobey, or obey out of fear, laws that are not consistent with social norms and to obey laws that reflect social norms.*

A good example of a water law that many farmers ignore is the law in Pakistan against trading canal water. In contrast, the 1981 Chilean water law that establishes private water use rights is widely obeyed because not only does Chile have a long record of private water development, but it also allocated the water rights based on past use (Hearne 1995). The one major problem Chile had was in granting water rights for nonconsumptive water use free of charge to power companies, who then changed the timing of water flows.

3. EXPERIENCE WITH WATER MARKETS

As discussed above, there are cases of both formal and informal water markets. The formal sales can be for a specified volume or share of water for a set period of time, or they can be for the transfer of permanent water rights. The more informal markets usually involve the sale of unmeasured flows of water in a canal for a set period of time or the number of hours water is delivered from a pump. Although the unit sold in the informal market may not be metered, both the buyer and the seller have good information on the volume transferred. The key difference between formal and informal markets is the enforcement of trade. If enforcement is by users with no recourse to the legal or administrative system, then the market is informal. Enforcement in informal markets is based on reputation of the individuals or personal trust within the irrigation community.

It is important that we use our experience with both formal and informal markets to help decide when and how water markets might be effectively introduced and how the transaction costs of trading can and should be reduced. Country experiences with water markets are wide-ranging, with most of the formal markets found in North and South America, while the informal markets are concentrated in the irrigated areas of South Asia. For example, in several areas of Chile, the legalization of water trading and recording of water use rights were sufficient to foster an active formal water market with relatively low transaction costs. The combination of effective water user associations and well-developed irrigation infrastructure allowed these farmers to easily trade water throughout the system, including trades with the urban sector (Hearne 1995).

3.1 Informal Water Markets

In the case of informal markets in Asia, Shah (1993) finds that in areas with dependable groundwater recharge, the selling of tubewell water has benefited a wide range of farmers by allowing them to increase crop production and income. Meinzen-Dick (1997, p. 2) finds that the economic gain from groundwater markets can come from

> *improving utilization of tubewell capacity, increasing access to irrigation water supplies (especially among farmers with small or fragmented holdings), and lowering water tables in areas of water logging. ... Access to groundwater makes it possible for those without wells to use groundwater for irrigation. This improves the equity of resource distribution because it is the smaller scale farmers who do not own tubewells. The opportunity to sell groundwater can make it profitable for farmers to invest in wells even if their own holdings are too small to use the full pumping capacity (see Shanker 1992). Even some farmers with wells purchase groundwater from others. This may be to irrigate fragments located away from their own well, or to take advantage of the more reliable water supply from wells located closer to the canals or in pockets of fresher groundwater.*

Groundwater markets are important for agricultural production and the distribution of water in much of the irrigated area of South Asia. Saleth (1997) estimates that 20 percent of the 14.2 million pumpsets are likely to be involved in water trading. This means that water markets are providing water for about 6 million hectares, or 15 percent of the total area irrigated by groundwater. In Pakistan, a survey of well owners found a similar percentage selling water, i.e., 21 percent (National Engineering Services of Pakistan, 1997).

In one of the few studies that have tried to estimate the economic returns from having access to water markets, Meinzen-Dick (1997) found that water markets increased the availability of water and the reliability of supplies. With greater water supplies and reliability, yields and income increased for those who purchased water, particularly for those who also had access to canal water. However, the highest yields and income were still found among farmers who owned tubewells and had access to canal water.

Even though groundwater markets appear to be growing rapidly and providing a more assured source of water to many farmers in South Asia, a number of concerns have been raised about the performance of informal groundwater markets. Given that markets for the sale of groundwater draw on an open access resource, it is not surprising that problems arise in areas with high water demands and limited groundwater supplies. Farmers have an incentive to ignore the scarcity value and buffer stock value of the groundwater and pump until their cost of pumping equals the market price of water (Ramasamy 1996). Over time, the cost and price rise as the groundwater level declines. For example, the overdraft in the Coimbatore district of India is almost 5 thousand cubic meters. Ramasamy (1996) estimates that if the overpumping continues, it will mean a drop in the total net returns to farmers of $42 to $69 million, due to higher energy costs and well investments. Here is a case where informal markets may exacerbate the over-drafting problem. Yet formal markets may not work any better unless water rights can be established in strict quantity terms. The problem is not the water markets but the lack of exclusive property rights. To establish exclusive water rights, the number of wells and the amount of pumping would have to be agreed upon and restricted. Such restrictions are probably unrealistic unless there is strong support in the irrigation community to implement pumping limitations and strict enforcement. However, if exclusive water rights can be established, then the water market should reflect the scarcity value of water and help restrain over-pumping.

Blomquist (1995) reports on one of the few cases where the demand for water is increasing and the community of water users has been able to stop the overdraft. In the Los Angeles–Orange County–Riverside metropolis, pumping is metered and taxed. He reports (1995, p. 43),

> *Water users have shifted at least some of their reliance on local groundwater to more expensive but more plentiful imported water. Surface and imported water are impounded for basin replenishment. In the coastal basins, saltwater intrusion from the ocean has been halted. In some basins, pumping rights have been defined, limited to the basin safe yield, and made transferable within the basin.*

A more typical case, in coastal Gujarat, is reported by Shah (1993). Here, the overdraft of coastal aquifers has caused declining groundwater supplies in some areas and saltwater intrusion in others. Shah argues that it is likely to be difficult to effectively reduce overdrafting without good local leadership and the involvement of water user groups. He argues that "legal, quasi-legal, and organization instruments of public policy will not, on their own, succeed in securing the compliance of farmers unless they are accompanied by measures aimed at affecting private returns to irrigation ... or unless the structure of property rights on the water resource itself is drastically reformed" (p. 47). Similarly, Meinzen-Dick (1997, pp. 18-19) doubts whether in Pakistan "government would have the institutional capacity to regulate sales among hundreds of thousands of private tubewells, and if it had such capacity, it is unclear what such direct intervention could achieve."

Furthermore, in both India and Pakistan, the impact of water markets on the overdrafting of groundwater is much smaller than the impact of the very low or zero per unit charges for electricity. The zero or near zero marginal cost of pumping in most states of India and Pakistan means that farmers have an incentive to use water to the point where the marginal value produced is close to zero. This, of course, encourages farmers who can sell water to use their wells at close to full capacity. The low power rates not only create overdrafting problems, but they also lead to an inefficient use of electricity in countries without adequate electricity supplies.

As noted above, water markets can actually help solve the overdrafting problem by increasing the incentives for efficient water use and making it possible to purchase water from areas not faced with an overdraft. The ability to find another source of water, but at a higher marginal cost, can help promote community action for self-regulation and demand management. Shah (1993) cites a case in coastal Gujarat where self-regulation was generated when additional new supplies were piped into the area.

In large areas of northern India and Pakistan, overdrafting is not a problem. The overdrafting problem tends to be concentrated in coastal areas and the hard rock areas of southern India. In many of the northern areas, pumping actually improves crop-growing conditions by lowering the water table (Shah 1993; Meinzen-Dick 1997). In cases where there are high water tables or rapid recharge rates, water markets are not likely to have negative externalities except for temporary well interference because of the closeness of neighboring wells.

In these cases, personal trust and repetition are likely to be enough to foster competitive informal water markets. This is particularly true where farmers own a number of separate plots that cannot be served by the same well. In such cases, most water sellers are also buyers, since most farmers have only one or two wells on their largest plots and purchase water to irrigate their other plots (Shah 1993; Saleth 1997; Meinzen-Dick, 1997). In addition, their one or two wells are likely to be underutilized unless they can sell water. Yet, because of conveyance costs and the need to get cooperation of neighboring farmers when water is to be transported any distance, transaction costs can be high. High transaction costs not only could limit trades, but they could also prevent water markets from being competitive.

This brings us to the other important concern that has been raised about water markets—the potential for monopoly pricing and discrimination. Groundwater markets are somewhat confined by the physical limits of the location and supply of groundwater. Still, pipelines can extend markets, as they have in parts of Gujarat, and the investment costs in wells should put a limit on monopoly power. An abusive monopolist who raises prices too high will find others investing in wells and undercutting her price. Shah (1993) notes a lack of balance between the numbers of buyers and sellers in areas with high capacity wells, where there may be as many as 70 or 80 buyers served by one seller. Yet, he does not say how many sellers an individual buyer can access. If the buyers can purchase water from four to five sellers, this may be enough to prevent monopoly pricing as long as the sellers do not collude.

The evidence on monopoly pricing is mixed. Meinzen-Dick (1997, pp. 9-10) found, in a 1991-92 survey, that the price of water was not much higher than the cost of pumping,

The two most common ways of charging for groundwater are a flat charge per hour of pumping (ranging from Rs 14 to Rs 80 per hour, depending on the pump type, capacity, and

location), and ... arrangements whereby the buyer supplies the diesel and motor oil for the pump, and pays an additional fee of Rs 4 to Rs 6 per hour to the well owner to cover the wear and tear on the engine. Water sellers with diesel pumps are apparently only recovering their own costs under either type of contract.

Saleth (1997) suggests that in India, monopoly rents may be extractive in some areas. He cites as evidence the variation in water charges compared to pumping costs in different areas. For example, water charges are 1.3 to 2.0 times higher than operating costs in the Indo-Gangetic region, but 2.5 to 3.5 times higher in the water-scarce hard rock regions of south India. Another explanation for the difference in rates is the difference in water scarcity. Since groundwater scarcity is greater in most hard rock areas than it is in the Indo-Gangetic plain, the rates may be directly related to the scarcity value of groundwater.

The degree of monopoly power may also be related to the terms of the transaction or contract for water. Not surprisingly, some of the contracts for water are quite similar to contracts for land. Contracts for water include crop-sharing, crop- and input-sharing, and labor arrangements. If the payment is cash-based, then the buyer has more freedom to take his/her business to another well owner anytime during the season. When the transaction is a contract in kind, especially if it is based on a crop-sharing or a crop- and input-sharing contract, then the buyer is tied to the seller for at least one season. Similarly, when compensation is in terms of labor, it may be difficult to change suppliers, depending on the length of time a buyer provides labor to the seller. Yet within the village setting of South Asia, it appears that informal markets do not face extreme cases of monopoly rent taking. In fact, if it restrains trading in areas with serious problems of declining groundwater, this may help reduce overextraction. When suppliers are taking advantage of their monopoly position and there are adequate groundwater supplies, the best strategy is to encourage (legalized) trading and increase competition through support of community or private well development (Palanisami and Easter 1991).

Thus, informal water markets seem to be a way to improve water use and incomes in irrigated areas where the water rights are not well-defined or recorded. They also may be a good option if third party challenges are common and result in high transaction costs for formal water markets. Finally, informal markets would work well in traditional irrigation systems where the farmers manage their own water allocation and are able to maintain a relatively modest level of transaction costs.

3.2 Formal Water Markets

In situations where informal markets can work well, it may not be necessary to incur the extra expense of establishing formal water markets. However, informal markets cannot be relied on for permanent water transfers or for interjurisdictional trades. Formal markets will be required to provide the certainty necessary for permanent water transfers or transactions between different sectors and jurisdictions. For example, formal markets in the Middle East may be one way to solve some of its critical interjurisdictional water allocation problems (Becker 1995; Becker et al. 1996).

Since the need for permanent trades and interjurisdictional water exchanges are likely to become more important with the rapid growth in nonagricultural water demands, we are likely to see a growth in formal markets. The growing demand in water-scarce regions has been a driving force behind the new interest in water markets. A number of studies have already illustrated the benefits that are possible from interjurisdictional trading.

Garrido (1997) shows that the economic gains, when trading is restricted to an individual water district or community, may be relatively modest. In contrast, if trades are allowed among communities that are subject to different supply constraints and drought conditions, the gains from trade can be substantial. In his analysis of intra-community trading in four irrigation communities, Garrido (1997) estimates total welfare gain would be no more than 10 percent over the current water allocation. However, with inter-community trading, the economic gain for one of the older irrigation communities could reach almost 50 percent. He also shows that both intercommunity and intracommunity trading are very sensitive to the level of transaction costs. If transaction costs exceed 8-12 percent of the market price, trading and the gains from trading would be too small to justify the expense of establishing formal markets. Yet Garrido (1997) may underestimate the potential gains since he only considers crops traditionally grown in the region, i.e., cotton, wheat, corn, oilseeds and sugar beets, and excludes any transfers to nonirrigation uses. In Chile, there was significant change in cropping that accompanied water trading (Hearne 1995).

In contrast, Horbulyk and Lo (1997) found in their study of potential gains from introducing water markets in the Alberta Province in Canada that most of the gains were likely to be from intrasub-basin trades. They considered four sub-basins and compared the current situation with the water allocation under four separate markets (one in each sub-basin), as well as with a market encompassing the total basin. The four separate market scenarios created 90 percent of the welfare gains that were obtained when unrestricted trading was allowed among the four sub-basins. The urban sectors purchased most of the water, except for the South Saskatchewan River, where the agricultural sector purchased additional water when market trading was allowed among the sub-basins.

In their analysis of selected water markets of Chile, Hearne and Easter (1997) found trading both within and between sectors. For permanent transactions either within or between sectors, well established water use rights that were recorded and recognized by government were critical. Several trades between farmers and the city of La Serena were not consummated because of the uncertainty regarding ownership of water rights. La Serena is a growing vacation destination located in a dry region on the coast some 400 km north of Santiago.

The rapid growth in demand has placed a strain on the city's water supply, particularly during the summer tourist season. The opening of water markets allowed the city to purchase water use rights and to save funds by delaying the need to develop new sources of water. Starting in 1992, the city's water company, ESSCO, purchased

292 shares, which added 28 percent to its water supply.[1] In addition, 63 shares were purchased by households for domestic uses upstream from La Serena, and 351 shares were purchased by farmers. The gains from trade averaged $846 per share with a transaction cost of $56 per share.

The gains from trading in La Serena and for sales to other domestic consumers were estimated based on the cost savings from delaying dam construction by 18 years compared to the value of water used in agriculture. Because of the purchase of water rights, La Serena will be able to meet its rapidly growing water demands without water from the proposed Poloma Dam until after the year 2010. The value of the cost savings, Φ, is calculated following

$$\Phi_t = \frac{C}{e^{r(T-t)}}$$

where

r = .12 (12 percent interest rate)
T = 2010
t = 1992
C = annualized payment by the city for the cost of the dam, CH $ 1 billion.[2]

The value of water in agriculture is based on farm budget analysis in which the net returns to water are estimated for both buyers and sellers of water in the agricultural sector. This is an upper bound to the farmers' willingness to pay; it is not a marginal analysis. Yet it provides a good basis for calculating the net gains from water trading, which are calculated as the difference between the buyers' and sellers' net return for water. In the case of La Serena, the opportunity cost of the water it purchased is its value in agricultural production.

The other study area in Chile with significant trading was the Limarí Valley, where trading was very active but restricted to only the agricultural sector. The local city already has adequate water supplies. A survey of 37 farmers selling water and 19 farmers buying water accounted for 9.2 million cubic meters being traded at an average annual price of $2.47 per cubic meter with transaction costs of $.07 per cubic meter. This was not a random sample or an attempt to obtain a complete sample. The numbers surveyed were large enough to indicate that the water market was very active and has created significant gains from trade. The largest gains were for water purchased by three large table grape producers who purchased 5.8 million cubic meters of water (63 percent of the total amount traded in the sample).

In the active water markets in Chile, the transaction costs were low and did not seem to constrain trading. However, in other areas in Chile, such as the upper section of the Maipo River that supplies the southwestern Santiago area and irrigates 100,000 hectares, transaction costs are high and trading is quite limited. The Maipo River is divided into three sections for management and water trading. Uncertain water rights and the lack of adjustable control structures both limit trading. Similar concerns with

[1] One share for ESSCO is 1 liter per second.
[2] For more details concerning the model and data, see Hearne and Easter, 1995.

transaction cost have been raised by Archibald and Renwick in their 1997 study of water transactions in the California. Using the same basic model as Hearne and Easter (1995), they found that a large number of potential trades had not occurred, even though substantial gains from trades were possible. The primary reason why these trades did not occur was largely the transaction costs related to possible third party effects and legal actions.

Archibald and Renwick's (1997) analysis of the California water market suggests that two types of transaction costs have limited potential for profitable trades. One, which is explicit and included in the sales price of water sold through the water bank, is administratively induced transaction costs (AIC). The other is policy-induced transaction costs (PIC), which they attribute to existing institutional requirements such as no injury to vested rights, no unreasonable effects on fish and wildlife, and no mitigation against third-party effects. The AICs include the costs of the search for buyers and sellers and negotiating quantities, timing, and other terms of transfer, and were \$50 per acre foot in 1991 and \$17.50 per acre foot in 1992 and 1994. In effect, what has happened is that when PICs were added, the supply curve (S_1) for water shifted up to above S_2 (Figure 1), which made many potential transfers of appropriative water rights in the Central Valley unprofitable. According to Archibald and Renwick (1997, p. 30),

> *Potential rents net of AICs for sellers ranged from \$18 to \$36 in the 'low' price market and from \$64 to \$100 in the 'high' price market. For buyers, potential rents net of AICs range from \$80 to \$149 in the 'low' price market and \$56 in the 'high' price market.*

Archibald and Renwick argue that these potential rents were too low to cover expected PICs and prevented trading in the Central Valley of California. Colby et al. (1990) found PICs in the West range from \$187 per acre foot, in Colorado, to \$54 per acre foot, in New Mexico, with all western states having less stringent state and federal transfer requirements than California. Policy-induced transaction costs in this range would be as much or more than the potential gains for trading in the California Water Bank.

Colby (1997) suggests that claims on water by Native Americans on many western rivers have the effect of imposing high transaction costs on water trading. The problem may be that markets will not be able to work under such high transaction costs. Conversely, she argues that when the transaction costs of water markets are compared with the cost of litigated solutions to the Native American water rights claims, water markets look much better.

4. CONCLUSIONS

As we consider the use of more formal water markets, the question of transaction costs must be addressed directly. What are the legitimate uses that need to be protected in water transactions? Clearly, return flows have to be considered, as do third party economic effects and instream uses, but should these interests be able to block all transfers because they might be affected adversely?

On the other hand, should third parties have to take into account the economic losses caused by not allowing the transfers? For example, the 1981 legislation establishing water markets in Chile did not include safeguards for possible third party or return flow impacts. In more recent legislation, the Chilean government has tried to correct for some of these potential problems by including provisions for minimum stream flows and other restrictions on water transfers. Is the 1981 Chile model a better one for developing countries to use as they approach the question of water reallocation? If they use this model, does it pose potential future problems, especially when they are dealing with return flows in international waters?

In countries with numerous small scale water users, it may be better to encourage the procreation of informal markets. If this can be done, many of the costs of establishing and enforcing water rights can be avoided. In addition, third party effects can be dealt with by the larger irrigation community. Yet, as markets expand into areas with limited groundwater supplies, regulations may be necessary to prevent serious market inefficiencies due to the lack of well-defined water rights. Furthermore, informal markets will not be adequate to take advantage of potential gains from trading among different jurisdictions. It is also likely that some instream uses and environmental damages will be ignored. Still, for many countries in the early stages of development, this may be a price they are willing to pay.

REFERENCES

Archibald, S., and M.E. Renwick. 1997. Market Potential and Performance in the California Water Industry. In K.W. Easter, M. Rosegrant, and A. Dinar, eds., *Markets for Water: Potential and Performance.* Boston, MA: Kluwer Academic Publishers.

Becker, N., N. Zeitouni, and M. Shechter. 1996. Reallocating Water Resources in the Middle East through Market Mechanisms, *International Journal of Water Resources Development* 12: 17-32.

Becker, N. 1995. Value of Moving from Central Planning to a Market System: Lessons for the Israeli Water Sector, *Agricultural Economics* 12: 11-21.

Blomquist, W. 1995. Institutions for Managing Groundwater Basins in Southern California. In A. Dinar and E.T. Loehman, eds., *Water Quantity/Quality Management and Conflict Resolution: Institutions, Processes, and Economic Analysis.* Westport, CT: Praeger Publications.

Colby, B., et al. 1990. Transaction Costs and Efficiency in Western Water, *American Journal of Agricultural Economics* 72: 1184-1192.

Colby, B. 1997. Negotiated Transactions as a Conflict Resolution Mechanism: Bargaining over Water in the American West. In K.W. Easter, M. Rosegrant, and A. Dinar, eds., *Markets for Water: Potential and Performance.* Boston, MA: Kluwer Academic Publishers.

Cooter, R.D. 1997. The Rule of State Law and the Rule-of-Law State: the Economic Analysis of the Legal Foundations of Development. In M. Bruno and B. Pleskovic, eds., *Annual World Bank Conference on Development Economics, 1996.* Washington, D.C.: The World Bank.

Garrido, A. 1997. An Economic Analysis of Water Markets within the Spanish Agricultural Sector: Can They Provide Substantial Benefits? In K.W. Easter, M. Rosegrant, and A. Dinar, eds., *Markets for Water: Potential and Performance.* Boston, MA: Kluwer Academic Publishers.

Greif, A. 1997. Contracting, Enforcement, and Efficiency: Economics beyond the Law. In M. Bruno and B. Pleskovic, eds., *Annual World Bank Conference on Development Economics, 1996.* Washington, D.C.: The World Bank.

Hearne, R.R., and K.W. Easter. 1997. The Economic and Financial Gains from Water Markets in Chile. In K.W. Easter, M. Rosegrant, and A. Dinar, eds., *Markets for Water: Potential and Performance.* Boston, MA: Kluwer Academic Publishers.

Hearne, R.R., and K.W. Easter. 1995. Water Allocation and Water Markets: An Analysis of Gains-from-

Trade in Chile. World Bank Technical Paper Number 315, Washington, D.C.

Hearne, R.R. 1995. The Market Allocation of Natural Resources: Transaction of Water-use in Chile. Ph.D. Dissertation, Department of Applied Economics, University of Minnesota.

Horbulyk, T.M., and L.J. Lo. 1997. Welfare Gains from Potential Water Markets in Alberta, Canada. In K.W. Easter, M. Rosegrant, and A. Dinar, eds., *Markets for Water: Potential and Performance.* Boston, MA: Kluwer Academic Publishers.

Meinzen-Dick, R. 1997. Groundwater Markets in Pakistan: Institutional Development and Productivity Impacts. In K.W. Easter, M. Rosegrant, and A. Dinar, eds., *Markets for Water: Potential and Performance.* Boston, MA: Kluwer Academic Publishers.

National Engineering Services of Pakistan (NESPAK). 1991. Contributions of Private Tubewells in the Development of Water Potential. Field Report, Ministry of Planning and Development, Government of Pakistan.

Palanisami, K., and K.W. Easter. 1991. Hydro-economic Interaction between Tank Storage and Groundwater Recharge, *Indian Journal of Agricultural Economics* 46: 174-179.

Ramasamy. 1996. Intertemporal Allocation of Groundwater Resources with Stochastic Surface Supplies: The Case of Southern India. Ph.D. Thesis, Department of Applied Economics, University of Minnesota.

Renfro, R.Z.H., and E.W. Sparling. 1986. Private Tubewells and Canal Water Trading of Pakistan Punjab Watercourses. In K.W. Easter, ed., *Irrigation Investment, Technology and Management Strategies for Development.* Boulder, CO: Westview Press.

Repetto, R. 1986. Skimming the Water: Rent Seeking and the Performance of Public Irrigation Systems. Research Report 4. Washington, D.C.: World Resources Institute.

Rogers, P. 1992. Comprehensive Water Resource Management: A Concept Paper. Working Paper Series 879, Infrastructure and Urban Development Department. Washington, D.C.: The World Bank.

Rosegrant, M., and H. Binswanger. 1994. Markets in Tradable Water Rights: Potential for Efficiency Gains in Developing Country Irrigation, *World Development* 22: 1613-1625.

Saleth, R.M. 1997. Water Markets in India: Economic and Institutional Aspects. In K.W. Easter, M. Rosegrant, and A. Dinar, eds., *Markets for Water: Potential and Performance.* Boston, MA: Kluwer Academic Publishers.

Shah, T. 1993. *Groundwater Markets and Irrigation Development: Political Economy and Practical Policy.* Bombay: Oxford University Press.

Shankar, K. 1992. *Dynamics of Groundwater Irrigation.* New Delhi: Segment Books.

Vaux, H., and R. Howitt. 1984. Managing Water Scarcity: An Evaluation of Inter-regional Transfers, *Water Resources Research* 20: 785-792.

18

OVERCOMING THE INTROSPECTIVE LEGACY OF TRADEABLE WATER ENTITLEMENT POLICIES IN SOUTH EASTERN AUSTRALIA

Henning Bjornlund
Jennifer McKay
University of South Australia, Australia

Tradeable water entitlement (TWE) policies and water laws in South Australia, Victoria, and New South Wales are very different, reflecting the local economic, geographical, and historical circumstances. Such introspectivity is not uncommon, but in resource development it creates particular problems. Prior to the involvement by the Council of Australian Governments in 1994, no attempts have been made to consider other nearby jurisdictions. Recently, the federal government as well as the Murray-Darling Basin Commission have been pushing for more interstate cooperation in these matters to create conditions for viable interstate water markets. The Murray-Darling Basin Ministerial Council has funded a limited trial of water transfers among the three states to take effect in 1997. This chapter will discuss this introspective legacy of TWE policies and water allocation laws and their implications for the wider application of water markets. Some ways of overcoming this legacy will also be considered, as well as the environmental and social ramifications of enlarged TWE.

1. INTRODUCTION

The development of water trade and water markets has taken place in several countries in arid and semi-arid regions of the world, among them Australia and the southwestern states of the USA. The economic, social, cultural, and environmental arguments for and against water trade have been well documented in the literature and will not be repeated in this chapter (for example, see Howe et al. 1986; Young 1986; and Nunn and Ingram 1988). Suffice it to say that, largely, water trade has been seen as a means to overcome the legacy of past water allocation laws and policies in order to satisfy increased demand from new water users from a finite resource without the use of draconian measures.

Past water policies reflect past intersectoral political values and power balances among different sectors of the economy. Intersectoral balances have changed over time and now result in policy changes facilitating a redistribution of resources reflecting these new realities. One of these, the riparian doctrine, was adopted by all states, but by the turn of the century all Australian states had discarded this system as too restrictive for future developments. Australian states also rejected the prior appropriation doctrine (which creates water as a personal property right) and instead retained control and ownership of water with the Crown. This allowed state governments to allocate water to individual users and projects in support of state policies. State water allocation policies, however, developed differently reflecting the different water uses and structures of the irrigation industry. Such water allocations were attached to specific parcels of land and not created as personal property.

Permanent water trade was first introduced in South Australia in 1983, followed by New South Wales in 1989 and Victoria in 1991. The new 1996 Water Resources Bill in South Australia is the only act formally removing the nexus between land and water, vesting the ownership of the water right in the licensee. In the other states, the ministers are given the power to remove the water right from one property and attach it to another. Water trade has been defined differently within each jurisdiction, reflecting the different ways in which the water entitlements have been defined, and the different economic, geographical, and historic circumstances within the three jurisdictions. As a consequence, very few interstate transfers have taken place. Water markets and interstate trade form an integral part of the water reform agenda presently being pushed by the Council of Australian Governments (COAG), as well as the Murray-Darling Basin Commission's (MDBC) Irrigation Management Strategy (MDBC 1996). In line with most other western countries, economic rationalism is the platform for contemporary Australian policy making in many areas, and the water industry is no exception. In this process, the COAG sees transferable water entitlements as an important tool. With TWEs, inefficient irrigators, not able to sustain full cost recovery water prices, will be able to fully or partly opt out of irrigation. Then more efficient water users will be able to buy water and, hence, inefficient water users will have a market for their water asset.

The COAG is, therefore, encouraging the states to amend and change legislation to remove impediments to trade such as restrictions regarding interstate trade. The federal government is using fiscal powers in this process by linking the implementation of water reforms to the payment of federal funds under the National Competition Policy.

The federal government has little power over water resources, apart from its general fiscal powers. Section 100 of the Constitution ensures states the full right to use the water of rivers for navigation and irrigation. On the other hand, Section 92 ensures absolute freedom of interstate trade and commerce. It has never been contested in any court whether restrictions on interstate water trade violate the rights under Section 92, if such trade does not infringe on the state rights under Section 100. Such court cases have taken place in the United States making it illegal for states to stop interstate water trade. See *Sporhase v Nebraska ex rel Douglas*, discussed in Shupe et al. (1989). See also Gould (1989).

The MDBC, the organization managing Australia's largest river system, expresses some concern about the consequences of water trade, if present water management practices are continued. The MDBC conducted a Water Audit in 1995. This resulted in water extraction from the river being capped to respect environmental concerns.

This chapter will first briefly discuss the differences in water management and allocation and transfer polices of each of the three states comprising Southeastern Australia, namely South Australia, Victoria, and New South Wales. Next we will discuss the Murray-Darling Basin. This Basin is the predominant feature in this region and has a history of shared management. This part will also include a discussion of the MDBC's capping initiative and its possible implication for future water markets and interstate trade. This will be followed by a brief discussion of the COAG and its water reform agenda. We will then briefly describe a pilot interstate water trading scheme which will be introduced by the MDBC in 1997 within a restricted area. Finally, we will discuss some possible environmental and social ramifications of trade and how these can be overcome. Throughout the chapter, reference will be made to some aspects of analysis of water trade in South Australia and Victoria. These are case studies of water markets along the Murray River in South Australia from 1987 to 1994 and within the Goulburn-Murray Irrigation District (GMID) in Victoria from 1991 to 1994. The two study areas are shown on the map in Figure 1.

2. STATE POLICIES

This section discusses water management and allocation and transfer policies in each of the three states in southeastern Australia.

2.1 South Australia

The Murray-Darling Basin Agreement (the Agreement) secures South Australia an annual minimum flow of 1,850 gigaliters (GL) specified on a monthly basis. New South Wales and Victoria each guarantee 50 percent of this quantity and can then share the rest of the flow in the River Murray and all the water in their respective tributaries. Of the allocation of 1,850 GL, 700 GL are specified for dilution and evaporation losses. Out of the total allocation, South Australia has only committed 870 GL for economic diversion for both domestic industrial and rural uses. Of this, 70 percent is for irrigation. This leaves a residual of 980 GL, or 280 GL more than required by the Agreement. This is widely accepted in South Australia as a de facto allocation to the environment. With the exception of severe droughts, South Australian irrigators are virtually 100 percent certain of receiving their annual entitlement

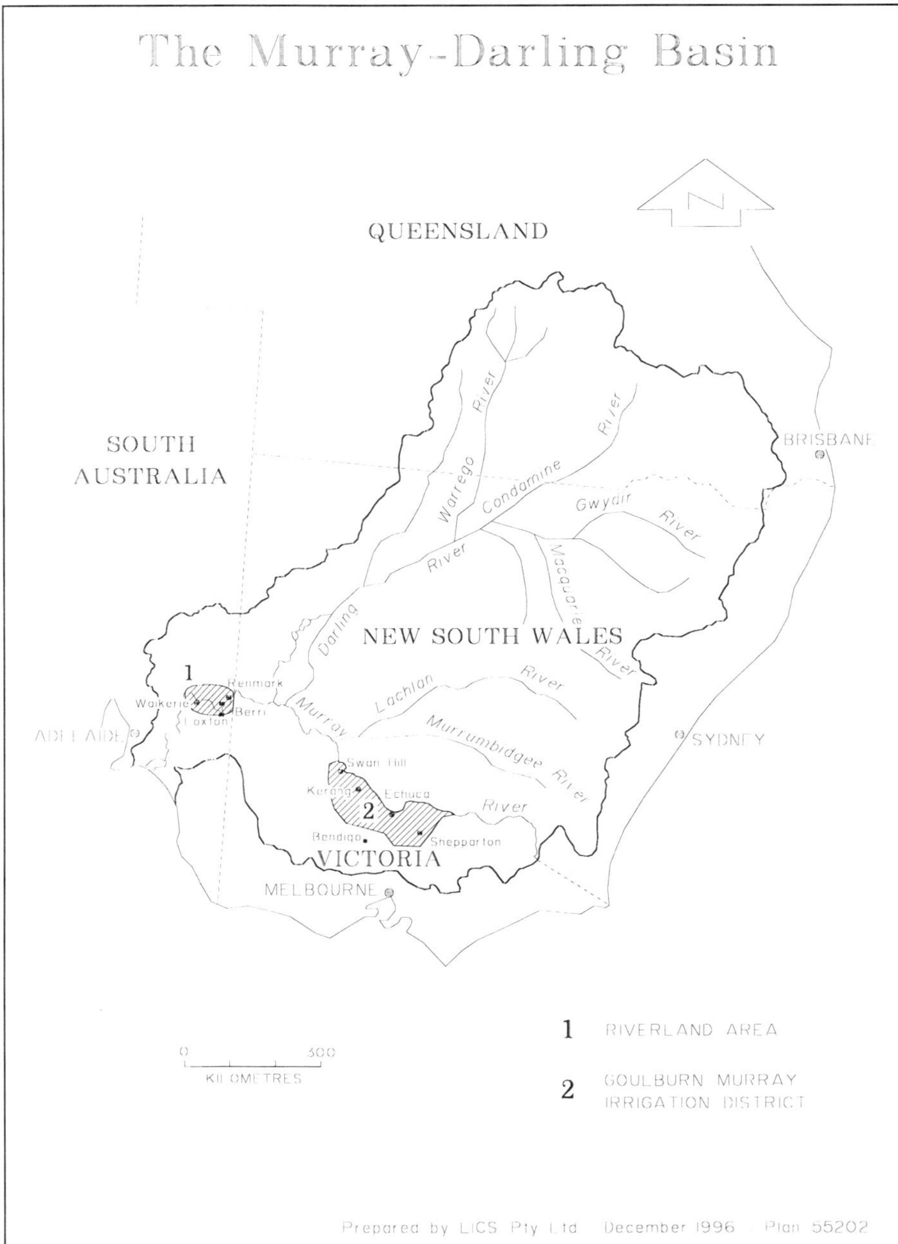

Figure 1. The Murray-Darling Basin
Prepared by Land Information Cartographic Services (LICS), Melbourne, Victoria, Australia

South Australia was conscious of the risks associated with over-commitment of the water resources of the River Murray as far back as 1969, and introduced a moratorium on all new allocations. With the 1976 Water Resources Act, the River Murray was declared a proclaimed water course. This allowed for close control of diversions. Volumetric allocations were introduced for all private irrigators based on their current and, in some cases, potential level of development. It soon became apparent that some allocations were excessive. In 1979 all allocations were revised based on actual consumption during the 1976-79 period, resulting in a reduction of nearly 10 percent. The South Australian government, at the time, was so concerned with overallocation in the Basin that the Minister for Water Resources raised objections with the New South Wales (NSW) government over its overallocation and likely downstream impact. When the NSW government refused to respond, the question was finally considered in the NSW Supreme Court (Smith 1994). Since then, the MDBC has raised grave concerns over water diversion from the river (Murray-Darling Basin Ministerial Council, 1995)

The South Australian legislation is silent on the issue of interstate trade, but the attitude is clear. South Australia's stand on the issue is that, since interstate trade is not restricted or prohibited in the Act, it is possible. South Australia does not find it necessary to include any special provision for interstate trade since it is explicitly possible to purchase and sell water under the Act. This flexible attitude is possibly caused by the expectation that, if interstate trade was widely used, water would flow toward South Australia due to the high value nature of the production. The horticultural and viticultural industries in the Riverland have, under the intrastate water trade program, proven their willingness to purchase water. As much as 91 percent of the water purchased goes into these industries, whereas 71 percent of the water sold goes out of properties with pastures or broad acre production such as grain and lucern (Bjornlund and McKay 1996). In the first annual report on the COAG's water policy reform implementation progress (The Progress Report), South Australia expressed strong support for interstate as well as intersectional transfers. It is, however, stressed that such transfers will have to be under certain constraints such as environmental consideration, uniform definitions of water rights among jurisdictions, and a uniform approach to sleeper and dozer licenses. Large scale up-stream trade to the eastern states should be avoided due to the loss of dilution flow, whereas down-stream trade should be encouraged due to both environmental and economic reasons (The Working Group 1995; see also Department of Environment and Natural Resources 1995).

2.2 Victoria

Victoria also has a system with relatively high supply security. The Victorian system is two tiered. Irrigators have water rights which have a delivery security of 97 years out of 100. In addition to the water rights, irrigators have access to sales water. Sales water allocations are announced annually based on the amount of water in the reservoirs and are expressed as a percentage of water rights. The high security level of the water rights is maintained by setting the annual sales water allocation at a level where next year's water rights are secured in the reservoirs.

The Victorian problem is that irrigators over many years have been accustomed to very high levels of sales water allocations normally in excess of 100 percent of water rights, and often considerably more. This has caused some irrigators to plant permanent pastures on the strength of these uncertain sales allocations. In recent years, the access to sales water allocations has been curtailed. Due to the 1994/95 drought, the water authority for the GMID moved to cap sales of water entitlements in 1995, at a maximum of 50 percent and 100 percent of water rights, respectively, within two different parts of the river system. This cap has since been made permanent in response to the MDBC's drive to limit diversion within the Basin. The amount of sales water is likely to decrease further under the new capping initiative. This development has resulted in increased activity on the transfer market as irrigators try to secure their permanent pastures with water rights rather than the more and more uncertain sales water. As a result, 57 percent of all water buyers rated security against drought as a very important reason for purchasing water. This contrasts to South Australia where only 13 percent is purchased for drought security, reflecting the difference in the two allocation policies (Bjornlund and McKay 1995, 1996).

In the case study area, most water is applied to pastures for animal production, predominantly dairy. This production requires a combination of permanent pastures, annual pastures, and some dry pastures. This combination is well suited for the allocation polices, apart from the fact that irrigators have overcommitted themselves with permanent pastures due to the traditional high level of annual sales of water allocations. The presence of annual pastures and dry pastures allows the irrigators to increase and decrease production on an annual basis depending on the availability of annual sales water. This production has also resulted in a high level of activities on the temporary transfer market.

The increased take-up of permanent water trade activating previously partly or fully unused water (dozer or sleeper water) will increase annual extractions of water from the river and thereby reduce annual allocations of sales water. This process will further increase the demand for water rights on the permanent transfer market.

The Victorian Water Act of 1989 with its 1995 amendment is the first Act to explicitly make provision for interstate trade. Trade is still only allowed on a temporary basis. Concern was raised over a possible increase in overall water use due to increased uptake of sales water entitlements if water was sold to New South Wales. The Minister has therefore been given the power to restrict transfer of the sales water component. As a consequence, transfers of private diverter licenses have been restricted to only the water right component. Also, under the 1995 amendment, the last steps were taken to bring all water diversion in Victoria under a bulk entitlement. Water allocation is thereby capped, as requested by the MDBC. In the Progress Report, Victoria is strongly supporting interstate trade, but again stressed the need for more consistency among jurisdictions, expressed the need to remove all cross subsidies for water crossing the border, and emphasized that reciprocity is a key issue in ensuring that water can move in all directions (The Working Group 1995).

2.3 New South Wales

New South Wales offers two levels of security of allocations. High security entitlements are offered for domestic and industrial users, for stock watering, and for irrigation of permanent plantings. All other water users have a normal security license. Holders of high security licenses receive 100 percent of their allocations except under severe drought. Holders of normal security licenses share what is left in the rivers and storages after the high security licenses have been supplied. Reservoirs in NSW are managed to supply the maximum water to irrigators in each season, carrying over a minimum to the next irrigation season (Alaouze and Whelan 1996). Sturgess and Wright (1993) call the quantity left in the reservoirs a 'prudent' carryover for future dry periods. Allocations for normal security licenses are announced annually based on water in storage. With little carryover, annual allocations fluctuate widely from year to year and from basin to basin depending on the basin's ability to collect and store water and the weather patterns. As an example, the Gwydir River Valley has averaged 45 percent of entitlement in recent years (Sturgess and Wright 1993). The official NSW policy is to achieve 70 percent reliability of supply for all the regulated rivers in the state (MDBC 1995). This variation in annual allocation creates uncertainty constituting an impediment to trade. The state has therefore assigned a reliability factor to water entitlements in an attempt to alleviate this uncertainty.

Water resources in NSW were overcommitted at an early stage as a part of the settlement of the state. This was due to, among other things, area-based entitlements and was without volumetric limitations. To overcome this, volumetric allocations were introduced in some areas in 1975/76 and officially introduced in the River Murray in 1981, effectively stopping the exploitation of existing users. To stop the introduction of new licenses, an embargo was put in place in 1981 for irrigation water licenses and in 1989 for industrial water licenses (Sturgess and Wright 1993).

Normal security entitlement holders are therefore exposed to a large level of uncertainty. Many of these are growing rice or using the water for improved pastures. Such farmers have a high degree of adaptability to absorb such fluctuations, but it does not allow the irrigators to maximize the outcome of their capital investments in machinery, etc. To help overcome these fluctuations, the NSW system allows intertemporal water transfers, allowing irrigators to draw on the following season's entitlements when such water is available (Alaouze and Whelan 1996). The present system also enables normal security holders to convert their license to a high security license against a reduction in their volumetric allocation.

New South Wales has introduced transfer of water among river valleys, provided they supply the same downstream demand (Pigram et al. 1992). A major transfer of 10,000 megaliters has taken place from the Murrumbidgee River to a large cotton grower on the lower Darling River. Such transfers are subject to significant local community opposition for reasons which have been discussed elsewhere (Sturgess and Wright 1993). Large numbers of temporary transfers were reported from the Murrumbidgee to the Murray Valley during the 1992/93 season (Hill 1992).

Like South Australia, New South Wales is silent on the issue of interstate trade. Considering the significant opposition to intravalley transfers, it must be expected that interstate transfers will attract an even larger opposition. Sturgess and Wright (1993) acknowledge that national efficiency might be enhanced if South Australia were able to buy additional water from New South Wales and Victoria. In 1992, the authorities in New South Wales and Victoria agreed to a cross-border temporary transfer of 50,000 megaliters. The official New South Wales position on interstate trade reported in the first Progress Report is positive (The Working Group 1995). The state's response to interstate trade is by far the least enthusiastic of the three southeastern states, and this positive statement might be under duress from the COAG Agreement and the financial ramifications of not implementing water reforms.

As it has been described, there are quite substantial differences in the way in which NSW and Victoria manage the same water resources of the River Murray. Since the two states share many of the same storages, this difference in carryover policy has created a lot of friction between the two states under the Agreement's water-sharing and accounting rules. Under the annual accounting rules, NSW benefited greatly from Victoria's conservative water management because the two states shared what was left in the reservoirs at the end of the year. This rule was changed in the 1992 Agreement to a continuous accounting rule, removing most of these benefits (Alaouze and Whelan 1996).

3. THE MURRAY-DARLING BASIN

The Murray-Darling Basin is by far Australia's largest river basin, covering approximately one-seventh of the continent. It accounts for about one-third of Australia's output from rural industries and contains almost 75 percent of the country's irrigated land (Murray-Darling Basin Ministerial Council, 1989a). To further complicate matters, the Basin covers four states: Queensland, New South Wales, Victoria, and South Australia (see Figure 1). Most of the water originates in three states, with South Australia as the recipient of the residual flow and at the mercy of the other states in the absence of intergovernmental agreements and cooperation. The water resources of the Basin are thus clearly of a trans-boundary nature with the associated problems.

The need for an intergovernmental agreement was evident right from federation in 1901. The Basin has been subject to a series of agreements with different emphases reflecting the changing political concerns and values as well as the changing economic powers within the Australian society. The first River Murray Water Agreement in 1914 was among the federal government, NSW, Victoria, and South Australia. It was mainly concerned with the management of the river system for navigation and irrigation. South Australia's main concern was to ensure adequate and controlled river flow, securing navigation to enable its inland rural produce to be transported to the coast for export. The emphasis of the Agreement has changed over the years to more and more concern with management of the catchment area. In the 1987 Murray-Darling Basin Agreement, the emphasis has moved to promote and coordinate the effective planning and management for equitable, efficient, and sustainable use of the water, land, and environmental resources of the catchment of the Murray-Darling river system. The latest 1992 Agreement, to which Queensland is also a party, has confirmed this emphasis which now is supported by the inclusion of the Salinity and Drainage Strategy. Thus, the area covered by the Agreement has changed from the river system to one that now includes land, water, and environmental resources (Fowler and Bates 1996).

As discussed previously, the MDBC and others have expressed great concern for the level of water diversion within the Basin. Due to allocation policies of the states, considerable quantities of unused but allocated water is in the system. Some water entitlements have never been activated (sleeper licenses), whereas others have been utilized only partly (dozer licenses). Fear was expressed for the possible consequences if such water were to be activated through water trade. The MDBC therefore commissioned a Water audit (MDBMC 1995) to investigate these issues.

The audit found that water diversions grew by 8 percent from 1984 to 1994. It also found that during the last five years only 63 percent of allocated water was actually used, leaving a potential increase of 37 percent. According to the audit, the reason for these unused allocations is not the allocation policies but limitations imposed by the irrigation infrastructure, limited river channel capacity, the number of on-farm storages, forms of irrigated agriculture in the southern part of the Basin which are unsuitable for opportunistic watering, low financial return from many irrigation activities, and the undeveloped market for trading water entitlements (MDBMC 1995, p. 3). It can be expected that, as the water market matures, many of these limitations will be reduced or removed. The audit estimates that even within the above limitations water diversion will increase by an additional 14.5 percent. The environmental and financial consequences for existing irrigators and for the general community is considerable.

The security of supply for existing irrigators is likely to be reduced. Both NSW and Victoria make large allowances for such unused entitlements when announcing their annual allocations. In many cases, the nominal volume of water allocated exceeds the water available. This has, in most instances, not created a problem due to the large proportion of unused allocations. If water trade activates sleeper and dozer licenses, the authorities will have to reduce annual allocations (MDBMC 1995). The audit also identified considerable environmental costs associated with increased water extraction from the river, especially in the form of increased salinity. Such an increase will potentially be at significant cost to the states. Under the Salinity and Drainage Strategy (MDBMC 1989b), the states are committed to maintain a certain salinity level at Morgan in South Australia and finance the elimination of any increased salinity in the river originating from activities within their jurisdiction.

3.1 Sleeper and Dozer Water in Case Study Areas: South Australia and Victoria

To determine the extent to which water trade has activated sleeper and dozer water, we have computed the proportion of their allocation that sellers and buyers actually used prior to the sale and how large a proportion of their allocation they sold. This should indicate to what extent sales are activating dozer and sleeper licenses. The findings can be seen in Tables 1 and 2.

Table 1 indicates that a substantial proportion of water sales could be sleeper and dozer water. As much as 87 percent of license holders in South Australia and 63 percent in Victoria used less than 50 percent of their water. As could be expected, buyers used a much higher proportion of their allocation. The trend in the two states is the same, but the findings also reflect different water allocation policies. Victoria has a much larger proportion among the buyers using more than 100 per cent of their water rights, indicating that they rely on annual sales water.

Table 1. Sleeper and Dozer Water: Percent of Allocation Used at Time of Transfer[1]

Percent of Allocation Used	Percent of Sellers		Percent of Buyers	
	South Australia	Victoria	South Australia	Victoria
No Allocation	0	0	31	10
0 - 25	74	39	13	1
26 - 50	13	24	6	1
51 - 75	10	24	8	4
75 - 100	3	8	19	5
100+	0	5	23	79

[1]In Victoria, allocation only refers to water rights and not the associated sales water.

Table 2. Sleeper or Dozer Water: Percent of Total Allocation Transacted[1]

Percent of Allocation Transacted	Percent of Sellers		Percent of Buyers	
	South Australia	Victoria	South Australia	Victoria
No Allocation	0	0	26	11
0 - 25	52	45	41	40
26 - 50	21	26	12	26
51 - 75	8	10	5	13
75 - 100	19	19	6	6
100+	0	0	10	4

[1]In Victoria, allocation only refers to water rights and not the associated sales water.

Table 2 confirms these conclusions. Most sellers are parting with a marginal proportion of their allocation, and only 19 percent of license holders in both states sell more than 75 percent of the water transacted. Bigger discrepancies among the states are apparent for the buyers, reflecting the difference in the reasons given for the purchases. In South Australia, more buyers are establishing their first irrigation farm. Therefore, a larger proportion had no allocation prior to purchase. Similarly, more farmers in South Australia indicated that they wanted to expand their operations. Hence, more irrigators increased their allocation by more than 100 percent. In Victoria, the main reasons for buying were security and applying more water on existing crops. Thus, more buyers purchase 26–75 percent, in essence, replacing reliance on sales water with water rights (Bjornlund and McKay 1995, 1996).

To establish the full extent of the sleeper and dozer issue, it is necessary to cross-tabulate the information in Tables 1 and 2 for the sellers. If the sum of the sellers allocation used and sold does not approach at least 75 percent, it can be concluded that the transfer has activated sleeper or dozer water. In South Australia, it was found that the proportion sold and used added up to more than 75 percent in only 28.8 percent of all transfers. In 71.2 percent of all transfers, sleeper and dozer water was activated. The survey of the sellers' reasons for selling confirms this because 28 percent of the sellers in South Australia indicated that the main reason for selling was to reduce or cease irrigation. The corresponding finding in Victoria was that, in 43.2 percent of all transfers, the percentage used and sold added up to 75 percent, indicating that in 56.8 percent of all transfers sleeper and dozer water was activated. The seller survey in Victoria showed that 47 percent of the respondents indicated that the main reason for selling was to reduce or cease irrigation (Bjornlund and McKay 1995, 1996).

Our survey of water sellers and buyers in Victoria and South Australia seems to confirm the concerns raised by the MDBC. Increased trade is activating sleeper and dozer licenses, thus increasing actual water use in the Basin. In response, the MDBC has initiated a capping initiative. The intention is to cap actual water use at the 1994 level. The process has created conflicts between the states due to the introspective past of their water management policies. Victoria and South Australia argue that they traditionally have showed restraint in their water management: Victoria, by maintaining very high water levels in the reservoirs, and South Australia, by a conservative allocation policy of capping water use at a very early stage and by not using their total allocation under the Agreement or utilizing opportunity flows. The outcome of the capping process so far is different for each state:

- In NSW, water use has been capped at the 1994 level. This will reduce the annual allocations. Increased trade will further reduce such allocations, putting increased pressure on the transfer market and the desire to convert to high security level allocations.

- In Victoria, water rights have been secured whereas sales water entitlements have to be kept at a level so that total water consumption does not exceed the

1994 level. The consequence of this policy is that, as trade activates more water rights, the annual allocations of sales water will decline and possibly disappear. This will further increase the demand for water rights in the transfer market.

- In South Australia, water allocations have been capped at the 1994 level.

As could be expected, irrigators in NSW are not impressed with this outcome.

4. COUNCIL OF AUSTRLIAN GOVERNMENTS

The Working Group on Water Resource Policy of the COAG (The Working Group), in its first report (1994), stressed that water resource policy should be seen as an integral part of the wider microeconomic reform and natural resource and environmental agendas. This clearly corresponds with the Agreement's increased emphasis under the 1987 and 1992 versions, changing the focus from the river system as such to managing the water, land, and environmental resources of the Basin. The Working Group also acknowledges that a number of water resource systems are overallocated. As the MDBC, they also see that this is not a problem where sleeper and dozer water exists and that trade in such areas could have a negative impact on the riverine environment. The Working Group recommends that in such instances dormant rights should be reviewed before the creation of tradeable water entitlements. This does not seem to be the case in any of the states except in South Australia, where it was initiated in 1979.

In its drive for full cost recovery within the water industry, the COAG sees TWEs as an important tool. The Working Group (1994, p. 18) states that

> *there would appear to be scope to offset the impact of rising water prices by allowing farmers to sell, in part or in whole, their water entitlements and use the proceeds to re-structure their properties or to leave the land all together.*

The Australian Bureau of Agriculture and Resource Economics conducted some modeling to estimate the impact of increased water prices on demand for irrigation water (see Hall et al. 1993). It found that increased prices will have very little effect on water demand. However, the modeling found that the impact on estimated regional gross margins and thereby on profit is likely to be substantial. Hall et al. (1993, p. 7) concludes

> *As a result, increases in water delivery charges are likely to induce substantial structural adjustment within irrigation regions. This is likely to lead to generally larger farms within regions.*

This clearly links in with the Working Group's expectations quoted above.

The Working Group also sees the potential for cross-border trading (especially cross-basin trading within the Murray-Darling Basin), but emphasizes that in such cases consistency in water pricing and entitlement systems is essential. In its second report, the Working Group (1995) points out that there is still a long way to go in order to achieve such consistency. The Working Group suggests close cooperation among the COAG, MDBC, and the Agriculture and Resource Management Council of Australia and New Zealand in order to achieve this consistency. To facilitate such consistency within water pricing and asset valuation, the COAG appointed an Expert Group on Asset Valuation Methods and Cost-Recovery Definitions for the Australian Water Industry (The Expert Group). This group produced its report and recommendations in February 1995.

5. THE MALLEE REGION TRIAL

The Murray-Darling Basin Ministerial Council is presently planning to test interstate water trading on a trial basis within a limited area of South Australia, Victoria, and New South Wales. The Region is the Mallee border region along the Murray River stretching from Nyah in Victoria to Mannum in South Australia. As of this writing, the trial is planned to start on July 1, 1997.

To overcome some of the controversies associated with interstate trade as well as some of the differences in water allocation laws among the three states, trade will be limited to the above area. This region has a relatively uniform production of horticulture and viticulture with high security licenses within all three jurisdictions. This prevents the lower value producing areas from fearing that they will lose all their water to the higher value producing regions. To overcome the differences in water allocation laws, the following restrictions have been imposed:

- Trade can only take place between private diverters. That is, members of government irrigation areas cannot participate.

- Only high security entitlements can be traded. These are water rights in Victoria, high security licenses in NSW, and all private diversion licenses in South Australia.

Community consultations have been ongoing for quite some time within the region, and negotiations are taking place to achieve consistency in water pricing and environmental standards. If the trial is successful, it will be extended to larger areas and other types of water entitlements.

6. SOME ENVIRONMENTAL AND SOCIAL RAMIFICATIONS OF A WIDER APPLICATION OF WATER MARKETS

Interstate as well as interregion or intervalley transfers are not without opposition because of environmental, social, and cultural concerns. Local communities along the Murrumbidgee in NSW have been quite vocal about possible transfers from that region to both the Darling River and the River Murray. The first such major transfer to the Darling River took place in 1992 after extensive negotiations, but only after the proposed permanent transfer of one million liters was changed to a five-year lease (Sturgess and Wright 1993). The social implications of major exports of water out of any specific region are quite clear:

- Net profits from the farming sector will be reduced, causing a decreased taxation base.

- Not only the farm sector will be affected, but the follow-on effect in the service and processing industry also will be considerable, with a reduced taxation base and increased unemployment and social cost.

- Property values will decline, reducing the base for local property taxes which support local infrastructure.

- The area will suffer a population decline, making a number of community facilities such as schools, child care, hospitals, and public transport unviable.

- The cost of maintaining the irrigation infrastructure will increase for the remaining irrigation properties.

- Land left uncultivated will form a haven for pests and weeds, affecting neighboring properties.

- Regions will lose their production base and thereby their way of life and local culture.

Such effects have been found in the United States and have been well-documented and discussed. For example, see Mumme and Ingram (1985), Ingram and Oggins (1992), DuMars and Minnis (1989), Colby (1988), and Colby et al. (1991).

International concerns for the environment and the global impact of national decisions when it comes to environmental and natural resource issues have been prominent for more than 25 years, especially within the auspices of the United Nations. The issue was extensively considered in the Stockholm Declaration in 1972; the Bruntland Commission's final report, *Our Common Future,* in 1987; and the United Nations General Assembly in 1989, leading up to the United Nations Conference on Environment and Development Earth Summit in Rio de Janeiro in June 1992. The latter conference produced four formal documents: the Rio Declaration, Agenda 21, and two conventions on biodiversity and climate change. The documents set out how sustainable development can be achieved on an international basis. Australia is a signatory to these documents (Raff 1995; Blay and Piotrowicz 1993; Bates 1994; and Wright 1995).

In response to these international developments, Australia has developed a framework for sustainable development. This process culminated in the Commonwealth, states, territories, and the Australian Local Government Association signing an Intergovernmental Agreement on the Environment (IGAE) in 1992 to pursue the effective integration of economic and environmental considerations in decision-making processes in order to improve community well-being and to benefit future generations. The IGAE commits all parties to ecologically sustainable development under the adoption of four main principles (Bates 1994):

- the precautionary principles,
- intergenerational equity,
- conservation of biological diversity and ecological integrity, and
- improved valuation, pricing, and incentive mechanisms.

With a clear connection to the IGAE, the COAG endorsed the Australian National Strategy for Sustainable Development on December 7, 1992. This document is expressly linked to the Bruntland Report and implementation of the initiatives toward global sustainable developments in the Rio Declaration and Agenda 21 (Raff 1995). Also relevant in this context are the COAG's water reform agenda, the MDBMC's Natural Resources Strategy (1989c), the Water Audit (1995), and the Capping Initiative, as well as the wider application of TWE in interstate and interregional transfers.

Locally, environmental concerns have been raised by the Australian Conservation Foundation and other community and environmental groups. Increased water diversion has a clear negative environmental impact due to decreased dilution flow and potentially increased drainage outflow containing additional salt and nutrients. The environmental impact of intervalley transfers tends to work opposite of the social impact. The exporting valley will experience environmental improvements due to an increased dilution effect and decreased drainage outflow, and the importing region will be at risk of environmental degradation due to decreased dilution and increased drainage outflow. The same is true with interstate and interregional transfers up and down a river. A downstream move will improve the environmental conditions, whereas an upstream move will worsen the same conditions.

Both the COAG and the MDBC are aware of these risks. For example, the COAG clearly states that trade must take place within the social, environmental, and physical constraints of catchments, and requires specific allocations for the environment before trade can be fully implemented. It also emphasizes that water resource policy must be seen as an integral part of microeconomic reforms and natural resource and environmental reforms. The Expert Group is recommending that water beneficiaries or impactors carry the full cost of providing and maintaining water for the environment (The Expert Group 1995).

The MDBC's Water Audit and Capping Initiative are also a clear acknowledgment of the environmental problems. The capping of water diversion at the 1994 level is a significant step. The Salinity and Drainage Strategy is also an important step, holding each jurisdiction financially responsible for the impact they cause on river salinity. The salinity impact can be computed for each transfer and entered on the salinity credit register for that state.

A clear commitment has been made by the Commonwealth, the MDBC, and the states to maintain and improve the environmental conditions of the river. The Working Group (1995) is confident that the tests they have recommended to be applied to transfers are sufficiently broad and robust to ensure that environmental amenities will not be significantly compromised by water transfers (Working Group 1995, p. 12). In South Australia, an irrigation and drainage management plan for the buyer's entire irrigation operation is required as a part of the transfer process to ensure that each transfer is at least salinity neutral to the river. In larger cases, or if uncertainty of future impacts exists, the buyers are required to establish a trust fund to finance future works to remedy any negative impacts. In Victoria, strict guidelines are in place governing water transfers. These guidelines specify the requirements as to soil salinity levels, channel capacity, as well as on- and off-farm drainage and recycling systems, and specify maximum water allocations per hectare depending on the conditions of these issues. The task ahead is to make sure that these estimates are correct, and to closely monitor the actual impact on the environment and ensure that the polluters/impactors are held responsible and remedy possible negative impacts. Both the COAG and the MDBC state that specific allocations to the environment need to be made. In reality, we have seen very few of these. Work needs to be done to specify what allocations will be adequate and ensure that these are actually put in place

7. CONCLUSIONS AND RECOMMENDATIONS

We have discussed state and federal policies related to water trade and water allocation and found a significant introspective legacy within these policies. This legacy has been created by state governments responding to different geographical, economic, and historical circumstances, and to changing political values and power balances within each jurisdiction.

We have also discussed how the federal government and the Murray-Darling Commission have recently worked with the states in order to achieve more consistency in policies as a prerequisite for wider application of water markets across boundaries, valleys, and sectors of the economy. The federal government as part of a wider microeconomic reform agenda, spurred on by economic rationalism, wants full-cost recovery within the water industry as well as other public or semi-public service providers by 2001. Within the water industry, TWEs are seen as a major tool in this process, allowing the inefficient water users to sell their water rights in full or in part and use the proceeds to restructure their remaining farm or establish themselves outside the farming community. It is expected that the efficient irrigators will be willing and capable of compensating the inefficient irrigators through this process. The federal government, through the COAG, is using its fiscal powers to coerce the less willing states to comply.

The MDBC also sees interstate trade as an important tool to achieve increased efficiency of the irrigation industry, and this concept forms an integral part of its

Irrigation Management Strategy. The Commission has recently moved to cap water consumption in the river as one of the conditions for increased trade.

Both the federal government and the MDBC see the environmental danger of increased trade as well as interstate and intervalley transfers. They require formal allocations for the environment and uniform water pricing and water supply security policies as a condition for the wider application of water trade. It is the opinion of the Working Group that the tests suggested for water trade are sufficient to ensure that the environmental concern is satisfied. We find that it is of great importance that a close monitoring of the actual social and environmental impact of trade is carried out to see whether the suggested tests and polices adequately protect the interest of the wider community.

REFERENCES

Alaouze, M.C., and S.P. Whelan. 1996. Economic Efficiency and Property Rights Issues in the Management of Rural Water. In Pigram, J.J., ed., *Security and Sustainability in a Mature Water Economy—A Global Perspective*. Proceedings of an International Workshop for the Water and Resource Economics Consortium. Centre for Water Policy Research, Armidale, New South Wales.

Bates, G. 1994. Implementing ESD, *Environmental and Planning Law Journal* 11: 251-253.

Bjornlund, H, and J. McKay. 1995. Can Water Trading Achieve Environmental Goals?, *Water* 22: 31-34.

Bjornlund, H., and J. McKay. 1996. Transferable Water Entitlements: Early Lessons from South Australia. *Water* 23: 39-43.

Blay, S.K.N., and R.W. Piotrowicz. 1993. Biodiversity and Conservation in the Twenty-First Century: A Critique of the Earth Summit 1992, *Environmental and Planning Law Journal* 10: 450-469.

Colby, B.G. 1988. Economic Impact of Water Law—State Law and Water Market Development in the Southwest, *Natural Resources Journal* 28: 721-749.

Colby, B.G., M.A. McGinnis, and K.A. Rait. 1991. Mitigating Environmental Externalities Through Voluntary and Involuntary Water Reallocation: Nevada's Truckee-Carson River Basin, *Natural Resources Journal* 31: 757-783.

Department of Environment and Natural Resources. 1995. *South Australia—Our Water Our Future—Providing for the Future*. Adelaide.

DuMars, C.T., and M. Minnis. 1989. New Mexico Water Law: Determining Public Welfare Values in Water Rights Allocation, *Arizona Law Review* 23: 817-839.

Expert Group on Asset Valuation Methods and Cost-Recovery Definitions for the Australian Industry. 1995.

Fowler, R.J., and G. Bates. 1996. Environment and Natural Resources. In J.A. Riordan, ed., *The Laws of Australia*. The Law Book Company.

Gould, G.A. 1989. Transfer of Water Rights, *Natural Resources Journal* 29: 457-477.

Hall, N., D. Poulter, and R. Curttoti. 1993. ABARE Model of Irrigation Farming in the Southern Murray-Darling Basin. ABARE Research Paper No. 93.

Hill, C. 1992. Property Rights in Australia—Their Status and Future Directions. Occasional Paper of the Australian Water Resources Council.

Howe, C.W., D.R. Schurmein, and W.D. Shaw. 1986. Innovative Approaches to Water Allocation, *Water Resources Research* 22: 439-445.

Ingram, H., and C.R. Oggins. 1992. The Public Trust Doctrine and Community Values in Water, *Natural Resources Journal* 32: 515-537.

Mumme, S.P., and H.M. Ingram. 1985. Community Values in Southwest Water Management, *Policy Studies Review* 5: 365-381.

Murray-Darling Basin Commission. 1996. *Interstate Water Trade. Free Trade Zone in the Mallee Border Region of the Murray-Darling Basin.*

Murray-Darling Basin Ministerial Council. 1989a. Murray-Darling Basin Natural Resources Management Strategy. Discussion Paper No. 2, Canberra, June.

Murray-Darling Basin Ministerial Council. 1989b. *Salinity and Drainage Strategy*. Canberra.
Murray-Darling Basin Ministerial Council. 1989c. *Murray-Darling Basin Natural Resources Strategy*. Canberra.
Murray-Darling Basin Ministerial Council. 1995. *An Audit of Water Use in the Murray-Darling Basin. Water Use and Healthy Rivers—Working Toward a Balance*. Canberra, June.
Nunn, S.C., and H.M. Ingram. 1988. Information, the Decision Forum, and Third-Party Effects in Water Transfers, *Water Resources Research* 24: 473-480.
Pigram, J.J., R.J. Delforce, M.L. Coelli, V. Norris, G. Antony, R.L. Anderson, and W.F. Musgrave. 1992. Transferable Water Entitlements in Australia. Report to the Land and Water Resources Research and Development Corporation. Canberra.
Raff, M. 1995. The Renewed Prominence of Environmental Impact Assessment: "The Tale of Two Cities," *Environmental and Planning Law Journal* 12: 241-263.
Shupe, S.J., G.D. Weatherford, and E. Checchio. 1989. Western Water Rights: The Era of Reallocation, *Natural Resources Journal* 29: 413-434.
Smith, M. 1994. River Murray Irrigation Water Allocations—Discussion Paper.
Sturgess, G.L., and M. Wright. 1993. *Water Rights In Rural New South Wales. The Evolution of a Property Rights System*. CIS Policy Monographs 26. St. Leonards, New South Wales: The Centre for Independent Studies.
Working Group on Water Resources Policy to the Council of Australian Governments. 1994. First Report.
Working Group on Water Resources Policy to the Council of Australian Governments. 1995. Second Report.
Wright, I. 1995. Implementation of Sustainable Development by Australian Local Governments, *Environmental and Planning Journal* 12: 54-61.
Young, R.A. 1986. Why Are There So Few Transactions Among Water Users, *American Journal of Agricultural Economics* 78: 1143-1151.

TABLE OF CASES

Sporhase v. Nebraska ex rel Douglas, 458 U.S. 941 (1982)

19 OPTIMAL ALLOCATION OF GROUND AND SURFACE WATER IN OAHU: WATER WARS IN PARADISE

James Moncur
University of Hawaii at Manoa, USA
Jim Roumasset
University of Hawaii at Manoa, USA
Rodney Smith
University of Minnesota, USA

Despite abundant rainfall and groundwater resources, the State of Hawaii has managed to erect so many barriers to efficient use that serious conflicts have erupted over use rights to water. In what is currently the most prominent battle, environmental and agricultural interests of windward Oahu are pitted against agricultural producers on the leeward side over the allocation of aqueduct water in the Koolau Mountains. To inform this debate, a model of conjunctive use is constructed and optimal allocation of ground, surface, and aqueduct water computed. Optimal use of the aqueduct water is as a substitute for groundwater. This postpones the eventual, expensive use of desalination, thus benefiting future users. By making use rights fully transferable and charging current users water's true opportunity cost, both windward and leeward users would be winners.

1. INTRODUCTION

Even before Western contact with Hawaii, development and use of the Islands' water resources were crucial factors in the well being of their people. More recently, water allocation has often generated great controversy. This chapter traces the evolution of one such controversy involving integrated development and management of ground and surface waters that occur in a somewhat unusual relationship. The controversy has arisen against profound changes in the institutional framework governing water use, some of it forced by hydrological factors and some by legal and legislative decisions. The question to be asked is whether the resulting institutions will improve or detract from the efficiency of use of the waters involved in this controversy. Similar controversies are brewing elsewhere in the state, not to mention in other states.

2. BACKGROUND AND PROBLEMS

Sugarcane has been grown successfully in Hawaii since 1835 (Hawaiian Sugar Planters Association 1973). Disruptions in U.S. sugar supply during the Civil War and later tariff agreements made sugar an enormously profitable crop for Hawaii (Mollett 1961). Early on, much of the crop was grown on rainy windward parts of the Islands, but planters soon realized the benefits of irrigation in raising yields and extending cultivation to dry leeward areas. By the late 1870s, planters had built several impressive irrigation works to exploit available surface water. However, these sources often dried up in the summer season and could not supply as much land as growers wanted to cultivate. In 1879, a driller brought in the state's first artesian well, on land owned by sugar planter James Campbell in the dry Ewa section of Oahu (D. Cox 1981). Engineers soon developed a practical understanding of groundwater hydrology principles then being formulated by contemporaneous European researchers (R. Cox 1981) and of the interconnectedness of surface and groundwater sources in the Islands.

2.1 Development of the Waiahole Ditch

Planters continued to seek more water. In 1913, work began on an aqueduct to bring water from the wet windward side of the island to the drier central and leeward parts. The resulting aqueduct system (Figure 1) includes 27 connected tunnels, 37 stream intakes, and four development tunnels, plus the Main Tunnel piercing the Koolau Mountains (Herschler 1966). Known as the Waiahole ditch, this early engineering feat added around 25 million gallons daily (mgd) to the amount of water available for irrigation in leeward parts of the island.[1] The principal beneficiary was Oahu Sugar Company, Ltd., but over the years, numerous small farmers and cattle growers have also come to depend on it. Further, these waters contribute recharge to the Pearl Harbor aquifer, which is the largest source of water for urban Honolulu: 40 percent or more of irrigation water (mostly drip irrigation) is thought to seep back into the aquifer and become available for pumping (Lao, Chester, Honolulu Board of Water Supply, personal communication, 1996).

In 1993, Oahu Sugar Company announced that it would phase out of business with its current crop. Since the shutdown, additional small farming operations have sprung up, growing such crops as watermelon, pineapple, bell peppers, herbs, and potatoes on small leasehold plots (Omandam 1995a). There may be potential for replacing imported food crops (Amaral 1995). Also, the leeward section of the island is the major locus of residential and resort growth, and landowners would presumably like

---

[1]By comparison, water usage for the entire island of Oahu, prior to the closing of the Oahu Sugar Company in 1995, was about 400 mgd (Wilson Okamoto and Associates, Inc., 1990).

to hang on to the water rights to supply future housing, hotels, golf courses and the like. This development is controversial in itself, even apart from water use implications.

The Waiahole Ditch enabled nearly a century of sugar and pineapple cultivation in leeward Oahu. At the same time, the ditch reduced the amount of water available to windward farmers, and dried up several small streams. Many also depended on the streams for fishing, not to mention cultural and recreational activity. The diversion may also have changed the character of parts of Kaneohe Bay, by reducing inflows of freshwater from the mountain streams emptying into the bay (Kresnak 1995).

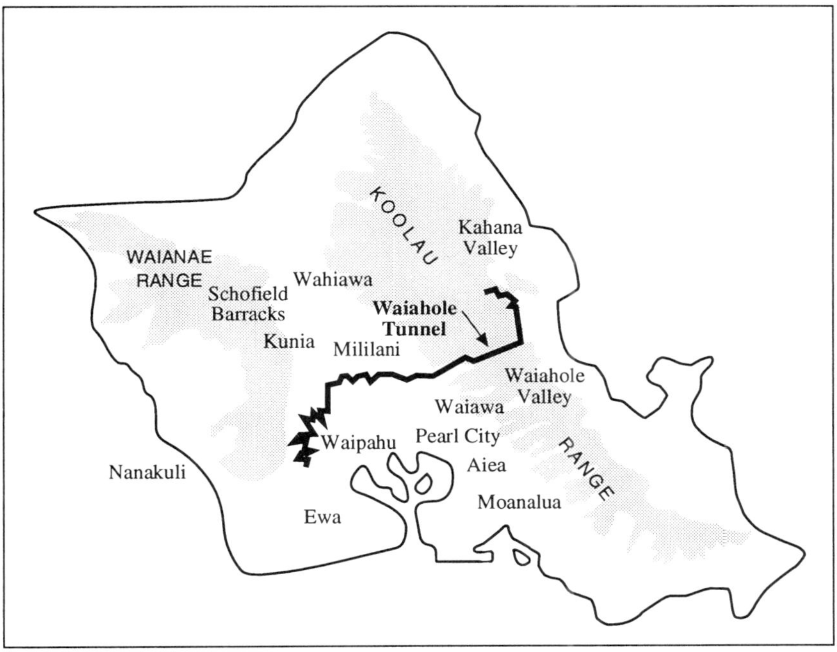

Figure 1. The Waiahole Ditch System, Oahu, Hawaii

Nevertheless, diversion of windward waters went more or less unchallenged until recent years. With the impending shutdown of Oahu Sugar Company, windward farmers have mounted a vigorous campaign for the return of streamflow to their side of the island. Indeed, they have asked for the return of *all* the water that currently flows through the ditch (Omandam 1995b). Moreover, agricultural use per se has become entwined with issues of preserving traditional Hawaiian culture, defending constitutionally guaranteed water rights to native Hawaiian lands, and protecting the aquatic environment.

The state Commission on Water Resources Management (CWRM) must decide who gets to use this water. Some 23 parties have applied for permits to continue existing uses, begin new uses, or reserve waters for future use (Waiahole Ditch Decision 1995). An interim decision by the state Commission on Water Resources Management left 12 mgd running through the tunnel while 16 mgd was returned to the windward side. This was to lead to a permanent allocation by June 1995, but instead began a contested case hearing, a quasi-judicial proceeding before the Commission (Omandam 1995b). As of this writing (April 1997), the Commission has finished taking testimony but has not yet made a final decision.

2.2 Water Sources and Tunnels

The tunnel's engineers planned to bore several short tunnels into the side of the mountain to capture waterfalls and emerging streams. This water was to be collected and sent through the Koolau mountains via a 2.76-mile Main Tunnel. The mountain through which the Main Tunnel was dug consists mostly of extremely porous lava formations, interrupted here and there by dikes. As D. Cox (1981, pp. 63-64) describes it,

> *In Hawaii lava ordinarily reaches the surface, in an eruption through long fissures. The dikes represent the lava left in the fissures after the eruption. Having cooled under pressure, the lava in the dike is much less permeable than the lava in the flows. Hence ... the dikes can serve as dams behind which water may be held to great heights in the ground.*

To the surprise of the tunnelers, construction activity punctured several of these dikes, releasing enormous flows. While still under construction, the yield of the main tunnel ran to 43 mgd (D. Cox 1981) with flashes reported at as much as 100 mgd in periods of heavy windward rainfall (Herschler 1966). The 40-year average flow has been about 27.3 mgd (Commission on Water Resource Management 1995). The initial storage of about 46 billion gallons has been long since depleted, but continuing rainfed flows from the dike compounds account for the bulk of the tunnel's flow. At present, most of the dike water is free-flowing, with few controlling works available to enable rational management. However, as Takasaki and Mink (1985, p. 32) note,

> *The vast initial storage that has been depleted could not likely be returned to its original state because of disruption to the integrity of the reservoir caused by tunnel construction. Significant portions of it, however, can be restored and manipulated economically in each of the development tunnels ... Correctly located and properly built bulkheads in the tunnels could provide the means for improved management of the water resource.*

Management of this resource clearly requires careful recognition of both ground and surface sources as well as their interconnections.[2] Under the best of circumstances, such an approach requires vast amounts of data and a clear understanding of economic

[2] The integrated nature of the surface and groundwater flows contributing to the Waiahole ditch system was shown graphically by newspaper photos of newly re-emergent waterfalls after the ditch flows were partially restored in 1994 (Lynch 1994).

and political goals and constraints. The central question of this chapter is whether the water law and administrative apparatus Hawaii has erected is up to the task posed by this case and others like it. Are Hawaii's water allocation institutions capable of re-allocating resources fairly and efficiently, and managing them efficiently once reallocated? Will they take appropriate cognizance of the interrelatedness of ground and surface water sources?

2.3 Institutional Background

The question of water rights has been an important and controversial one in Hawaii, especially since the 1973 state Supreme Court ruling in *McBryde v. Robinson*. Traditional Hawaiians basically shared the available supplies as a common property resource, an allocation principle presumably adequate for the agricultural and domestic purposes of their times.[3] The rise of sugar and expansion of taro cultivation in the nineteenth century, however, greatly increased the demand for water and brought conflict not resolvable within the traditional institutions. These conflicts were resolved first by alienating land and water from the monarchy to private individuals in the "Great Mahele" of 1848. Subsequently, a series of court decisions stretching well into the twentieth century refined the right to use water. By 1920, water rights were apparently well defined and clearly transferable between uses, users, and locations (Anderson 1985, p. 903).

It is notable that no formal legislative direction was involved in any of this activity in spite of crises from time to time. Over-exploitation of groundwater, however, was a recurring problem in the early twentieth century. In response to declining heads, a system of regulations was put in place aimed at stopping leakage and obvious waste, rather than cutting final demand for water (see D. Cox, 1981, p. 65).

While these regulations satisfied most parties, the system was not without controversy. Expansion of economy and population continued to strain the available water sources (see, for example, Board of Water Supply, 1982, p. 62). Many interconnections between ground and surface water and between different users' water rights remained contentious and unresolved.

One such argument arose between two sugar planters on Kauai. In the late 1940s, Gay & Robinson replaced an old and leaky diversion channel with a new tunnel, in the process reducing water available to nearby McBryde Sugar Company. After a decade of negotiations, McBryde sued in 1959 to clarify the water right. Eventually, the state Supreme Court ruled, quite surprisingly,[4] that while both companies had appurtenant or riparian rights to use "surplus" water, neither had a property right in it. Instead, the court declared the state to be the owner.

---

[3]The following paragraph draws from the account of Anderson (1983, especially pp. 901-906). See also D. Cox (1981).

[4] "The court flabbergasted everybody involved in a fight already 50 years old by declaring the water didn't belong to either plantation, it was public property. No one expected this outcome to what seemed to be a purely private fight" (Smyser 1989).

Gay & Robinson then sued the State of Hawaii in federal court (*Robinson v. Ariyoshi*), claiming their property had been taken without compensation. After years of going from court to court, the Ninth Circuit Court ruled in 1989 that even though *McBryde v. Robinson* had "placed a cloud on the title of the various private owners, this inchoate and speculative cloud is insufficient to make this controversy ripe for review." Thus the court did not actually endorse *McBryde v. Robinson*, but it failed to sustain the private property concept of water rights that Hawaii had evolved prior to this decision. In doing so, the court left open the door for the State government to continue its tendency to exercise substantial control over water resources—a tendency already well under way.

Uncertainty over the status of water rights generated by this lengthy court fight left a large opening for rent seeking activity. This uncertainty, along with other factors, led the 1978 Hawaii Constitutional Convention to propose an amendment directing the state legislature to establish an agency to protect the state's water resources and to regulate water use.[5] Voters approved the amendment in 1978. Almost a decade of studies, proposals, hearings and discussion followed. Major proposals from official study commissions (Governor's Advisory Commission 1979; Legislative Reference Bureau 1986) would have required all users to obtain permits of limited duration from a state authority with powers to deny renewal and to disapprove transfers. The Advisory Study Commission went so far as to denounce market institutions: "Indeed, the commission strongly feels that water and the right to use it should not be the subject of purchase and sale on the open market." This sentiment appeared in early drafts of the legislative act creating the code, although it was deleted before final passage.

The resulting State Water Code (State Legislature of Hawaii, Act 45 1987; Hawaii Revised Statutes, ch. 174C) has been called a masterpiece of compromise (Law 1988). What remains unclear is whether the compromise melded the best or the worst of competing paradigms.

2.4 Provisions of the Water Code

The State Water Code establishes an administrative mechanism for Hawaii's water resources with certain planning and managerial functions covering both ground and surface water, supposedly in an integrated fashion. The six-person Commission on Water Resource Management has a full time Deputy and staff (the Division of Water Resource Management, Department of Land and Natural Resources). The code

---

[5]The Hawaii State Constitution, Article XI, section 7, states: "The State has an obligation to protect, control and regulate the use of Hawaii's water resources for the benefit of its people. The legislature shall provide for a water resources agency which, as provided by law, shall set overall water conservation, quality and use policies; define beneficial and reasonable uses; protect ground and surface water resources, watersheds and natural stream environments, establish existing correlative and riparian uses and establish procedures for regulating all uses of Hawaii's water resources."

empowers the Commission to designate water management areas wherever water resources are threatened as to quantity or quality. Water outside these areas is not subject to regulation, although all water users must file a declaration of their uses. The code also requires registration of existing wells and stream diversion works, whether or not in a water management area, as well as permits for constructing new wells or installing pumps.

The Commission is to establish administrative rules and has jurisdiction over disputes involving water uses. It may also acquire land for flood control, water management, or water-related resource conservation, including streams and other watercourses, parks and recreation areas, beaches, submerged lands, and other open areas. A water management plan, with objectives specified in the law, is to be developed as guidance for allocation decisions. The Commission is also to provide for gathering extensive hydrologic data, and must establish a sustainable yield for each hydrologic unit in the state. The Commission may declare a water shortage and determine classes of permits allowing water use during the drought period.

Within Designated Management areas, any water use except domestic applications requires a permit. Permits will specify the water source, quantity, use, location, and other information requested by the commission. Applications for permits must establish that the proposed use of water (1) can be accommodated within the capacity of the source; (2) is a reasonable-beneficial use; (3) will not cause negative third-party effects; and (4) is consistent with "the public interest" and with state and county plans. The Commission may allow the permittee to transport the water outside the overlying land or watershed, but must approve almost all modifications of permit terms. County water utilities, however, need approval only for changes in quantity or changes involving third-party effects.

A major responsibility of the Commission is protecting water quality and in-stream uses. Similarly, the code specifies protection of Native Hawaiian water rights, a problem as difficult for the Islands as some Native American water rights issues have been on the mainland.

The 1987 code mandated that after five years' experience, a review commission would study its effects and recommend changes. That Commission held extensive hearings beginning in July 1993, and issued its report in December 1994 (Review Commission 1994). The report included extensive proposals for changes to the water code. These changes are intended, among other things, to (1) essentially make the entire state a designated water area and therefore subject to regulation by the CWRM; (2) clarify water rights and establish a hierarchy of water uses; and (3) push the CWRM in the direction of a planning and rule-making body and away from case-by-case decision making. The state legislature, however, has so far declined to take up these proposals. While legislators have not announced their reasons for failing to discuss the proposals, their reluctance is not surprising in the midst of highly controversial hearings and decisions currently on the CWRM agenda, including first and foremost the Waiahole Ditch water question.

2.5 Problems

Water is a scarce commodity in many places. The goal of an allocation mechanism such as the Hawaii State Water Code should be to encourage a degree of conservation befitting this scarcity, i.e., to encourage the *broadly conceived* economic efficiency of water allocation and use. An efficient resource allocation system needs to accommodate changes in underlying economic, demographic, and climatic factors by fostering the timely transfer of the resource from low- to higher-valued uses. The Waiahole Ditch case illustrates the lack of such flexibility. The Oahu Sugar Company ceased using the water gradually, as its current crop was harvested. By late 1994, the ditch's full flow was available for reallocation. The CWRM, chaired by the departing director of the state Department of Land and Natural Resources, got all parties to agree to a six-month temporary solution dividing the water with 16 mgd for windward, 12 mgd for leeward (The Honolulu Advertiser 1994), thus handing the problem over to the newly elected administration and providing time for formal hearings.

Once a "permanent" Waiahole decision has been made, will users (and potential users) have appropriate incentives to seek out, accept, or allow transfers? The code allows transfers of the permits if the place, quantity, and purpose of the use, and possibly other conditions as well, remain unchanged. The Commission may require changes in other aspects of the permit as a condition for approving a requested modification. An application for transfer is to be considered on a par with any competing applications. All these factors reduce the likelihood that any permit holder will apply for a transfer or modification, unless forced as in the Waiahole case.

Permits are issued without an expiration date but the CWRM must review all permits every twenty years. As a political and administrative body, the Commission has only weak incentives and limited capabilities to mandate efficiency-enhancing changes based on these reviews (see, for example, Trelease 1974, or Anderson 1983). Permits may also be revoked for nonuse over a continuous period of four years or more, thus reducing the flexibility of rights holders to plan and implement water-using projects.

The code does not permit, let alone encourage, temporary reallocations such as might be undertaken under water "banking" or "leasing" arrangements (see, for example, Saliba and Bush 1987, pp. 46-47, 156.) Similarly, there is no incentive to install bulkheads, revive groundwater storage capacity, and rationalize its use. Water allocated to instream flow standards, once established, will be extremely difficult to transfer.

The Review Commission's proposed hierarchy of uses promises further ossification in the uses of water. The hierarchy, for example, gives highest priority to certain native Hawaiian water rights. These rights are non-transferable, particularly to non-Hawaiians, thus limiting the flexibility of Hawaiians to make the most of their own resources.[6]

---

[6]A recent court ruling on the subject of native Hawaiian land rights indicates that certain lands reserved for Hawaiians cannot even be sublet to non-Hawaiians (Tanji 1996). If applied to native Hawaiian water rights as well, this would further limit transferability and thus economic efficiency.

3. TOWARDS A SOLUTION

The decision to allocate 16 out of 28 mgd of water from the Koolau aqueduct to windward Oahu was not based on any explicit efficiency calculations but reflected an opinion that since the sugar industry was going out of business in Hawaii, the Oahu Sugar Company was less likely to put the entire 28 mgd to "beneficial use." The possibility that some of this quantity could be allocated to the Board of Water Supply, which supplies most of urban Honolulu, while leaving considerable amounts to leeward agriculture, was apparently not seriously considered. The leeward service area of the Koolau aqueduct overlaps with that of the Pearl Harbor Aquifer, which is the primary source for the entire Pearl Harbor Water Management Area (WMA). The aqueduct also supplements other sources that service the windward side, especially in the Koolaupoko district, containing the Kailua-Kaneohe population center. Since the windward and leeward districts are, in effect, connected via the fungibility of aqueduct water, there are likely to be efficiency gains available from unified management of the two use areas and three water sources. An illustrative model for this problem is discussed below.

3.1 A Model of Integrated Management

Demand in the Pearl Harbor WMA is given by the sum of urban and agricultural uses, both of which are represented by the function $D_{it} = \alpha_i e^{g_i t}(p_{it} + d_i)^{-\eta}$, where t is the number of years after the benchmark year (1991), g_i is the growth rate of demand (due to population and income growth), p_{it} is the wholesale water price, and d_i is the per unit cost of distribution, i = 1,2. Demand in the Koolaupoko district, usually called "windward," is given by the same functional form, where i = 3. In the benchmark year, D_1 = 88 mgd is leeward urban demand, D_2 = 101 mgd is leeward agricultural demand, D_3 = 38 is windward demand, and $p + d$ = \$1.23. Using these benchmark values yields α_1 = 93.63, α_2 = 107.47, and α_3 = 42.56 as the leeward urban, leeward agricultural, and windward demand coefficients, respectively.[7] Windward demand is initially assumed to grow at 0.5 percent per year, reflecting the relatively stringent zoning restrictions and the fact that urban and agricultural demands are integrated in our model.

The sources of supply are the Pearl Harbor Aquifer, 28 mgd from the Koolau aqueduct, and 38 mgd of surface water and sustainable groundwater yield from the Koolaupoko district (Wilson Okamoto Associates, Inc., 1990), and desalination at a cost of $\bar{p}$ = \$3.00 per thousand gallons.

Optimal use is achieved by maximizing discounted net social welfare subject to constraints on aquifer head behavior, available aqueduct water, and windward surface

[7]The part of windward Oahu serviced by the Koolau aqueduct is the Koolaupoko district, which consumed 38 mgd in 1988 (Wilson Okamota & Associates, Inc., 1990). This figure was adjusted upwards to conform to the benchmark year.

water use,

$$\max_{\{q_{1t},q_{2t},b_{1t},b_{2t}\}} \int_0^\infty e^{-rt}\left\{\sum_{i=1}^{2}\left[\int_0^{q_{it}+b_{it}+s_{it}} D_{it}^{-1}(x)\,dx\right] - [c(h_t)+d](q_{1t}+q_{2t}) - [\bar{p}+d](b_{1t}+b_{2t}) - d(s_{1t}+s_{2t})\right\}dt$$

$$+ \int_0^\infty e^{-rt}\left\{\int_0^{S^T+\bar{S}-s_{1t}-s_{2t}} D_{3t}^{-1}(x)\,dx - d\,(\bar{S}-s_{1t}+s_{2t})\right\}dt, \tag{1}$$

subject to

$$\dot{h}_t = w - l(h_t) - q_{1t} - q_{2t} + s_{1t} + s_{2t},$$

$$\bar{S}_t = s_{1t} + s_{2t} + s_{3t},$$

where $D_{it}^{-1}(x)$ are the inverse demand functions, $q_{1t} + q_{2t} \geq 0$ is aquifer extraction at time t, $b_{1t} + b_{2t}$ is the amount of the backstop resource (desalination) used, $l(h)$ is aquifer leakage as a function of aquifer head h, and h_0 is the exogenous head at time 0. As explained elsewhere (Krulce et al., 1997, present a more formal derivation), coastal groundwater is a renewable resource since leakage $l(h_t)$ is an increasing function of head, such that net recharge, $w - l(h_t)$, increases as the water head falls.

Social benefit is represented by leeward urban gross surplus, $\int D_{1t}^{-1}(x)\,dx$, agricultural gross surplus, $\int D_{2t}^{-1}(x)\,dx$, and windward gross surplus. Social costs have three components: the total cost of extracting water from the aquifer $c(h_t)(q_{1t} + q_{2t})$; the total cost of producing desalinated water $\bar{p}(b_{1t} + b_{2t})$; and water distribution costs dW_t. Here, $c(h_t)$ is the per unit cost of water extraction given that the head level is h_t, and $q_{1t} \geq 0$ and $q_{2t} \geq 0$ represent the quantity of aquifer water supplied to the urban and agricultural markets, respectively. The exogenous parameter $\bar{p}$ is the per unit cost of desalination, while $b_{1t} \geq 0$ and $b_{1t} \geq 0$ represent the quantity of desalinated water supplied to the urban and agricultural markets. Per-unit water distribution costs are denoted d. Finally, S^T is the constant windward flow of surface water and sustainable groundwater, $\bar{S}_t$ represents the time t flow of water through the Koolau aqueduct, with s_{1t}, s_{2t}, and s_{3t} representing the exogenously supplied levels of Koolau aqueduct water supplied to the leeward urban, leeward agricultural, and windward markets respectively.

The solution to (1) is governed by the constraints and the necessary conditions for efficient groundwater extraction and optimal allocation of aqueduct water between the two sides of the mountains. The condition for efficient groundwater extraction is

$$\lambda_t = \frac{\lambda_t^*}{\lambda} - \frac{l'(h_t)\lambda}{r} - \frac{c'(h_t)(q_{1t}+q_{2t})}{r} \tag{2}$$

where λ_t is the in situ shadow price of water, $D^{-1}(x) - c(h)$. The left-hand side of (2) represents the marginal benefit of extracting an extra unit of water, and the right-hand side is marginal user cost (MUC), i.e., the decline in present value of the remaining groundwater. It is instructive to note that ignoring the last two terms of MUC yields the familiar Hotelling condition for non-renewable resources. The last term represents the present value lost by facing higher extraction costs in the future due to the extra unit of extraction. The second to last term captures the increase in present value due to the decrease in leakage from withdrawing more water.

Efficient allocation of aqueduct water requires that the ex situ shadow price (value of the inverse demand function) of leeward and windward water be equal or that the value on one side be higher than on the other even when it gets all aqueduct water (corner solution). To obtain a numerical solution for particular parameters, an algorithm can be devised that treats aqueduct allocation as exogenous for the purposes of determining the optimal path of groundwater extraction and then searching for a vector of aqueduct allocations that satisfies the efficiency condition regarding the ex situ shadow prices of leeward and windward water.

3.2 The Application

Following Mink (1980) and Krulce et al. (forthcoming), the leakage function is estimated as $l(h) = .2497h^2 + .022h$, the extraction cost function as $c(h) = c_0\,(h_0/h)^2$, and the initial head as 15 feet. Figure 2 illustrates the ex situ shadow prices of windward and leeward water based on the assumption that all aqueduct water is allocated to the leeward side. As shown, this (corner solution) allocation is correct until the year 1999 when the windward price path crosses from below, indicating that some positive amount of water should be allocated to the windward side thereafter.

Figure 3 gives the leeward and windward ex situ shadow prices (alternatively "wholesale" prices) for an allocation giving 24.5 mgd to leeward and only 3.5 mgd to windward. Note that this allocation meets the efficiency requirement of equal prices in the year 2035. This suggests that the optimal allocation of water to the windward side is zero from the benchmark year until 1999 and then gradually increases up to a maximum allocation of roughly 3.5 mgd. Since Figure 3 also shows the windward price falling below the leeward path after 2035, the allocation to windward should be decreased below 3.5 mgd after 2035. To investigate how much this reduction should be, we also found that if 2.05 mgd is given to the windward side, both price paths hit the backstop in the same year (2071).

In summary, optimality requires that aqueduct water allocated to windward be gradually increased to about 4 mgd and then even more gradually decreased to 2 mgd.

Figures 4 and 5 illustrate the sensitivity of results to the growth of windward demand. Note that with zero demand growth (Figure 4), optimality is achieved by allocating all aqueduct water to the leeward side. Even so doing, water is more valuable on the leeward side in all time periods.

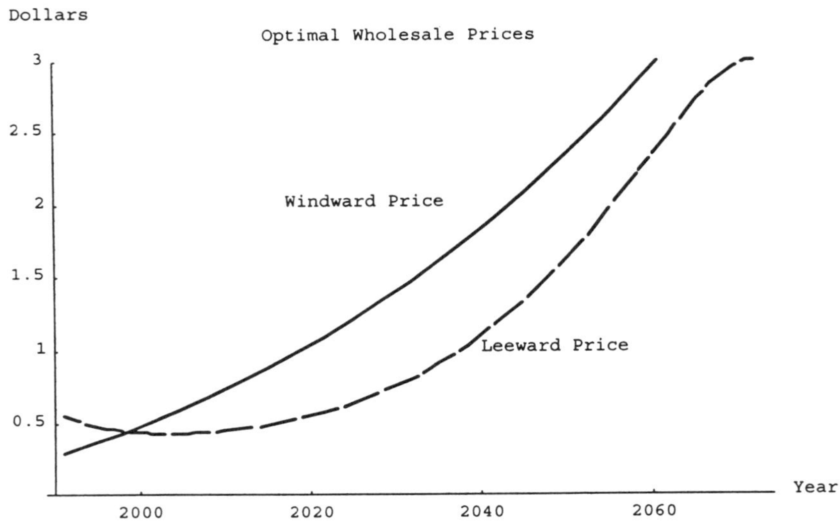

Figure 2. Optimal Wholesale Prices of Water if Windward Demand Grows at 0.5 Percent Per Year and Leeward Gets All Tunnel Water

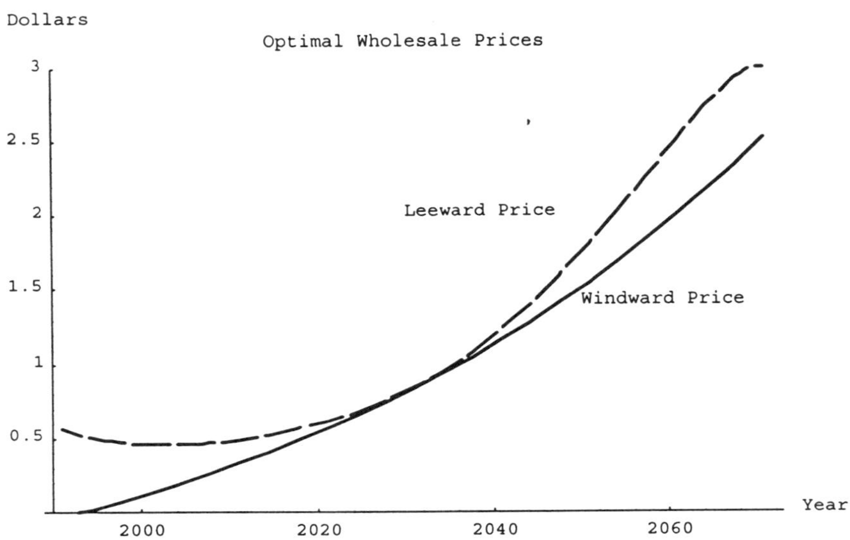

Figure 3. Optimal Wholesale Prices if Windward Demand Grows at 0.5 Percent Per Year and Leeward Gets 24.5 mgd of Tunnel Water

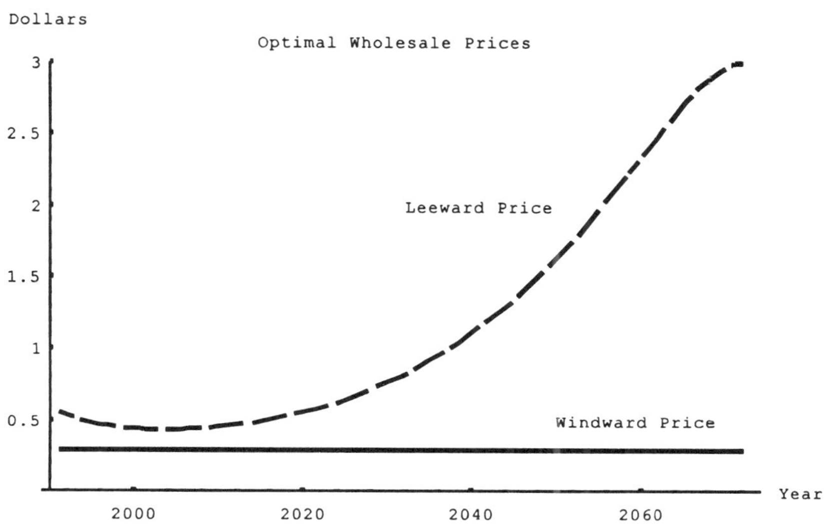

Figure 4. Optimal Wholesale Prices if Windward Demand Is Constant Over Time and Leeward Gets All Tunnel Water

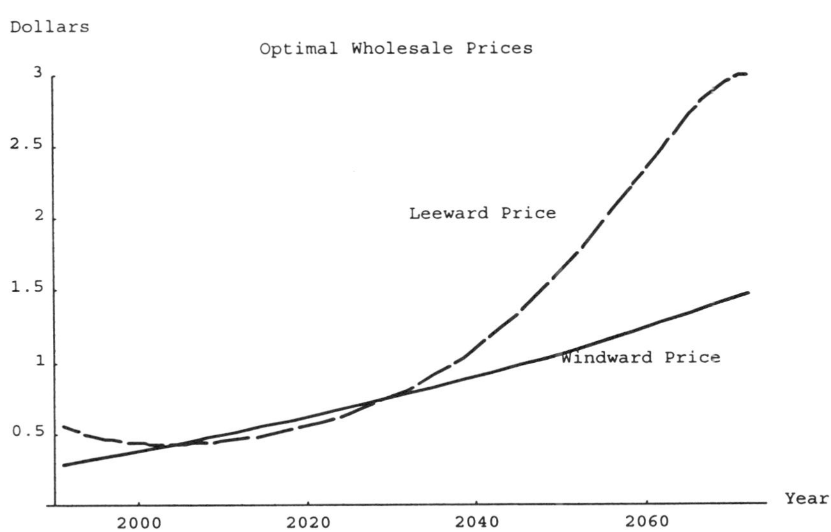

Figure 5. Optimal Wholesale Prices if Windward Demand Grows at 0.25 Percent Per Year and Leeward Gets All Tunnel Water

Figure 5 illustrates the intermediate case wherein windward demand grows at 0.25 percent annually. In this case, allocating all the aqueduct water to the leeward side is optimal during the initial and final phases and is nearly optimal during an intermediate phase from about 2005 to 2035.

Figure 6 shows the optimal profile of head in the Pearl Harbor Aquifer. Note that optimal extraction calls for an initial phase of conservation for the first 20 years or so, followed by a period of drawdown until a steady state head level of about 10 feet is reached slightly after the year 2070. Recalling that the optimal allocation of aqueduct water is almost all to leeward, we now see that the highest and best use of the contested water is neither to current leeward and windward users but to future generations, primarily of leeward users. That is, the water in question can be used to facilitate conservation of groundwater, thus substantially postponing the eventual need to desalinate. Conserving groundwater also requires that current overuse be reversed. This is most easily accomplished by setting wholesale prices equal to the corresponding ex situ values.

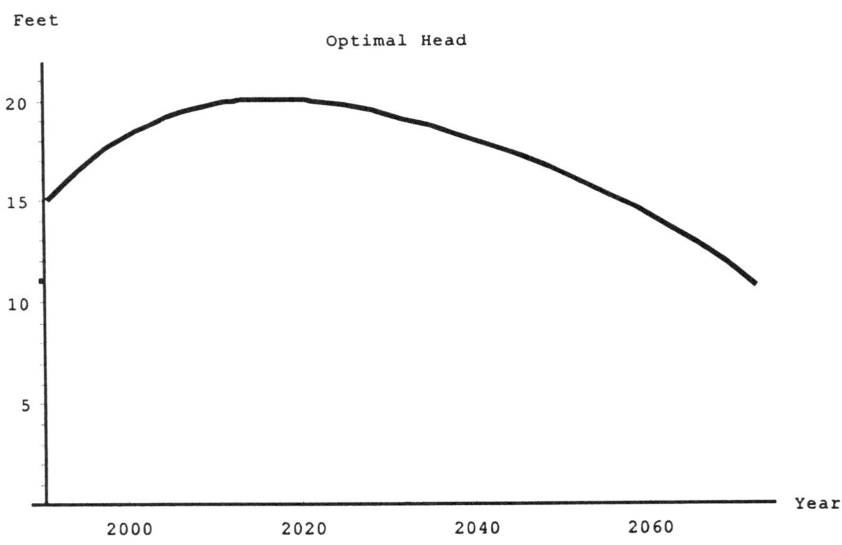

Figure 6. Optimal Head for the Pearl Harbor Aquifer if Leeward Gets All the Tunnel Water

Figure 7 shows the optimal paths of ex situ groundwater values (wholesale prices) and extraction costs. The efficient wholesale price falls during the initial phase of rising head, then rises to its steady state level of \$3. The big difference between wholesale price and extraction cost of groundwater is due to the last term in equation (2). By refraining from extracting water until the extraction cost is equal to the backstop price (the appropriate rule for a non-renewable resource), the water authority is able to garner a stream of "dividends" in the form of lower extraction costs in future years, and this increases the present value of the resource. Note too that the wholesale price of leeward water reaches a minimum of approximately two times the extraction cost and a maximum of almost six times. Since the Board of Water Supply bases water fees on extraction costs (and *historical* extraction costs at that), water on the leeward side is being seriously underpriced and correspondingly wasted.

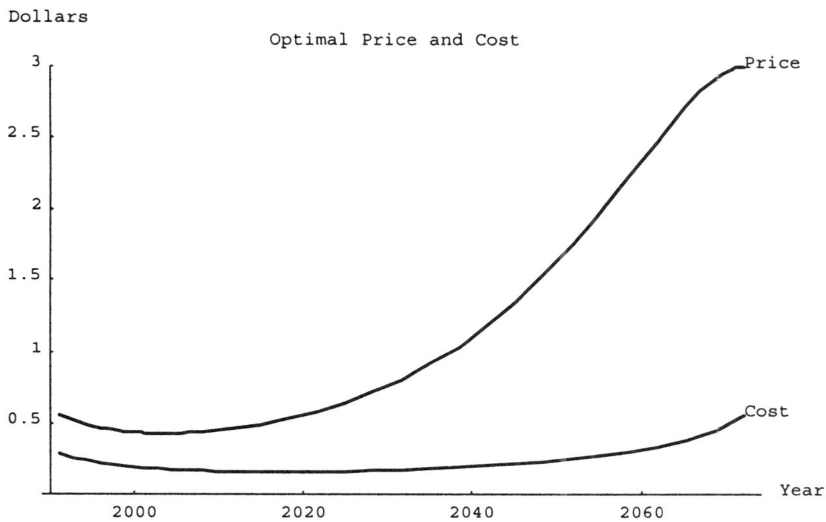

Figure 7. Optimal Price and Cost for Pearl Harbor Aquifer Water if Leeward Gets All the Tunnel Water

4. CONCLUSIONS

The Water Commission could stick to its announced interim allocation of 16 mgd for windward users and still not interfere with efficient use so long as water rights were made transferable. Revenue from the sales could be used for preservation of Hawaiian culture and to further environmental and other objectives that motivated its decision to reallocate most of the aqueduct water to the windward side.

If aqueduct water rights were made transferable and BWS pricing rules were reformed to allow the inclusion of marginal user cost of groundwater, the result would be primarily, in effect, a reallocation to future users, not a reallocation among current users. This could delay the very costly construction and operation of desalination facilities for most of the twenty-first century.

Hawaii has very substantial water resources. But by underpricing water, misallocating fungible water, and restricting transfers, the governing agencies have created the specter of impending shortages and unnecessary conflicts. The moral of the story is that when governments ignore efficiency in setting economic policies, they create artificial scarcity from the abundance of creation. The key to cooperation is refocusing on the win-win opportunities that removing barriers to efficiency always affords.

REFERENCES

Amaral, A.C. 1995. Letter to the Editor. *The Honolulu Advertiser*, June 30, p. A.

Anderson, T.L., ed. 1983. *Water Rights: Scarce Resource Allocation, Bureaucracy, and the Environment.* Cambridge, MA: Ballinger Publishing.

Anderson. T.L. 1985. The Market Alternative for Hawaiian Water, *Natural Resources Journal* 25: 893-910.

Board of Water Supply. 1982. *Oahu Water Plan.* City and County of Honolulu, Hawaii.

Commission on Water Resources Management. 1995. Waiahole Ditch Informational Packet. Honolulu, Hawaii (July).

Cox, D. 1981. A Century of Water in Hawaii. In F.N. Fujimura and W.B.C. Chang, eds., *Groundwater in Hawaii: A Century of Progress.* Water Resources Research Center, University of Hawaii at Mānoa, Honolulu.

Cox, R.H. 1981. Groundwater Technology in Hawaii. In F.N. Fujimura and W.B.C. Chang, eds., *Groundwater in Hawaii: A Century of Progress.* Water Resources Research Center, University of Hawaii at Mānoa, Honolulu.

Governor's Advisory Commission. 1979. *Hawai'i's Water Resources: Directions for the Future* (January).

Hawaiian Sugar Planters Association. 1973. Sugar Manual. Honolulu.

Herschler, L.H. 1966. The Waiahole Water Company: Fifty Years of Water Service. Reprinted in Commission on Water Resources Management, 1995.

Honolulu Advertiser, The. 1994. Full Flow Resumes in Waiahole Stream. December 21, p. A3.

Honolulu Advertiser, The. 1995. Waiahole Ditch Decision Anticipated in February. December 23, p. A3.

Kresnak, W. 1995. Waiahole Water Dispute Simmers. *The Honolulu Advertiser*, June 20, p. A1.

Krulce, D.L., J.A. Roumasset, and T. Wilson. Forthcoming. Optimal Management of a Renewable and Replaceable Resource: The Case of Coastal Groundwater, *American Journal of Agricultural Economics.*

Law, L.S. 1988. State Water Code—A Masterpiece of Compromise. *Wiliki o Hawai'i* 23: 1-3; 23: 2.

Legislative Reference Bureau. 1986. Advisory Commission on Water Resources to the Thirteenth State Legislature (January).

Lynch, R. 1994. On Tap: Sticky Battle for Old Waiahole's Water. *Honolulu Star Bulletin*, May 20, pp. A1, A4.

Mink, J.F. 1980. *State of the Groundwater Resources of Southern O'ahu.* Honolulu: City and County of Honolulu Board of Water Supply.

Mollett, J.A. 1961. *Capital in Hawaiian Sugar: Its Formation and Relation to Labor and Output, 1870-1957.* Honolulu: Hawaii Agricultural Experiment Station, Agricultural Economics Bulletin 21 (June).

Omandam, P. 1995. Waiahole Water Dispute Takes Center Stage. *Honolulu Star-Bulletin* June 29, p. A3.

Omandam, P. 1995b. Testimony Under Way over Waiahole Water Control. *Honolulu Star-Bulletin*, November 10, p. A3.

Review Commission on the State Water Code. 1994. Final Report to the State Legislature. Honolulu, December.

Saliba, B.C., and D.B. Bush. 1987. *Water Markets in Theory and Practice.* Boulder, Co: Westview Press.

Smyser, A.A. 1989. Richardson Court Bent Rules in Public's Favor. *Honolulu Star-Bulletin*, October 17, p. A14.

State Legislature of Hawaii. 1987. Act 45, State Water Code.

Takasaki, K.J., and J.F. Mink. 1985. *Evaluation of Major Dike-Impounded Ground-water Reservoirs, Island of Oahu.* Washington, D.C.: U.S. Government Printing Office.

Tanji, E. 1996. Use of Homestead Land by Non-Hawaiians Illegal. *Honolulu Advertiser,* May 25, p. A1.

Trelease, F.J. 1974. The Model Water Code, the Wise Administrator and the Goddam Bureaucrat, *Natural Resources Journal* 14: 207-229.

Wilson Okamoto Associates, Inc. 1990. Oahu Water Management Plan: Technical Reference Document. Paper prepared for Department of General Planning, City and County of Honolulu.

TABLE OF CASES

McBryde Sugar Co. v. Robinson, 54 Haw. 174, 504 P.2d 1330, *aff'd on rehearing,* 55 Haw. 260, 517 P.2d 26 (1973).

Robinson v. Ariyoshi, 887 F.2d 215,217 (9th Cir. 1989) *vacating* 676 F. Supp. 1002 (D. Haw. 1987).

20

THE EFFICIENT SHARING OF AN UNCERTAIN NATURAL RESOURCE: A CONTRACT THEORY APPROACH

Richard Barrett
University of Birmingham, U.K.

This chapter examines the efficient sharing between two parties of a natural resource, such as water from a river, when information on the benefits which will accrue is private. In order to maximize these benefits, the two parties agreed to a contract. In this chapter, various forms of contracting are considered, the form adopted determining how the division of the natural resource between the parties is controlled. The main object of this chapter is to investigate efficient pricing and the conditions under which each of these forms is best. The theory is illustrated by an application to the Lesotho Highlands Water Project.

1. INTRODUCTION

Throughout we are concerned with two parties, S (the "seller") and B (the "buyer"). In applications they are typically countries. Available for S and B to "share" is an amount, W, of a divisible natural resource. For concreteness, we assume that S is the owner, but we could allow any initial allocation of W without much effect on the analysis.

An analogous situation is where S and B are children, invited to share a cake. If they cannot agree (the "threat point"), adults will confiscate the cake and eat it themselves. An equitable solution might be that each receives half the cake. In general, however, this is not efficient. Perhaps B likes cake more than S, and the total benefit from the cake can be increased by B receiving the larger share (which may in compensation require a payment from B to S).

The really difficult question in such bilateral monopoly situations is how the "cake" will, or should, be divided. Pigou (1932) thought the matter intractable, or at any rate beyond the scope of economics. Subsequent "solutions," provided for example by

Nash (1950) and Rubinstein (1982), are in various ways open to objection.[1] In the two-country context, the division of the cake might be determined by an international court, and an investigation of the principles which in practice have governed such court decisions could be very interesting. We do not, however, venture into this arena. Instead, the focus of this chapter is on the efficiency of the contract to which S and B might agree.

It is assumed that, at the time the contract is agreed, W is uncertain, and for each party the benefits which derive from the consumption of any given quantity of the resource are also uncertain. When the uncertainty about W is resolved and W becomes known, this is public knowledge, but when the uncertainty about benefits is resolved, they are known only to the party concerned.

The model could be applied, either positively or normatively, to sharing (a) water from a river or from underground sources or (b) reserves of natural gas or oil. Such resources often straddle national boundaries, and are sure to be uncertain in extent and in the long-term benefits they provide. These benefits will to a degree be private information.

In Section 2, the main assumptions of the model are set out, and several contract forms are described. The contract forms typically involve prices, and in Section 3 efficient pricing is examined, both when prices are linear and when they are non-linear. Then, in Section 4, the advantages of the different contract forms are discussed. Section 5 presents an application to the Lesotho Highlands Water Project, and Section 6 concludes.

2. THE MODEL

Suppose S and B are to share an amount, W, of a natural resource owned by S. At t=0, they sign a contract and at t=1, accordingly, S receives the allocation W_S and B receives W_B. Assume:

(A1) At $t = 0$, W has a known probability distribution, and a value for W is realized at $t = 1$.

(A2) Both S and B are risk neutral.

(A3) The costs of provision of the natural resource are fixed.

(A4) In monetary terms, the marginal benefits obtained from the division of W are, respectively,

---

[1]Nash's assumptions can be criticized. For example, they allow independent affine transformations of the two parties' utility functions. The result is that symmetry is invoked in situations in which there is a lack of symmetry once one takes account of interpersonal comparisons of utility. Rubinstein's "alternating offers" model has the unattractive feature that "last mover advantage" can only be removed by extending the sequence of offers to infinity or, analogously, "shrinking" the prize to zero.

$$MB_S = MB_S(W_S),\ (W_S \geq 0)$$
$$MB_B = MB_B(W_B),\ (W_B \geq 0).$$

At t=0, these marginal benefit schedules have known probability distributions. A schedule for each party is, at t=1, realized as private information—observed only by the party concerned. For each party, marginal benefits are a continuous and decreasing function of consumption of the natural resource.

(A5) The probability distributions for W, $MB_S(\cdot)$ and $MB_B(\cdot)$ are mutually independent.

(A6) Unless otherwise stated, all information is public knowledge.

(A7) Side-payments are permitted.

(A8) Any contract agreed between S and B at t = 0 is costless to negotiate and execute, and can only take account of public knowledge. Negotiations at t = 1 are prohibitively costly.

According to the Coase Theorem (Coase 1960), the parties will choose a contract on the Pareto efficiency frontier—a contract which in relation to public knowledge is ex-ante Pareto efficient. We may prefer to say that they *should* choose such a contract, in order to leave economists something to do! By assumptions (A2) and (A7), choosing a contract on the Pareto efficiency frontier is equivalent to maximizing the expected joint benefits. However, the optimization needs to take account of the fact that, as far as each exercises control, S and B will be governed by their private information concerning benefits. The nature of this control will be discussed shortly.

Denote the parties' joint benefits from consumption of the natural resource by B, and the expectations operator by E where E denotes averaging with respect to variation in each of W, $MB_S(\cdot)$, and $MB_B(\cdot)$. The expected joint benefits are

$$EB = E\int_0^{W_S} MB_S(x)dx + E\int_0^{W_B} MB_B(x)dx. \tag{1}$$

The form of contract which S and B agree is important in determining how W is divided between W_S and W_B. Five such forms are worth distinguishing, the first four each involving a price. In general, the price can be non-linear, as will be discussed in the next section.

A price presupposes a buyer. We have let S be the seller and B the buyer, so B pays S for B's allocation of the natural resource. However, as indicated in the introduction, this characterization of ownership is inessential. A different configuration of property rights would not change the analysis, since assumption (A7) allows a side-payment to affect any division of the expected joint benefits, EB. Thus, there is no loss of generality in letting S be the seller and B the buyer. As we will now see, this specification allows us to make use of some existing terminology.

Subject to price, one form of contract (*seller control*) allows S to choose how W is shared, while another form (*buyer control*) gives this choice to B. A third form of contract (*non-integration*) allows S to choose how much of W is offered to B, and B to choose how much of the amount offered to accept.[2] A fourth form (*reverse*) interchanges the roles of S and B in *non-integration*; in this case, B makes an initial decision on how much to buy, but must also buy whatever additional amount (if any) S chooses to sell. A fifth form (*command*) determines directly the division of W between S and B. One can think of more complex contract forms, but these five are sufficiently varied and simple to yield insights.

2.1 Incommensurability

Assumption (A4) incorporates the notion that the benefits which S and B derive from the natural resource are "commensurable" with money. The effect is twofold: one party's benefit can be compared with the other's and, should one party's allocation be reduced, compensation in terms of money is feasible. However, it is interesting to note some expressions of doubt concerning the commensurability of such benefits. Sunstein (1994), for example, cites instances where legal practice reflects a belief in incommensurability. Consider the degradation of a natural resource (such as the Alaskan coastline). He shows that, in the event of this kind of environmental damage, a court may require restoration to the "status-quo ante" rather than compensation, even in situations where compensation would be the cheaper option.[3] (Thus Sunstein distinguishes commensurability from the existence of "trade-offs.")

It can be argued that, in the case of water, lifestyles may be so intricately bound up with the resource that commensurability is in doubt, but we do not pursue the matter further.

2.2 Complete versus Incomplete Contracts

Ayres and Gertner (1992, p. 730) define a contract as "contingently complete" if all the potential gains from trade have been realized. In the real world, a contract may never be complete in this sense—a fully contingent contract may be too complex and costly an instrument to be a practical possibility. But in the simplified framework of this chapter, the parties will rationally choose completeness when future negotiations

---

[2]The terms "seller control," "buyer control," and "non-integration" derive from Grossman and Hart (1986, 1987). Grossman and Hart consider a contract between a buyer and a seller, who may at some fixed point in the future trade an indivisible good. Future valuations will be private information. They find that, optimally, the right to enforce trade should be allocated (if to either party) to the party whose valuation of the good is the more uncertain.

[3]Another instance is that, in response to an action for breach of contract, a court may insist on "specific performance" rather than damages. "What the plaintiff wants, and what she is entitled to get, is a good that she values in the way that she values the object for which she has contracted" (Sunstein 1994, p. 12).

are excessively costly.[4]

The choice of an incomplete contract may be desirable when the future costs of negotiation are not excessive. In this case, any of the five forms of contract could be combined with the resumption of negotiations in the future, at a date when the two parties' private observations of the marginal benefit schedules is open to exploitation. In the resumed negotiations, one side or the other may suggest additional or reduced sales, at higher or lower prices, in an attempt to realize additional gains from trade.

Since the interests of S and B conflict with respect to price, it is reasonable to expect that future negotiations will be rather difficult and protracted, and that a complete contract will be chosen.[5] (Current negotiations may also be rather difficult and protracted, providing an incentive to settle the difficult questions now rather than postpone them to the future.) Thus, in this chapter we rule out additional negotiations at $t = 1$. The private nature of the two parties' observations of the marginal benefit schedules plausibly makes such negotiations prohibitively costly.

3. EFFICIENT PRICING

We now examine each of the five regimes in turn. The immediate object is to characterize efficient pricing rules both when prices are constrained to be linear and when they are allowed to be non-linear. For the moment, let $p = p(W_S)$ denote the marginal price schedule faced by both S and B. The schedule is horizontal when prices are constrained to be linear. Since the realization of W is public knowledge, $p = p(W_S)$ can be conditioned on W. (This dependence is suppressed in our notation.) At one point, we will relax the assumption that S and B face the same prices.

We assume that S and B choose optimally in exercising control, and in this section we also assume that all solutions are interior solutions. Let E_S and E_E denote taking expectations with respect to $MB_S(\cdot)$ and $MB_B(\cdot)$, respectively.

3.1 Seller Control

Under *seller control*, S chooses W_S after observing the realization of $MB_S(\cdot)$. Thus,

$$MB_S(W_S) = p(W_S). \tag{2}$$

---

[4]Ayres and Gertner's other sense in which contracts are complete is "obligational completeness" (1992, p. 730). This requires the obligations of the parties to be specified in all possible states of the world. For instance, there must be an agreed procedure by which any dispute concerning the implementation of the contract is subjected to the decision of the courts. Contingent completeness is the more relevant concept in this chapter.

[5]For an attempt in a multiperiod context at modeling wage bargaining and strikes when information about firm profitability is private to the firm, see Tracy (1987).

Equation (1) becomes

$$EB = E\int_0^{W_S} MB_S(x)dx + E\int_0^{W-W_S} MB_B(x)dx, \tag{3}$$

where, from (2), W_S depends on the realization of $MB_S(\cdot)$ and on the marginal price schedule.

Assuming linear pricing, we have $W_S = W_S(p)$ where p is conditioned on W. Differentiating with respect to p, equating to zero and applying (2), the efficient level for price (for any realization of W) is characterized by

$$p = [E_S\, W_S']^{-1}\, E_S\, E_B\, W_S' MB_B(W\text{-}W_S). \tag{4}$$

On the right-hand side, E_S describes weighted averaging with respect to variation in $W\text{-}W_S$ and MB_B as the $MB_S(\cdot)$ schedule varies.

Consider now non-linear pricing. The marginal price schedule can be chosen to achieve a target value for W_S for each $MB_S(\cdot)$ schedule. Differentiating with respect to W_S, equating to zero and applying (2), the efficient level for marginal price [for any realizations of W and $MB_S(\cdot)$] is characterized by

$$p = E_B\, MB_B(W - W_S). \tag{5}$$

As $MB_S(\cdot)$ varies, (5) implicitly maps out an upward sloping marginal price schedule, $p = p(W_S)$.

Equation (5) differs from equation (4) in that on the right-hand side of (5) averaging is for a given level of $W_B = W - W_S$. Intuitively, this is because, as $MB_S(\cdot)$ varies, p varies accordingly. A higher $MB_S(\cdot)$ schedule means a higher value for p, in order to maintain equality between $MB_S(W_S)$ and the average value for $MB_B(W - W_S)$ as the $MB_B.(\cdot)$ schedule varies. Thus, averaging on the right-hand side of (5) is for a given $MB_S(\cdot)$ schedule, and so involves variation in only the MB_B schedule. A consequence is that variation in MB_B in (5) is likely to be much less than in (4).

3.2 Buyer Control

Buyer control interchanges the roles of S and B in *seller control*, giving analogous results. Thus S and B are interchanged in equations (2), (4), and (5).

3.3 Non-Integration

Non-integration is a hybrid of the previous two regimes where S retains some of W and offers the rest to B. If B accepts all that is offered, S has effective control, but effective control otherwise passes to B. Call these alternate events T and T′. Let π be the probability of T and $1 - \pi$ the corresponding probability of T′, where π depends on W and prices.

Under linear pricing, we have $W_S = W_S(p)$ given T, and $W_B = W_B(p)$ given T′. Applying the weights, π and $1 - \pi$, it can be shown that the efficient level for price (for any realization of W) is characterized by

$$\pi\, E_T\, W_S'[p - MB_B(W_B)] + (1 - \pi)\, E_{T'}\, W_B'[p - MB_S(W_S)] = \lambda. \tag{6}$$

For given W, E_T and $E_{T'}$ denote averaging with respect to variation in both $MB_S(\cdot)$ and $MB_B(\cdot)$, conditional on the events T and T′, respectively. Here λ is an adjustment factor which reflects the effect of variation in p on π (the probability of T). Thus, for example, it might be optimal to increase π by lowering p because the fit is better under T than under T′. Because event T (B accepts all of W that is offered) is defined by $MB_B(W - W_S) \geq p$, the first term in (6) is positive and the second term negative.

Consider non-linear pricing. Given that S and B face the same marginal price schedule, the effect of variation in $MB_S(\cdot)$ can be "eliminated" from the first term as in *seller control*, but since the marginal price schedule will be upward sloping, the improvement in "fit" for the first term is at the cost of worsening that for the second. Suppose, however, it is arranged that S and B face different price schedules. Then, in (6), W_S and W_B can be separately targeted, and (6) is replaced by

$$p = E_{BT}\, MB_B(W_B) - \lambda_S \quad \text{(when T occurs)} \tag{7a}$$

$$p = E_{ST'}\, MB_S(W_S) - \lambda_B \quad \text{(when T}' \text{ occurs).} \tag{7b}$$

For given W, E_{BT} denotes averaging with respect to variation in $MB_B(\cdot)$ conditional on event T, and $E_{ST'}$ denotes averaging with respect to variation in $MB_S(\cdot)$ conditional on event T′. Both λ_S and λ_B are adjustment factors which reflect the effect of variation in the marginal price schedules on π (the probability of T). Again, the explanation for raising or lowering π concerns the relative fit under T and T′.

A special arrangement will, however, be needed for S and B to face different price schedules. An example is where S and B each buy the natural resource from a third party and the third party returns to them, as lump-sums, the expected total value of the resource. The third party must be content to break even on average (otherwise the arrangement incurs a cost for S and B).

3.4 Reverse

Reverse interchanges the roles of S and B in *non-integration*, giving analogous results. Thus S and B are interchanged in equations (6) and (7), and in the definitions of T and T′.

3.5 Command

Under *command*, W_S and $W_B = W - W_S$ are determined directly as functions of W.

Differentiating with respect to W_S in equation (3), and equating to zero, we obtain (for any realization of W)

$$E_S\, MB_S(W_S) = E_B\, MB_B(W_B). \tag{8}$$

4. RANKING CONTRACT FORMS

Our aim now is to rank contract forms in terms of efficiency. For this purpose we assume efficient pricing. For simplicity, we also assume that the marginal benefit schedules are linear. First suppose that, for any realizations of W, $MB_S(\cdot)$, and $MB_B(\cdot)$, the first-best outcome has an interior solution. The first-best outcome is of course characterized by

$$MB_S(W_S) = MB_B(W_B). \tag{9}$$

Let $d = MB_B(W_B) - MB_S(W_S)$. From (9), for any configuration of the parameters concerned and rules about pricing, $E\, d^2$ provides an inverse ranking of the five contract forms in terms of efficiency.[6]

Denote the induced weak ordering by "$\succeq$," and its asymmetric and symmetric factors by "$\succ$" and "$\sim$," respectively. Let S̲ denote *seller control*, B̲ *buyer control*, N̲ *non-integration*, R̲ *reverse,* and C̲ *command.* If, for example, *seller control* is more efficient than *buyer control* (that is, gives a lower value for $E\, d^2$), then S̲ $\succ$ B̲. Our results from the previous section allow us to draw conclusions about the ranking of the different regimes under different pricing rules.

First, linear pricing cannot be more efficient than non-linear pricing, which subsumes it. However, in the case of either S̲ or B̲, non-linear pricing is likely to be much more efficient. On the right-hand side of equation (4), MB_B varies as the $MB_B(\cdot)$ schedule varies and also as W_S varies; on the right-hand side of equation (5), MB_B only varies as the $MB_B(\cdot)$ schedule varies. Thus, the variation in B's marginal benefit is likely to be much greater for linear than for non-linear pricing, and this will be true also of $E\, d^2$. Equations (4) and (5) concern S̲ under linear and non-linear pricing, respectively; analogous results hold for B̲.

Given non-linear pricing, it follows from equations (5) and (8) that S̲ $\succ$ C̲ and B̲ $\succ$ C̲. This is because, according to equation (8), $E\, d^2$ is the sum of the average variances

---

[6]Assume that the marginal benefit schedules are linear with slopes $-b_S$ and $-b_B$, respectively. It is easy to show that the deadweight loss in relation to the first-best outcome is given by

$$D = d^2/(b_S + b_B).$$

Thus, because the denominator is constant, $E\, d^2$ ranks contract forms according to deadweight loss. Also, $E\, d^2$ is an approximate measure of deadweight loss when the marginal benefit schedules are non-linear.

for $MB_S(W_S)$ and $MB_B(W_B)$ under C̲, whereas according to equation (5) we have only the average variance of one of these two marginal benefits to worry about under S̲ or B̲. The choice between S̲ and B̲ depends on the relative variability of the two schedules. If $MB_S(\cdot)$ is the more variable, then S̲ ≻ B̲. If $MB_B(\cdot)$ is more variable, the result is B̲ ≻ S̲.

With linear pricing in the example of the appendix, we have S̲ ≻ C̲ ≻ B̲ if S has the steeper marginal benefit schedule, and B̲ ≻ C̲ ≻ S̲ if B has the steeper schedule.

Assume different price schedules for S and B. It follows that N̲ ⪰ S̲ and R̲ ⪰ B̲. This is because S̲ is now a special case of N̲ and, analogously, B̲ a special case of R̲. Clearly, with different price schedules for S and B, W_B in the case of N̲ (respectively, W_S in the case of R̲) can be set equal to W independently of the level of W_S (respectively, W_B). When under N̲ we have W_B=W, this is equivalent to S̲ and, analogously, W_S=W under R̲ is equivalent to B̲.

Up to this point, all our results depend on the assumption of interior solutions. A sufficient condition to have interior solutions is that equation (9) always holds. We now relax this assumption and assume instead that the marginal benefit schedules are horizontal (marginal benefits no longer depend on the amounts of the natural resource allocated). Apart from this, we make no change in assumptions. In a sense this is opposite to the case of interior solutions, because now with probability one all solutions are corner solutions.

Assume that each party's marginal benefit from the resource has a probability distribution which is uniform over an interval. Let the means of the distributions be μ_S and μ_B, and the standard deviations σ_S and σ_B, respectively. In Barrett and Senaoana (1996), with these assumptions and linear pricing, we derive closed form expressions for prices and expected joint benefits under the various regimes. We then prove the following results:

(a) S̲ ≻ C̲ and B̲ ≻ C̲.
(b) If $\mu_S = \mu_B$ and $\sigma_S = \sigma_B$, then S̲ ~ B̲ ~ N̲ ~ R̲.
(c) If $\mu_S > \mu_B$ and $\sigma_S = \sigma_B$, then N̲ ≻ S̲ ~ B̲ ≻ R̲.
(d) If $\mu_S = \mu_B$ and $\sigma_S > \sigma_B$, then S̲ ≻ N̲ ~ R̲ ≻ B̲.
(c′) If $\mu_B > \mu_S$ and $\sigma_S = \sigma_B$, then R̲ ≻ S̲ ~ B̲ ≻ N̲.
(d′) If $\mu_S = \mu_B$ and $\sigma_B > \sigma_S$, then B̲ ≻ N̲ ~ R̲ ≻ S̲.

Note that (c′) and (d′) are analogous to (c) and (d), and can be derived by interchanging S and B, S̲ and B̲, and N̲ and R̲ in (c) and (d).

5. THE LESOTHO HIGHLANDS WATER PROJECT

In 1986, Lesotho and the Republic of South Africa signed a long-term contract (The Treaty, 1987) for the transfer of water from Lesotho to South Africa under the Lesotho Highlands Water Project (LHWP). With the help of World Bank loans, Lesotho bears the costs (construction and operational) of the project. Thus, Lesotho is the "upstream producer" and South Africa the "downstream buyer." With the first stage of the project expected to be completed in 1995, the Treaty relates to the period 1995 to 2045. It specifies that in exchange for "royalties," South Africa is to receive all the "reserved supply" (water which is "98 percent sure"), while what is left, "excess water," will be shared. In addition to royalties, the Treaty sets a price which South Africa pays for any excess water which it receives, and a further benefit for Lesotho is the use of the water to generate hydroelectric power.

In Barrett and Senaoana (1996), we suggest that the shareout relating to the LHWP, as far as it can be evaluated, is decidedly in favor of Lesotho. We show there that this result is not explicable in terms of the conventional Nash solution, and go on to conclude that there seems to be a need for a more "ethical" explanation—Lesotho is the poorer of the two countries—or else a more "political" explanation—South Africa needs to have a friend on its borders.

The Treaty (not the most lucid instrument) could be interpreted as incomplete. It specifies that the agreement of both countries is required before the excess water is allocated in any year. In Barrett and Senaoana (1996), we infer that the form of contract adopted for the division of excess water is *non-integration*. There are two reasons for our conclusion. First is the belief that, in the absence of an agreement, a "default rule" will assign the excess water to Lesotho, since ownership of the LHWP gives Lesotho residual property rights. Second is the fact of the advance setting by the Treaty of a price (subject to indexation) for excess water. The price is determined for the entire fifty-year period that the contract is due to run. This fixed price seems to rule out any possibility that the two countries intend to conduct serious bargaining in the future over the allocation of excess water.

Because the excess water is not as significant as the reserved supply, Barrett and Senaoana (1996) argue that the marginal benefit schedules for excess water are approximately horizontal, and apply the results cited at the end of the last section. These suggest that Lesotho's marginal benefits from excess water will normally be higher than South Africa's.

How can this outcome be interpreted? All of the reserved supply is delivered to South Africa, which is less richly endowed with water resources than Lesotho, and it might thus be anticipated that the benefits from excess water will also be greater in South Africa than in Lesotho. A key consideration is, we feel, Lesotho's geographic proximity to the LHWP. This has a specific effect in raising its benefits from excess water in comparison with those of South Africa. The reason is that costly installations required to transfer water large distances are likely to be a more important factor in the case of excess water than in the case of the reserved supply. The reserved supply (98 percent sure and suitable for municipal uses) is both more highly valued and more manageable than excess water.

6. CONCLUSIONS

The main results of this chapter were presented in Section 4. They show the importance of non-linear pricing in increasing efficiency. For non-linear pricing, they also show that, in choosing *seller control* or *buyer control*, control should be given to the party with the more variable marginal benefits. When, however, there is not much basis for choosing between the parties in this respect, and corner solutions (the allocation of the whole of W to one party) are a significant feature (in probabilistic terms), it may then be optimal to choose a *non-integrated* or *reverse* framework. In this case, initial control should be assigned to the party whose marginal benefits are likely to be the higher of the two. *Non-integration* or *reverse* approaches should also be chosen when the two parties can be given different price schedules, but this may not be a practical possibility.

Since storage is such an important aspect in the case of water, future research should extend the model to a multiperiod framework. Because good storage facilities will reduce the variability of a country's marginal benefits (from the current allocation of the natural resource), one might conjecture that, other things equal, control should be given to the country with worse storage facilities.

It will also be worth relaxing the assumption that W, $MB_S(\cdot)$ and $MB_B(\cdot)$ have mutually independent probability distributions. As an example of stochastic dependence in the case of water, low rainfall or a hot summer may simultaneously lower W and raise the marginal benefit schedules. However, as long as there exists important residual information which is private, it appears that these correlations will not affect the basic theory much. They should, of course, be incorporated in any practical application of the theory.

APPENDIX

Let $MB_S = a_S + \epsilon_S - b_S W_S$ and $MB_B = a_B + \epsilon_B - b_B W_B$, where ϵ_S and ϵ_B are independent with zero means and variances σ_S^2 and σ_B^2, respectively.

A.1 Seller Control

Assume *seller control*. Under linear pricing, from equation (4),

$$p = E_S\, a_B - b_B(W - W_S)$$

where, from equation (2),

$$W_S = (a_S + \epsilon_S - p)/b_S.$$

Thus,

$$p = a_B - b_B[W - (a_S - p)/b_S],$$

$$d = MB_B - p = \epsilon_B + (b_B/b_S)\epsilon_S,$$

and

$$E\,d^2 = \sigma_B^{\,2} + (b_B/b_S)^2\sigma_S^{\,2}. \tag{10}$$

Under non-linear pricing, from equation (5),

$$p = a_B - b_B W_B,$$

$$d = MB_B - p = \epsilon_B,$$

and so now

$$E\,d^2 = \sigma_B^{\,2}. \tag{11}$$

A.2 Buyer Control

Interchanging the roles of S and B in *seller control* under linear pricing we have

$$E\,d^2 = \sigma_S^{\,2} + (b_S/b_B)^2\sigma_B^{\,2}, \tag{12}$$

and under non-linear pricing,

$$E\,d^2 = \sigma_S^{\,2}. \tag{13}$$

A.3 Command

Under *command*, from (8),

$$d = [MB_B(W_B) - E_B\,MB_B(W_B)] - [MB_S(W_S) - E_S\,MB_S(W_S)] = \epsilon_B - \epsilon_S.$$

Consequently,

$$E\,d^2 = \sigma_B^{\,2} + \sigma_S^{\,2}. \tag{14}$$

From equations (10)–(14), we conclude under linear pricing that $\underline{S} \succ \underline{C} \succ \underline{B}$ if $b_S > b_B$, and $\underline{B} \succ \underline{C} \succ \underline{S}$ if $b_B > b_S$; and under non-linear pricing that $\underline{S} \succ \underline{C}$ and $\underline{B} \succ \underline{C}$; $\underline{S} \succ \underline{B}$ if $\sigma_S > \sigma_B$, and $\underline{B} \succ \underline{S}$ if $\sigma_B > \sigma_S$.

REFERENCES

Ayres, I., and R. Gertner. 1992. Strategic Contractual Inefficiency and the Optimal Choice of Legal Rules, *The Yale Law Journal* 101: 729-773.

Barrett, C.R., and M.P. Senaoana. Forthcoming. The Price of Water under the Lesotho Highlands Water Project, *Bulletin of Economic Research*.

Coase, R.H. 1960. The Problem of Social Cost, *Journal of Law and Economics* 3: 1-44.

Grossman, S.J., and O.D. Hart. 1986. The Costs and Benefits of Ownership: A Theory of Vertical and Lateral Integration, *Journal of Political Economy* 94: 691-719.

Grossman, S.J., and O.D. Hart. 1987. Vertical Integration and the Distribution of Property Rights. In *Economic Policy in Theory and Practice*, Sapier Conference Volume. London: Macmillan.

Nash, J.F. 1950. The Bargaining Problem, *Econometrica* 18: 155-162.

Pigou, A.C. 1932. *The Economics of Welfare*. London: Macmillan.

Rubinstein, A. 1982. Perfect Equilibrium in a Bargaining Model, *Econometrica* 50: 97-109.

Sunstein, C.R. 1994. Incommensurability and Kinds of Valuation: Some Applications in Law. Paper presented at the conference, "Incommensurability and Value," Chateau du Baffy, France, April 1994.

Tracy, J.S. 1987. An Empirical Test of an Asymmetric Information Model of Strikes, *Journal of Labour Economics* 5: 149-173.

The Treaty between the Kingdom of Lesotho and the Republic of South Africa on the Lesotho Highlands Water Project. 1987. Ministry of Water, Energy and Mining, Government of Lesotho. Maseru: Government Printing.

21

BILATERAL POLICY COORDINATION UNDER UNCERTAINTY

Sinaia Netanyahu*
University of Maryland, College Park, USA

Difficulties in exclusion and a high degree of exhaustibility of trans-boundary natural resources naturally lead players to behave strategically. Strategic behavior imposes externalities. The resulting outcome of unilateral management of a resource is suboptimal from the perspective of a social planner. Alternatively, the management of trans-boundary natural resources common to economically interdependent political entities can become more efficient when entities coordinate trans-boundary resource policies rather than each entity acting alone. However, governments must make decisions regarding domestic and trans-boundary resource issues under uncertainty. The effects of various sources of uncertainty on the incentives of governments to coordinate trans-boundary water policies are discussed by comparing the coordinated outcome to the outcome in the absence of coordination.

1. INTRODUCTION

There are more than 200 major international river basins around the world about one quarter of which are shared by more than two countries (United Nations 1975). In addition, numerous aquifers lie under the borders of neighboring countries. Consumption of trans-boundary water resources which is associated with limited exclusion and a high degree of rivalry often leads nations to behave strategically with respect to those resources. As a result, a significant quantitative and qualitative degradation of those resources may occur. Externalities resulting from the use of trans-boundary resources induce conflicts among the political entities which are located within the resource's geographic area (Kneese 1971). However, sovereign principals and ineffective international enforcement mechanisms suggest that conventional economic approaches dictated by a social planner such as regulation, taxes, and tradeable permits are not likely to be effective in attaining an efficient use of common resources or resolving trans-boundary conflicts.

---

*This research was funded in part by the U.S – Israeli Binational Agricultural Research and Development Fund, BARD Grant No. US-2637-95R.

Alternatively, political entities can internalize externalities by coordinating resource-related policies. Policy coordination can be further complicated by uncertainty. Designing the model upon which gains from coordination can be analyzed is subject to uncertainty with respect to the economic environment of the countries involved and other random effects.

This chapter focuses on the impact of uncertainty on gains from coordination of policy related to trans-boundary natural resources. The organization of the chapter is as follows. Section 2 discusses the management of trans-boundary water resources. Section 3 discusses incentives for policy coordination. Section 4 discusses sources of uncertainty. Section 5 reviews the literature on trans-boundary policy coordination. Section 6 presents the model and discusses policy coordination under deterministic and uncertain settings. Section 7 contains concluding remarks.

2. MANAGEMENT OF TRANS-BOUNDARY WATER RESOURCES

Increasing demand for and deteriorating supply of fresh water resources often lead governments to restructure domestic water sectors. Such economic restructuring is designed subject to an increasing number of conflicting domestic targets (e.g., promoting the agricultural sector or protecting it in drought while maintaining sufficient quantities and quality standards of water resources used by other sectors). Domestic concerns over the availability of water are exacerbated when the water resources are common to two or more countries. Water policies of one riparian country affect the ability of the neighboring country to achieve its targets.

The magnitude of the policy spillover can be altered if governments agree to coordinate their water policies. For such agreements to be effective, an enforcement mechanism that insures compliance should be in place. In the absence of a super-government that can effectively enforce agreements, efficient exploitation of trans-boundary natural resources can be managed by multilateral cooperation (Biswas 1993).

Successful coordination is likely to require transmission of accurate information with respect to the state of the resources and the respective economies. However, such information is often limited. This limitation implies that the management of a trans-boundary water resource involves various aspects of uncertainty. Uncertainty exists with respect to available quantities of common water, alternative water sources, hydrological interaction between water resources, the initial position of the agricultural sector, data quality, the regeneration rate of renewable water resources, the impact of policy tools on a government's targets, and the implicit weights given by the neighboring country to objectives such as agricultural output and water reserves. Constructing models becomes a difficult task under such uncertainties.

3. INCENTIVE FOR POLICY COORDINATION

Policy coordination is defined as the agreement by two or more countries to a cooperative set of policy changes where neither country would wish to undertake the policy changes on its own but where each expects the package to leave it better off relative to a non-cooperative equilibrium.

It is assumed that each government tries to target more objectives than it has policy tools to achieve them. Therefore, government must inevitably trade off between objectives. In a certain world, governments are better able to determine the trade off between objectives so policy coordination is likely not to have as much added value. However, in reality, governments must make decisions regarding domestic and trans-boundary issues under uncertainties. Uncertainties can originate from limited knowledge of the physical nature of the natural resources and the state of the respective economies that have access to such resources. Restricting the environment of policy coordination to one policy issue (management of trans-boundary water resources) and two players, this chapter focuses on the impact of uncertainties on gains from coordination of policies related to trans-boundary natural resources.

Cooper (1985) identifies four reasons for coordination among governments. Three of these reasons are associated with standard microeconomic "market failure" analysis: existence of public goods (the free-rider problem) that span international boundaries, presence of externalities (spillovers), and existence of a limited number of heterogeneous governments in size and influence (monopoly and relative power). The fourth argument is associated with time lags and discontinuities in the process of decision making.

Inefficiencies in the use of trans-boundary resources are present when one country's resource exploitation policies result in externalities on the resources available to the neighboring countries. To overcome the externality effect that is created by a unilateral action, microeconomic analysis suggests that externalities should be internalized. Internalizing the externalities requires passing the authority to a central planner that can collect taxes, issue permits, set regulations, or at least regulate and enforce markets and property rights. However, due to sovereignty issues, it is unlikely that trans-boundary water exploitation policies will be determined or effectively enforced by a central authority (Barrett 1990). In addition, the problem of allocating property rights arises. A bargaining mechanism is not likely to solve the international property rights problem due to a lack of appropriate institutions that have the power to allocate trans-boundary resources. The failure of the market mechanism and bargaining leaves a third possibility for internalizing the spillover effect, i.e., policy coordination (Folmer and Musu 1992).

The multi-country use of common resources may result in one of four different types of externalities.

- Unidirectional externalities—where the exploitation of the resource by an upstream country affects a downstream country.
- Complex unidirectional externalities—which are similar to unidirectional externalities but the damage occurs also in the upstream country.

- Regional reciprocal externalities—when a group of countries contribute to and suffer from the damages.
- Global environmental externalities—when most countries, worldwide, are affected in one way or another.

The framework presented in this chapter investigates regional coordination assuming a regional reciprocal transmission of externalities via water policies. The nature of common resources dictates that exclusion of other parties from using the resource is difficult and that subtractability, as a result of common use, is high (Ostrom et al. 1994). That is, not only is it difficult to prevent access to the resource, but the exploitation of the resource by one entity leaves the others with less of the resource. The exclusion problem may not be eased by the existence of hard-to-define property rights when monitoring and enforcement are costly and/or infeasible (Barrett 1990). Additionally, when the externality is reciprocal, subtractability is likely to be high.

Resources common to two or more political entities can be managed with various degrees of cooperation such as full harmonization of policies, joint expenditures, a rule-based framework, continuous exchange of information, and continuous joint decision-making (i.e., policy coordination). Some degrees of cooperation threaten sovereignty and some require voluntary and self-enforced participation. In the case of trans-boundary water resources, the degree of cooperation is influenced by hydro-geological, economic, cultural, and political factors. Incentives to cooperate and the magnitude of cooperation are further enhanced by the definition of property rights in an aquifer, externalities experienced in the absence of joint management, external water resource opportunities, monitoring, and effectiveness of enforcement. Political stability and international image can also encourage cooperation. This chapter considers binary coordination, i.e., countries either fully coordinate trans-boundary water policies or not.

While the advantages of coordination are easy to detect, the obstacles to coordination should not be ignored. In addition to uncertainties, coordination faces obstacles related to compliance and credibility. Fieleki (1988) notes some unfavorable consequences of coordination. Following the argument against centralized government, Fieleki suggests that it is risky to encourage a high degree of collaboration among governments because that promotes their own power rather than serving their constituencies. The second argument is related to the feasibility of coordination. Two aspects are noted. The first aspect concerns intra- and inter-domestic interest groups and government conflicts over the goals that should be coordinated. The second aspect relates to adequacy of economic know-how, suggesting that many possible models of the economy are available, each of which yields a different outcome. The latter aspect describes best the main focus of this paper. Uncertainty associated with the structure of economic models influences the incentives for coordination. Therefore, policy modeling of trans-boundary resource coordination should be studied in this context.

When governments choose domestic water related policies, they face a trade off between the cost of dwindling water resources and economic gains in water intensive sectors, especially the agricultural sector. If policies are not bilaterally (or regionally) coordinated, then neither government takes account of the spillovers of its policies onto the other country. For example, when a government values water reserves more heavily than agricultural output, it may desire to set certain levels of water quality-quantity. Hence, the government may introduce stringent water policies. As water use restrictions are introduced in, say, an importing country, the marginal value of the water in that country must increase relative to the other country. The increase in water prices in one country leads to an increase in the price of its agricultural products and to a reduction in consumption of its agricultural products. If water intensive goods are traded between the entities that share water resources, then an increase in agricultural prices will increase the relative prices of the traded crops in the importing country. Alternatively, the importing country's desire to protect itself from future price increases and to increase its food self-sufficiency can result in a policy that encourages an increase in its domestic agricultural output. This policy, in turn, induces an increase in water pumping and may raise extraction costs. Also, in the case of common aquifers, increased pumping results in a decline of water tables and often in a deterioration of the quality of water making both countries worse off.[1] Therefore, with uncoordinated (Nash) equilibrium, governments may set water related policies while ignoring the externalities such policies can impose on a neighboring country. However, when governments pursue common regional resource policies, all can potentially attain a higher level of welfare relative to the uncoordinated case.

4. SOURCES OF UNCERTAINTIES

We begin by introducing two types of uncertainty: additive and multiplicative. As an example, consider the following model:

$$
\begin{aligned}
y &= \alpha_1 w + \beta_1 w^* + z \\
y^* &= \alpha_1^* w^* + \beta_1^* w + z^* \\
s &= \alpha_2 w + \beta_2 w^* + q \\
s^* &= \alpha_2^* w^* + \beta_2^* w + q^* .
\end{aligned}
$$

Domestic and foreign (denoted by asterisks) countries each have two targets: agricultural output (y and y^*) and water reserves (s and s^*). Each country attempts to achieve two targets by setting water policies (w and w^*). The economy of each country is exposed to various shocks and uncertainties. Additive uncertainty (noted by z's and q's) refers to shocks (e.g., exogenous agricultural demand or water reserve shocks) that affect government targets but do not change the effectiveness of its policy instruments. Multiplicative uncertainty (α's and β's) refers to possible changes in government targets (e.g., due to technological innovation) which effect

---

1 For example, damage can occur by saline intrusion if the aquifer is located close to an ocean or sea.

the policy instrument and, in turn, the government ability to achieve the economy's goals. Multiplicative uncertainty may originate from the structure of the model and from parameters. Model uncertainty refers to viewing the parameters of the policy instrument as an estimation error that results from lack of knowledge of the fundamental structure of the model. Parameter uncertainty refers to viewing the parameters of the policy instrument as random variables. Parameter uncertainty is attributed to domestic uncertainty (α's) and foreign uncertainty (β's) that transmits across the borders. Domestic uncertainty refers to the impact of the parameter of the domestic policy instrument on achieving domestic targets. Foreign uncertainty refers to the impact of the parameter of a foreign policy instrument on achieving domestic targets.

5. LITERATURE REVIEW

The analytical framework in this chapter relies on the framework established in the literature on international macroeconomic policy coordination analysis. In the case presented in this chapter, trans-boundary policies regarding common resources affect nations' ability to achieve conflicting economic targets. This fact and the fact that nations' economies are interdependent especially due to externalities that result from the use of a common resource make the framework of international policy coordination fits the discussion in this chapter. Ghosh nad Masson (1994) present an overview of the formal literature on international macro-economic policy coordination. Canzoneri and Henderson (1991) show that potential welfare gains from coordination are much larger when there are on-going conflicts between the two countries.

Uncertainty was explicitly introduced in models of inteınational macroeconomic policy coordination only recently. In a series of articles, Ghosh and Masson (see review in Ghosh and Masson 1994) analyze the impact of uncertainty on international coordination. Considering additive and multiplicative uncertainty, they examine the impact of such uncertainties on the incentives for policy coordination. They find gains from coordination to be larger in the uncertainty case than in the certainty case. This general framework is adopted in this chapter to analyze trans-boundary resource policy coordination under uncertainty.

Recent calls by Fieleki (1988), the World Commission on Environment and Development (1987), and Caldwell (1990) suggest that increasing regional and global economic interdependency require coordination of resource policies.[2] Just et al. (1991), Heal (1991), Safadi and Low 1991, and Beltratti (1995) all note that the implications of regional and global resource externalities should be further studied in the contexts of uncertainty and trans-national policy.

---

[2] In Caldwell, see especially the discussions on implementation of international environmental policy on regional bases with respect to international river basins, European regional cooperation, and Oceans and Coastal Programs of the United Nations Environmental Program.

Although trans-boundary resource problems cover a broad spectrum of natural resources, the economic literature concentrates mainly on economic analysis of fisheries and air pollution. Anderson (1975), Munro (1979), and Kaitala (1993) discuss optimal management of trans-boundary fisheries. Trans-boundary air pollution problems are discussed, for example, by Maler (1989, 1991) and Carraro and Filar (1995).

Joint (reciprocal) externalities are analyzed by, for example, Van der Ploeg and De Zeeuw (1992), Hoel (1992) and Kaitala, et al. (1992). Incentives for free-riding in trans-boundary resource policy coordination are discussed by Maler (1991). Models of linkages of environmental policies to other forms of cooperation are discussed by Folmer and Howe 1991.

Elements of uncertainty and risk stand as obstacles in reaching regional/ international agreements. The sources of uncertainty and risk that are often noted in the environmental literature are scientific (Nicolaisen et al. 1991). In addition, policy makers often fail to understand ecological systems and the long run effects of resource and environmental mismanagement on the economy. Unpredictable development of abatement technologies adds uncertainty regarding abatement costs. Uncertainty, irreversibility, and learning in combination with strategic behavior are analyzed by Ulph and Ulph (1995) in the context of a global warming dynamic game.

Finally, water basin models are surveyed by Spulber and Sabbaghi (1994). In particular, the authors discuss organizational arrangements, authorities' objectives and instruments, investment planning, and water markets mostly in relation to intrastate and interstate basin problems and, to a lesser degree, to international basins. Deterministic and stochastic dynamic programming applications for optimal management of domestic water resources are offered in Esogbue (1989). An analysis of dispute resolution mechanisms for solving intrastate, interstate, and international water disputes is given in Dinar and Loehman (1995). However, the studies outlined above do not account for the impact of uncertainty on the gains from cooperation over the management of trans-boundary resources given countries' conflicting domestic objectives.

6. ANALYTICAL FRAMEWORK FOR POLICY COORDINATION

For the purpose of simplification, this chapter divides the world into two regions identified as the home country and the foreign country. In both countries, water is mainly allocated to the agricultural sector. In each country, policy makers target several conflicting objectives such as agricultural production and water reserves. In this case, increasing agricultural production, holding all other factors constant, requires large allocations of water to agriculture. Given additional non-agricultural demands, such allocation poses a threat to national water reserves. By increasing agricultural production, a government increases the country's level of food self-sufficiency. On the other hand, by maintaining a high level of water in an aquifer, policy makers achieve several purposes: (i) leaving water for non-agricultural sectors and for future generations, (ii) reducing agricultural welfare loss in case of bad weather, and (iii) preventing saline intrusion and contamination.

When the number of targets is the same as the number of policy instruments, a government does not face trade offs between targets (i.e., neither instrument has an advantage over the other instruments in affecting the targets). However, when a government faces a number of targets larger than the number of available instruments, then trade offs are inevitable. Since each government assigns relative weights to targets, the trade offs among targets are assumed known. That is, in the absence of shocks, policy makers aware of resource constraints should not experience unpredictable trade offs between agricultural output and water reserves. However, in the presence of uncertainties, governments face unpredictable trade offs between competing targets.

Suppose the two countries share the same water sources. Uncertainties, reciprocal externalities, and strategic behavior associated with shared resources imply that these conflicting goals cannot be easily attained (Ward 1993). Policy makers are induced to make choices among targets at the margin. Analyzing conflicting targets and trade offs among them requires optimization of a national welfare function. Such a framework was introduced by Theil (1964) for analyzing government policy and extended by Niehans (1968) and Hamada (1979) for analyzing government policy in an international framework. Karp and Sachetti (1996) discuss the issue of conflicting targets in the context of environmental problems.

For example, in the context of shared resources, an economy which experiences a negative water reserve shock induces its government to adjust water policies in the agricultural sector. If the government neglects doing so, then the country's water resources will experience an increasing rate of deterioration, lowering the country's welfare. Alternatively, governments can tighten their water supply to the agricultural sector and achieve a lower rate of degradation or avoid it altogether. Such a step buys the government a welfare gain from controlling degradation and water reserves for non-agricultural sectors at the cost of a welfare loss from reducing agricultural output. Whether governments choose low water reserves with high agricultural output or high water reserves with low agricultural output depends on its preferences.

6.1 Model

Consider an agricultural production function $y = f(w,x)$ where w is water and x is an indicator for all other inputs with prices p_w and p_x, respectively. If a government supplies water up to quantity w^s at price p_w, then the agricultural sector's maximization problem is max $p\,y - p_w w - p_x x$ such that $w \leq w^s$. This results in two possible regimes for agriculture (where p represents the agricultural producer price index):

Case 1: If the water constraint binds, then the profit function is $\pi_1(p, w^s, p_x)$.
Case 2: If the water constraint does not bind, then the profit function is $\pi_2(p, p_w, p_x)$.

The corresponding agricultural supply, y, and derived demand for water, w, are $y = y(p, w^s, p_x)$ and $w = w^s$ for Case 1 and $y = y(p, p_w, p_x)$ and $w = w(p, p_w, p_x)$ for Case 2.

These two cases represent two possible regimes. Case 1 applies to a situation where there is an exogenous water constraint. The constraint might be attributed to (i) an aquifer's safe yield over which both entities bargain and agree or (ii) the physical/natural characteristics of the aquifer where the maximum water quantity that can be pumped from the aquifer is w^s. Case 2 applies to a user-fee regime. While in Case 1 the aggregate amount of water is a constraint, in Case 2 the price of water determines use. The same framework can represent the agricultural sector in the foreign country (variables and parameters associated with the foreign country are noted with an asterisk).

The two economies can be modeled through four sets of home and foreign equations: agricultural output, water demand, water cost, and water reserve.

(i) Agricultural output functions. Assuming that the agricultural sector is constrained by water supply but that the water constraint is not binding, the agricultural output functions are

$$y = y(p, p_w, p_x)$$
$$y^* = y^*(p^*, p_w^*, p_x^*).$$

(ii) Water demand functions. The derived demand functions when the water constraint is not binding are

$$w = w(p, p_w, p_x)$$
$$w^* = w^*(p^*, p_w^*, p_x^*).$$

(iii) Water cost functions. Water extraction cost, c, depends on the level of the water table in the aquifer, s,

$$c_w = c_w(s)$$
$$c_w^* = c_w^*(s^*).$$

(iv) Water reserve functions. Water stock is affected by the water movement between the two political entities and, hence, depends on the relative water table level at the beginning of each period. Since the aquifer is a renewable resource, the recharge level, r, is considered along with water consumed by both entities, w and w^*. The water reserve equations are

$$s = s(s_0, r, w, w^*)$$
$$s^* = s(s_0^*, r, w, w^*).$$

Finally, functions that govern terms of trade for small countries with large agricultural sectors are

$$p = p(y, y^*)$$
$$p^* = p^*(y, y^*).$$

The above framework yields equations for home and foreign agricultural and water sectors of the form

$$y = y(\alpha_1, \beta_1, w, w^*)$$
$$y^* = y^*(\alpha_1^*, \beta_1^*, w, w^*)$$
$$s = s(\alpha_2, \beta_2, w, w^*)$$
$$s^* = s^*(\alpha_2^*, \beta_2^*, w, w^*),$$

where each government has two targets—agricultural output and water reserves—and one policy instruments—water. Both economies are exposed to water reserve shocks and aggregate agricultural demand shocks. Stochastic shocks are assumed to be independent and identically distributed with mean zero and constant variance.

The economic framework of the following analysis of coordination under uncertainty is based on the framework introduced by Ghosh and Masson (1994). In their analysis of interdependent economies of two countries who attempt to coordinate macroeconomic policies, they use a linear-quadratic model. In our case, we further generalize their framework to model the economic interdependence between two countries who share common water resources.

Consider the general framework of a two-country model that reflects multiple targets and policy tools:

$$Y = Y(\alpha w, \beta w^*)$$
$$Y^* = Y^*(\alpha^* w, \beta^* w^*),$$

where Y is an N-vector of targets, w is a K-vector of policy instruments, and α's and β's are N x K matrices of parameters. Targets may include agricultural output, environmental standards, water reserves, etc. Policy instruments may include quotas, taxes, subsidies, permits, etc.

In a static model where governments target goals periodically, the objective functions can be written as

$$V = \max U(Y) = \max U[(y_1(\alpha w, \beta w^*), \ldots, y_N(\alpha w, \beta w^*)] \tag{1}$$

$$V^* = \max U^*(Y^*) = \max U^*[y_1^*(\alpha^* w, \beta^* w^*), \ldots, y_N^*(\alpha^* w, \beta^* w^*)]. \qquad (2)$$

This framework can be used conceptually to discuss the outcome of a unilateral policy decision where the Nash bargaining solution is used to discuss the outcome of coordinated action.

6.2 Modeling Policy Coordination under Certainty

Governments' actions (policies) transmit across boundaries and affect the welfare of other countries. In the uncoordinated (Nash) regime, an individual country sets policies by maximizing its own welfare taking the actions of the other government as given. Thus the welfare spillovers of its actions on the other country are ignored. That is, the home government chooses its own water policy setting w to maximize its own objective function, V, taking w^* as given.

In this section, policy makers are assumed to choose their optimal targets in a deterministic setting. Shocks are observed before the optimal policies are chosen. Hence, policy makers do not face unpredicted trade off between targets. All trade offs between the objectives are known to the government and reflected through relative weights distributed among targets. It is in each country's best interest to set policies that increase its welfare. Therefore the α's are assumed to be positive. The effect of foreign policies on domestic targets is not clear a-priori (i.e., β's can be either positive or negative). However, the foreign effect, β, on domestic targets, Y, is likely to be dominated by the domestic effect of policies, α, on domestic targets.

We first look at the uncoordinated outcome and then at the coordinated outcome. The optimality conditions for each government can be derived using equations (1) and (2). The optimality condition of the home country is:

$$\frac{\partial U}{\partial w} = \frac{\partial U[y_1(\alpha w, \beta w^*), \ldots, y_N(\alpha w, \beta w^*)]}{\partial w} = AU_y(\cdot) = 0.$$

When the number of targets and policies coincides, A is a full rank matrix and therefore $U_y(\cdot) = 0$.

The effect of foreign policy on home country welfare is

$$\frac{\partial U}{\partial w^*} = \frac{\partial U[y_1(\alpha w, \beta w^*), \ldots, y_N(\alpha w, \beta w^*)]}{\partial w^*} = BU_y(\cdot).$$

Since $U_y(\cdot) = 0$, $\partial U / \partial w^* = 0$ as well. This result indicates that, at the Nash equilibrium, the home country's welfare is not effected by the foreign country's policies. The above derivation is similar in the case of the foreign country.

In order to evaluate the coordinated regime, Ghosh and Masson (1994, p.51) suggest modeling the coordinated regime by assuming a single global social planner who chooses policies for both countries that maximize a weighted average of both countries' objective functions. However, because sovereignty principles are so important for countries, governments are likely to avoid a regime that is coordinated by an arbitrator. Instead, governments are likely to be engaged in bargaining over policy issues and the distribution of total welfare among their countries. A useful approach to search for the cooperative policy is to apply the Nash bargaining solution where the product of each government's welfare gain over the noncooperative equilibrium (U^N or U^{N*}, respectively) is maximized

$$V^c = \max_{w^c, w^{c^*}} \; [U^c(y_1(\alpha w^c, \beta w^{c^*}),\ldots, y_N(\alpha w^c, \beta w^{c^*})) - U^N]$$
$$[U^{c^*}(y_1^*(\alpha^* w^c, \beta^* w^{c^*}),\ldots, y_N^*(\alpha^* w^c, \beta^* w^{c^*})) - U^{N^*}].$$

First order conditions, $\partial V^c / \partial w^c$ and $\partial V^c / \partial w^{c^*}$, yield

$$AU_y^c(\cdot)U^* + A^* U_{y^*}^{c^*}(\cdot)U = 0$$
$$BU_y^c(\cdot)U^* + B^* U_{y^*}^{c^*}(\cdot)U = 0,$$

where U and U^* represent the surplus obtained by the home and foreign country, respectively, from cooperation relative to the noncooperation outcome and A and B are K x N matrices of α's and β's, respectively. If $A = A^*$ and $B = B^*$ then the first order conditions can be rewritten as

$$[U_y^c(\cdot)U^* + U_{y^*}^{c^*}(\cdot)U][A - B] = 0.$$

Since home and foreign policies are not likely to affect a country in a similar way, we assume that $A - B \neq 0$ and therefore

$$U_y^c(\cdot)U^* + U_{y^*}^{c^*}(\cdot)U = 0.$$

We also know that U, $U^* > 0$. Hence,

$$U_{y^*}^{c^*} / U_y^c \leq 0.$$

This condition implies that a Pareto improving move from the coordinated equilibrium requires that a unilateral change in any policy setting (and therefore changes in one country's targets) will necessarily induce a welfare gain in one country but simultaneously reduce welfare in the other country. Moreover, because U, $U^* > 0$, the ratio between U_{y*}^{c*} and U_y^c must be equivalent in magnitude (but opposite in direction) to the ratio between U and U^*. That is, the absolute value of the ratio between welfare responses of both countries to a change in each country's own policy setting must be equal to the ratio between the countries' levels of surplus from cooperation.

6.3 Modeling Policy Coordination Under Uncertainty

Often physical/natural processes and their impacts on the economy are not well understood by researchers and policy makers who are forced to make decisions under uncertainty. Insufficient knowledge concerning various factors leads to an imperfect description of the economy in economic modeling. That is, the modeled environment is subject to random shifts (additive or multiplicative shocks). This observation suggests that instead of maximizing the preference function, it is necessary to maximize the *expectation* of the preference function.

6.3.1 *Additive Uncertainty*

We first introduce additive uncertainty in the form of shocks that affect the level of targets but that do not change the multipliers of the policies. Common sources of such uncertainty are lags in data collection and policy implementation, imperfect knowledge about the structure of the economy, and other exogenous shocks. In this section, it is assumed that there is an exogenous shock, s, with zero mean and variance σ^2.

The objective functions of the home and foreign governments are

$$V = \max EU(Y) = \max EU[y_1(\alpha w, \beta w^*) + s, \ldots, y_N(\alpha w, \beta w^*) + s]$$

$$V^* = \max EU^*(Y^*) = \max EU^*[y_1^*(\alpha^* w, \beta^* w^*) + s, \ldots, y_N^*(\alpha^* w, \beta^* w^*) + s],$$

where E is the expectations operator.

The first order condition of the home country with respect to its policy change is

$$\frac{\partial U}{\partial w} = \frac{\partial E[U(y_1(\alpha w, \beta w^*) + s, \ldots, y_N(\alpha w, \beta w^*) + s)]}{\partial w} = AE[U_y(\cdot)] = 0.$$

The derivative of the home country utility function with respect to the foreign country policy change is

$$\frac{\partial U}{\partial w^*} = BE[U_y(\cdot)].$$

From the first order condition we know that $E[U_y(\cdot)] = 0$. Hence $\partial U/\partial w^* = 0$ as well. Additive uncertainty does not cause one country's policy instrument to effect the other country's welfare.

Accounting for an exogenous shock in the coordinated regime yields a result similar to that obtained under certainty. Hence, we conclude that additive uncertainty does not create additional incentives for government to coordinate trans-boundary policies.

6.3.2 *Multiplier Uncertainty*

Structural parameters of a model cannot be estimated with complete accuracy. Uncertainty regarding these parameters is modeled as uncertainty in the domestic and foreign policy multipliers where the policy multipliers are assumed to be random variables.

The objective functions of the home and foreign countries are

$$V = \max EU(Y) = \max EU[y_1(\alpha w, \beta w^*), \ldots, y_N(\alpha w, \beta w^*)]$$

$$V^* = \max EU^*(Y^*) = \max EU^*[y_1^*(\alpha^* w, \beta^* w^*), \ldots, y_N^*(\alpha^* w, \beta^* w^*)].$$

At the uncoordinated regime, the first order condition of the home country is

$$\frac{\partial U}{\partial w} = \frac{\partial E[U(y_1(\alpha w, \beta w^*), \ldots, y_N(\alpha w, \beta w^*)]}{\partial w}$$
$$= E[AU_y(\alpha w, \beta w^*)] = 0,$$

where A and $U_y(\cdot)$ are correlated and $U_y(\cdot) \neq 0$. This also implies that $\partial U/\partial w^* \neq 0$. At the Nash equilibrium, the foreign country policy setting affects the home country welfare. This suggests a possibility for improvement of both countries' policy settings such that both countries gain from coordination of policy.

At the coordinated regime, parties maximize

$$V^c = \max_{w^c, w^{c^*}} \left[E\{U^c[y_1(\alpha w^c, \beta w^{c^*}), \ldots, y_N(\alpha w^c, \beta w^{c^*})]\} - U^N \right]$$

$$\times \left[E\{U^{c^*}[y_1^*(\alpha^* w^c, \beta^* w^{c^*}), \ldots, y_N^*(\alpha^* w^c, \beta^* w^{c^*})]\} - U^{N^*} \right].$$

The first order conditions are

$$\frac{\partial V^c}{\partial w^c} = E(AU_y^c)U^* + E(A^* U_{y^*}^{c^*})U = 0$$

$$\frac{\partial V^c}{\partial w^{c^*}} = E(BU_y^c)U^* + E(B^* U_{y^*}^{c^*})U = 0.$$

Solving first order conditions yields

$$\frac{U^*}{U} = -\frac{E(A^*U^{c^*}_{y^*}) - E(B^*U^{c^*}_{y^*})}{E(AU^c_y) - E(BU^c_y)}.$$

The left hand side of this equation is positive. To obtain a positive sign on the right hand side implies that a unilateral change in the policy setting of one country will increase its welfare but necessarily reduce welfare for the other country.

Note that since the coordinated equilibrium is the unique Pareto efficient outcome, it must be welfare superior to the uncoordinated equilibrium. This is in sharp contrast to the deterministic case where there is no incentive to coordinate policies. Hence, one can conclude that multiplier uncertainty gives rise to welfare benefits from coordination.

Finding the optimal policy settings for the w's for the coordinated and uncoordinated regimes, comparing them, and investigating their effects on water reserves requires a model with explicit functional form. However, this generalized framework introduces the importance of uncertainty in identifying incentives for countries to coordinate their activities over common resources.

7. CONCLUDING REMARKS AND FUTURE RESEARCH

The literature on trans-boundary natural resources is mostly developed in the framework of a deterministic environment. Only recently has the issue of uncertainty been introduced in the literature on international policy coordination. However, this development has been only in the area of macroeconomics which naturally focuses on different aspects and has different implications. This chapter discusses the effect of uncertainty on potential gains from trans-boundary water policy coordination. It finds that existence of multiplicative uncertainty is a key condition for gains from coordination while under either certainty or additive shocks there is no incentive to coordinate trans-boundary policies. Further analysis with a more explicit functional form is needed.

REFERENCES

Anderson, L.G. 1975. "Criteria for Maximizing the Economic Yield of an Internationally Exploited Fishery." In H.G. Knight, ed., *The Future of International Fisheries Management*. St. Paul: West.

Barrett, S. 1990. The Problem of Global Environmental Protection, *Oxford Review of Economic Policy* 6: 68-79.

Beltratti, A. 1995. "Consumption of Renewable Environmental Assets, International Coordination and Time Preference." In C. Carraro and J.A. Filar, eds., *Control and Game-Theoretic Models of the Environment,* Annals of the International Society of Dynamic Games. Boston: Birkhauser.

Biswas, A.K. 1993. Management of international waters: problems and perspective, *Water Resources Development* 9: 167-188.

Caldwell, L.K. 1990. *International Environmental Policy; Emergence and Dimensions.* Second Edition. Durham: Duke Press Policy Studies and Duke University Press.

Canzoneri, M. and D. Henderson. 1991. *Monetary Policy in Interdependent Economy: A Game Theoretic Approach.* Cambridge: MIT Press.

Carraro, C. and J.A. Filar, eds. 1995. *Control and Game-Theoretic Models of the Environment,* Annals of the International Society of Dynamic Games. Boston: Birkhauser.

Cooper, R.N. 1985. "Panel Discussion: The Prospects for International Economic Policy Coordination." In W.H. Buiter and R.C. Marston, eds., *International Economic Policy Coordination.* Cambridge: Cambridge University Press.

Dinar A. and E.T. Loehman, eds. 1995. *Water Quantity/Quality Management and Conflict Resolution.* Westport: Praeger.

Esogbue, A.O., ed. 1989. *Dynamic Programming for Optimal Water Resources System Analysis.* Englewood Cliffs: Prentice Hall.

Fieleki, N.S. 1988. *The International Economy under Stress.* Boston:Ballinger Publishing Company. Fedearl Reserve Bank of Boston.

Folmer, H. and C. Howe. 1991. Environmental Problems and Policy in the Single European Market, *Environmental and Resource Economics* 1: 17-41.

Folmer, H. and I. Musu. 1992. Trans-Boundary Pollution Problems, Environmental Policy and International Cooperation: An Introduction, *Environmental and Resource Economics* 2: 107-16.

Ghosh, A.R. and P.R. Masson. 1994. *Economic Cooperation in an Uncertain World.* Oxford: Blackwell.

Helm, D. ed. 1991. *Economic Policy Towards the Environment.* Oxford: Blackwell.

Hoel, M. 1992. "Emission Taxes in a Dynamic International Game of CO_2 Emissions." In R. Pethig, ed., *Conflicts and Cooperation in Managing Environmental Resources* Berlin: Springer-Verlag.

Just, R.E., N. Bockstael, R.G. Cummings, J. Miranowski, and D. Zilberman. 1991. "Problems Confronting the Joint Formulation of Commercial Agricultural and Resource Policies." In R.E. Just and N. Bockstael, eds., *Commodity and Resource Policies in Agricultural Systems* Berlin: Springer-Verlag.

Kaitala, V. 1993. Equilibrium in a Stochastic Resource Management Game under ImperfectInformation, *European Journal of Operational Research* 71: 439-53.

Kaitala, V., M. Pohjola, and O. Tahvonen. 1992. Trans-Boundary Air Pollution and Soil Acidification: A dynamic Analysis of an Acid Rain Gain Between Finland and the USSR. *Environmental and Resource Economics* 2: 161-81.

Karp, L. and S. Sachetti. 1996. Limited Cooperation in International Environmental Agreements. Working Paper No. 763, Department of Agricultural and Resource Economics, University of California, Berkeley.

Kneese, A.V. 1971. *"Background for Economic Analysis of Environmental Pollution"* In P. Bohm and A.V. Kneese, eds., *The Economics of Environment.* London: Macmillan.

Maler, K.-G. 1989. "The Acid Rain Game." In H. Folmer and E. vanIerland, eds.. *Valuation Methods and Policy Making in Environmental Economics* Amsterdam: Elsevier.

Maler, K.-G. 1991. "Incentives in International Environmental Problem." In H.Siebert, ed., *Environmental Scarcity: The International Dimension* Kiel, FGR: Institute fur Weltwirtschaft an der Universitate Kiel.

Munro, G.R. 1979. The optimal Management of Trans-boundary Renewable Resources, *Canadian Journal of Economics* 12: 355-76.

Niehans, J. 1968. Monetary and Fiscal Policies in Open Economies Under Fixed Exchange Rates: An Optimizing Approach, *Journal of Political Economy* 76: 893-920.

Nicolaisen, J., A. Dean, and P. Hoeller. 1991. Economic and the Environment: A Survey of Issues and Policy Option, *OECD Economic Studies* 16: 7-43.

Ostrom, E., R. Gardner, and J. Walker. 1994. *Rules, Games, and Common-Pool Resources* Ann Arbor: The University of Michigan Press.

Safadi, R. and P. Low. 1991. International Policy Coordination and Environmental Quality Symposium on International Trade and the Environment. Washington, D.C: The World Bank.

Spulber, N. and A. Sabbaghi. 1994. *Economics of Water Resources: From Regulation to Privatization* Boston: Kluwer Academic Publishers.

Theil, H. 1964. *Optimal Decision Rules for Government and Industry,* Studies in Mathematical and Managerial Economics. Amsterdam: North-Holland Publishing Company.

Ulph, A. and D. Ulph. 1995. *Who Gains from Learning about Global Warming?* Italy: Fondazione Eni Enrico Mattei.

United Nations. 1975. Management of International Water Resources: Institutional and Legal Aspects. Natural Resources/Water Series No. 1. New-York: The United Nations.

Van der Ploeg, F. and A.J. De Zeeuw. 1992. International Aspects of Pollution Control, *Environmental and Resource Economics* 2: 117-139.

Ward, H. 1993. Game Theory and the Politics of the Global Commons, *Journal of Conflict Resolution* 37: 203-235.

World Commission on Environment and Development. *1987. Our Common Future*. Oxford: Oxford University Press.

22 TRANS-BOUNDARY WATER ALLOCATION BETWEEN ISRAEL AND THE GAZA STRIP: DESALINATION, RECYCLING, AND FRESH WATER

Eyal Brill*
Hebrew University of Jerusalem, Israel
Ujjayant Chakravorty
University of Hawaii at Manoa, USA
Eithan Hochman
Hebrew University of Jerusalem, Israel

The availability of water in the Gaza Strip in relation to its population size and climatic conditions results in a water scarcity problem almost unparalleled in the world. Population density in the Gaza Strip is estimated at 2,000 people per square kilometer (compared to 200 hundred in Israel). In 1991, average annual water consumption was 35 cubic meters per capita (compared to the Israeli urban consumption of about 100 cubic meters per capita). Israeli-Palestinian cooperation over integrated water management addresses multiple regional objectives. It captures the utilization of multiple water resources such as fresh, desalinated, and recycled water, by multiple users in the agricultural, urban, and industrial sectors. Conveying water from various sources to users depends on spatial distribution. In order to address the problem, a conceptual model based primarily on the theory of optimal control is used to derive optimal solutions. Empirical data and computer simulation assist in the estimation of several solution scenarios.

1. INTRODUCTION

Water is scarce both for Israel and for its neighbors. With the recent establishment of the Palestinian entity in parts of the West Bank and the Gaza

* The authors are grateful to the Department of Agricultural and Resource Economics, University of Maryland, for support that led to the completion of this paper. At the time of this study, Eyal Brill held a post doctoral appointment and Eithan Hochman was visiting professor, both at the University of Maryland, College Park.

Strip, the problem of water allocation between Israel and its Palestinian neighbor and of alternative uses within Israel has become severe, influencing the ongoing Middle East peace negotiations. However, while the water problem in Israel currently affects only agricultural users, the availability of water in the Gaza Strip in relation to its population size and climatic conditions results in a water scarcity problem almost unparalleled in the world. Population density in the Gaza Strip is around 2,000 people per square kilometer and annual water consumption is 35 cubic meters per capita.[1] The water constraint as it is interpreted in its quality and quantity dimensions affects public health and environmental issues. Its solution is a major element in the success of any peace agreement in the Middle East.

There is a host of important technological, economic, and political issues that need to be considered explicitly in any systematic analysis of the problem. First, Israel is currently using all of its potential fresh water supply. Current rates of use far exceed estimates of aquifer "safe yields," leading to a continuous rise in salinity levels in most regions of the country (Yaron 1994). High land prices both in the center and in the north of Israel, combined with financial crises in the agricultural sector (see Zusman, Kislev, and Lerman 1988), have resulted in a reduction in the amount of land allocated to agriculture in these areas. During the 1990s Israeli agriculture has been characterized by a move southward in search of alternative agricultural sites. Some major crops, such as citrus, have already moved to the north of the Negev, an arid area, mostly unpopulated. Ensuring future food security will require Israel to make significant water quantities available to agriculture in the arid Negev, using mainly recycled water.

Second, both Gaza and the West Bank rely on Israel for access to underground aquifers that historically have been used mostly by Israel for meeting its domestic needs. Both Gaza and the West Bank are densely populated urban centers located in an arid environment. Rising demand in Israel and in Palestinian regions as a result of population growth and scarcity of fresh water suggests the need to look into the economics of two non-exclusive technological options: (i) desalination of water for household use, and (ii) recycling of urban water for agriculture use.

Water sharing has been a major issue in the ongoing Middle East peace negotiations, and there are a large number of policy studies available on the subject. This chapter examines potential cooperation between Israel and the Gaza Strip in solving water problems on both sides of the border. The analysis uses a stylized model of allocation over space that incorporates the following features:

- The longitudinal shape of the area, with fresh water sources located in the watersheds in the north, such as Lake Kinneret and the Mountain Aquifer, and water carried from the north to the south via a national aqueduct, the National Water Carrier (NWC);
- The location of agricultural areas around the NWC and the location of urban centers around Tel Aviv and Gaza at the southern end of the NWC;

[1] Israel's population density is around 200 people per square kilometer and annual water consumption is 100 cubic meters per capita.

- The generation of supplemental water in the south through desalination of brackish water utilizing Gaza's saline water sources; and
- The recycling of water from urban center uses to agriculture.

These issues are well known to policy makers in Israel and the Palestinian entity, as well as to those directly involved in the bilateral negotiations. However, no integrative analysis that links the spatial dimension to the broader question of alternative supply options and intersectional and intercountry water allocation has been done so far. Important policy questions include the following:

- Are desalination and recycling economically feasible?
- What proportion of agricultural needs must come from recycled urban use?
- How much water would become available for sharing with the Palestinian entity and Jordan under alternative institutional arrangements?

These policy issues were succinctly summarized in a recent editorial in the conservative Jerusalem Post (1997): "The question is not whether Israel will get the water it needs but from where: aquifers, recycling, waste treatment or desalination."

This chapter addresses the above problem by developing a stylized, spatial model of water allocation between agricultural and urban uses. Spatial models which determine water prices over space under distribution losses, and which compute optimal investment in loss reduction at each location have been developed by, for example, Chakravorty, Hochman, and Zilberman (1995). We modify this approach by incorporating the cost of pumping water over distance, the generation of water through alternative technologies such as desalination, and the reuse of recycled urban water in agriculture. For the sake of simplicity, costs of water distribution from the NWC to end-users were not included. This approach is based on the fact that conveyance facilities are already in place in Israel so conveyance costs within regions may be considered small compared to the cost of pumping water over relatively large distances.

The chapter is structured as follows. Section 2 provides the background for the water problem in Israel and the Gaza Strip. Section 3 develops a simple spatial optimal control model of water allocation. Section 4 discusses the characteristics of a cooperative solution between the parties. Section 5 describes and analyzes three scenarios of cooperation between Israel and the Gaza Strip. Section 6 concludes the chapter by highlighting policy implications of particular water allocation patterns on water sharing between Israel and Gaza.

2. BACKGROUND AND MOTIVATION

Both Israel and the Palestinian entity face severe water shortage problems. In Israel, all the fresh water potential is currently being utilized. By the year 2010, the annual deficit is estimated to be 200-600 million cubic meters (MCM), depending on whether costly investment in generating new water supplies is made (Tahal 1988; Nevo 1992). Urban demand in Israel has steadily increased in recent years, partly because of the large migration of Jews from the former Soviet Union. In the near future, the Israeli urban population in the north and along the coastal area will consume most of the annual fresh water generated by sources north of the Sea of Galilee. Under such circumstances, the likelihood of maintaining the southern part of the NWC is diminished unless there is regional agreement among Israel, Jordan, and the Gaza Strip to utilize the NWC to convey water from the Sea of Galilee to Gaza. Future increases in demand for irrigation water can only be met through large-scale use of recycled water in agriculture. Current predictions suggest that in the years 2000-2010 about 35 to 50 percent of all water used by the agricultural sector will be treated wastewater from the urban sector. Israel's recycled water sources can provide only a portion of its irrigation needs. The Negev population is not expected to grow dramatically over the next ten years. Therefore, the quantities of recycled sewage water Israel can be expected to produce would be insufficient to satisfy the demand of the agricultural sector in the Negev.

The gap between supply and demand is even more acute in the Palestinian territories. The safe yield of the coastal sub-aquifer, which supplies water to the Gaza Strip, is estimated at 60 MCM per year. Current usage rates are in the range of 120-150 MCM per year (Awartani 1990). This excessive use has resulted in seawater intrusion, leading to salinity levels of around 400 ppm of chlorides (Goldberger 1992). Salinity levels are rising annually by almost 10-20 milligrams per liter of chloride and, in most parts of Gaza, water is not fit for drinking. Increased per capita consumption together with population growth is expected to increase water demand in the West Bank and Gaza to about 450-500 MCM per year by the year 2010. The Palestinian water problem is amplified with prices of desalinated water ranging between $0.75 to $1.20 per cubic meter. Given these prices, desalinated water will be beyond the reach of the vast majority of the population. However, if desalinated water can be consumed sequentially, first as fresh water for households and second as recycled sewage for agriculture, its cost can become affordable.

Sharing of water resources is one of the main issues under discussion in negotiations under the framework of the Oslo Agreement. Under the agreement, Israel recognizes water rights of the Palestinians and both parties agree to undertake follow-up negotiations. However, most Middle East water analysts concur that, in an environment of acute water scarcity, an agreement to share the major aquifers and water basins will be difficult to achieve (Cohen 1996). Thus, it is highly likely that sustainable cooperation between the two parties will include the generation of new water resources through desalination and recycling.

In order to make additional water supplies available to the population in Gaza, desalination of brackish water seems to be the only viable solution in the long run. Several other large-scale projects are under active consideration. Some, such as (i) moving Mediterranean or Red Sea water to the Dead Sea (which is 400m below sea level) to generate energy for desalination, and (ii) importing water from other regions such as Turkey, may have little chance of realization in the near future. At current prices, desalinated water is not economical even for higher-valued domestic uses. However, if desalinated water is first used in the domestic urban sector and then recycled for agriculture, it could become affordable. This chapter develops a stylized model of water allocation in Israel with fresh water generated in the north (Sea of Galilee). The water is pumped to the south via the National Aqueduct (see Figure 1). Some of the water is used for agriculture along the NWC. In the south, the residual fresh water from the north is supplemented with desalinated water. This water is used by the urban sector of the Gaza Strip. Next, the urban sewage is recycled and pumped back for reuse by the agricultural sector. This model, although highly stylized, is used to examine key technological and policy options that will affect water allocation between Israel and the Palestinian territories.

3. THE MODEL

The schematic framework of the model is shown in Figure 1. A fresh water source is located at point A. Water is pumped from A via the aqueduct to an urban center located at B. Water is withdrawn from the canal for use on agricultural land along AB. Residual water from A is supplemented by desalinated water for use by households at location B. Urban area sewage is then recycled. A fraction of the water is lost during recycling, but the remaining is pumped back in the direction BA for agricultural use. Thus farmers located closer to A are likely to use fresh water while those located closer to B may use recycled water for irrigation. The model will determine the exact boundary between the two groups.

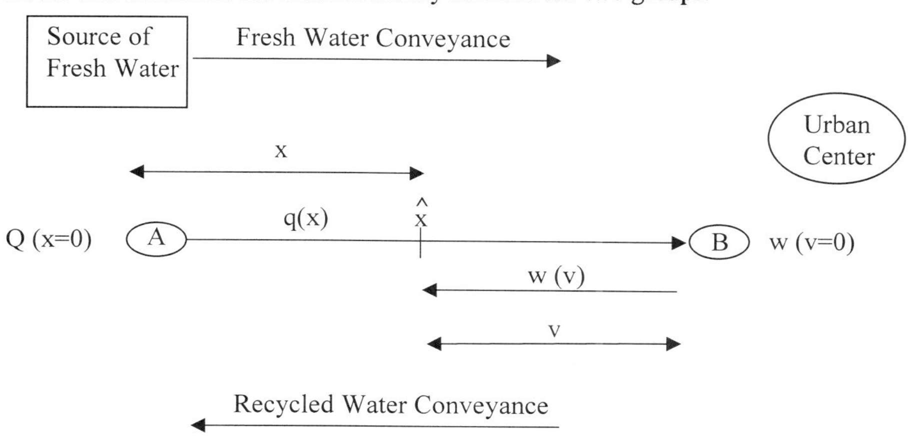

Figure 1. The Schematic Model

Model notation is as follows. Let the amount of water generated at A be denoted by z_0 which is endogenously determined. Agricultural users are located along a continuum from A to B and withdraw q(x) units of water at each location x, where x is the distance measured from A along AB. The production function for agriculture is given by f(q), where f has the usual neoclassical properties: $f' > 0$, $f'' < 0$.

The width of the farmed area is assumed to be a constant α. The cost function for generating z_0 units of water is given by $C(z_0)$, and has the usual properties, i.e., $C'(z_0) > 0$, $C''(z_0) > 0$. Let the residual amount of water flowing at any location along AB be denoted by $z(x)$. Then the cost of pumping water at each location is $mz(x)$, where m is the constant unit cost of pumping.

At the urban center B, S units of water are desalinated at a unit cost of a, where S is a choice variable to be determined. The residual water from farming, $z(X)$, where X is the fixed length between A and B, and the desalinated water, S, are for urban and municipal uses at a marginal value given by an inverse demand function $D_u^{-1}[S + z(X)]$. The water from the urban center, $W_0 = \beta[S + z(X)]$, where $0 < \beta < 1$ is the coefficient of return, is recycled at a unit cost of b, and the recycled water, W_0, is pumped back into the canal in the direction of A. Let v denote distance from B along BA. Users located along v withdraw $w(v)$ units of recycled water from the canal. Let $\hat{X}$ be the location which divides the agricultural region (see Figure 1). That is, farmers to the left of $\hat{X}$ use fresh water and farmers to the right of $\hat{X}$ use recycled water. Let Y be the aggregate agricultural output from utilizing water for irrigation along the distance x, i.e.,

$$Y = \int_0^X f(q(x))\alpha dx .$$

Then the inverse demand function for agricultural output is given by $D_a^{-1}(Y)$. Using the change of variables defined by $v = X - x$ implies $dv = - dx$, so that $v = 0$ implies $x = X$, and $v = X - \hat{X}$ implies $x = X - (X + \hat{X}) = \hat{X}$. The social planner's net benefit (NB) optimization problem can be written as

$$\max_{q(x),z(X),z_0,S,\hat{X},w(v)} NB = \int_0^{\hat{X}} \{D_a^{-1}(Y)f[q(x)]\alpha - mz(x)\}\, dx - C(z_0)$$

$$+ \int_0^{X-\hat{X}} \{D_a^{-1}(Y)f[w(v)]\alpha - mw(v)\}dv$$

$$+ \int_0^{S+z(X)} D_u^{-1}(\tau)d\tau - aS - b\beta[S + z(X)] - mz(\hat{X})(X - \hat{X}) \quad (1)$$

where the first integral represents the area under the demand curve for agricultural products less the cost of pumping water in the region that uses fresh water. Similarly, the second integral denotes a corresponding measure for the region between X and $\hat{X}$ that uses recycled water. The last integral represents benefit to urban consumers. The remaining terms denote the cost of water generation, the cost of desalination and recycling, and finally the cost of pumping the residual fresh water from $\hat{X}$ to X to be used by the city.

The objective function in (1) is maximized subject to the following constraints: (i) the amount of recycled water available,

$$\int_{X}^{\hat{X}} q(X-x)\alpha dx \le W\,, \tag{2}$$

which describes the total amount of fresh water used for agriculture along AB, (ii) the equation of motion,

$$\dot{z}(x) = -q(x)\alpha, \;\; 0 \le x \le \hat{X}\,, \tag{3}$$

which describes the residual amount of fresh water along AB, and (iii) the equation of motion,

$$\dot{w}(v) = -q(v)\alpha, \;\; 0 \le v \le X-\hat{X}\,, \tag{4}$$

which describes the residual amount of recycled water along BA. Alternatively, (4) can be expressed as

$$\dot{w}(X-x) = -q(X-x)\alpha, \quad \hat{X} \le x \le X\,. \tag{5}$$

The Hamiltonian for the above problem is

$$H = D_a^{-1}(Y)f(q)\alpha - mz(x) + D_a^{-1}(Y)f[q(X-x)]\alpha - mw(X-x)$$

$$-\lambda(x)q(x)\alpha - \theta(X-x)q(X-x)\alpha\,, \tag{6}$$

where θ and λ are the usual co-state variables attached to the equations of motion in (3) and (5). The first order conditions are

$$D_a^{-1}f'(q) = \lambda(x), \quad 0 \le x \le \hat{X}\,, \tag{7}$$

$$\dot{\lambda}(x) = m, \quad 0 \le x \le \hat{X}\,, \tag{8}$$

$$D_a^{-1}(Y)f'[q(X-x)] = \theta(X-x), \quad X \le x \le \hat{X}\,, \tag{9}$$

$$\dot{\theta}(X-x) = m, \quad X \le x \le \hat{X}\,, \tag{10}$$

$$D_u^{-1}[S+z(X)] = \alpha + b\beta\,. \tag{11}$$

Solving equations (8) and (10) yields

$$\lambda(x) = \lambda(0) + mx\,, \tag{12}$$

$$\theta(X - x) = mx + \theta(X)\,. \tag{13}$$

Conditions (12) and (13) suggest that both shadow prices increase linearly with distance as shown in Figure 2. Thus, it follows that fresh water allocated to agriculture will decrease as it moves away from its source, and recycled water allocation will also decrease as it moves away from the city. The latter condition implies that $x = 0$, $\theta(X) = \theta(X)$, $\theta[X - (X - \hat{X})] = m(X - \hat{X}) + \theta(X)$. Finally, the transversality conditions are obtained by differentiating the objective function in (1),

$$\theta(X) = -\frac{\partial(NB)}{\partial(z_0)} = C'(z_0)\,, \tag{14}$$

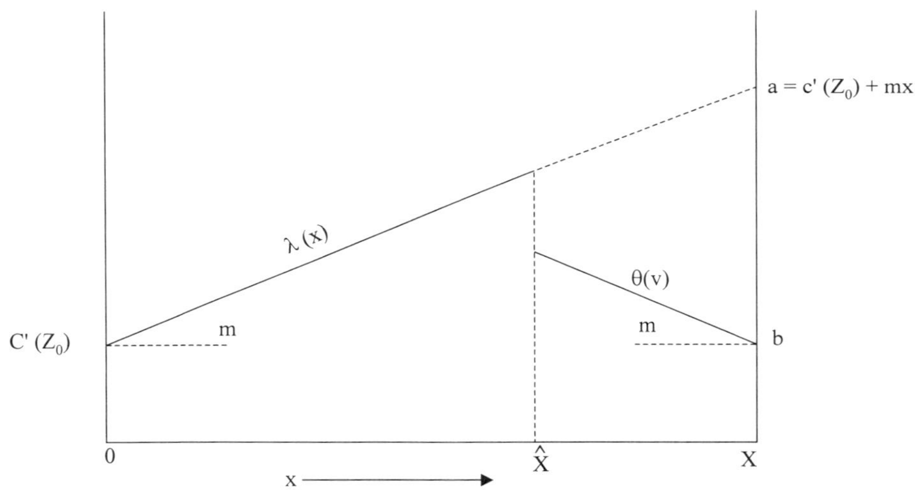

Figure 2. Distribution of Shadow Prices Over Space

$$\theta(X) = -\frac{\partial(NB)}{\partial W} = b\,, \tag{15}$$

$$\lambda(X) = \frac{\partial(NB)}{\partial z(X)} = D_u^{-1}[S + x(X)] - b\beta\,.$$

Using (11), (15) becomes

$$\lambda(X) = a + b\beta - b\beta = a. \tag{16}$$

Using (14) and (16) in (12) yields

$$\lambda(X) = C'(z_0) + mX = a \Rightarrow C'(z_0) = a - mX.$$

Therefore,

$$\lambda(\hat{X}) = C'(z_0) + m\hat{X} = a - m(X - \hat{X}).$$

Similarly,

$$\theta(\hat{X}) = m(X - \hat{X}) + \theta(X) = m(X - \hat{X}) + b.$$

The boundary between farmers using fresh water and those using recycled water is given by $\hat{X}$, which is determined by differentiating (1) and noting that at $x = \hat{X}$ and $v = X - \hat{X}$,

$$\frac{\partial(NB)}{\partial\hat{X}} = D_a^{-1}(Y)f[q(\hat{X})]\alpha - mz(\hat{X})$$

$$-D_a^{-1}(Y)f[q(X - \hat{X})] - mw(X - \hat{X}) + mz(\hat{X}) = 0. \tag{17}$$

The first two terms in (17) represent rents to fresh water at $x = \hat{X}$. The second two terms denote rents to recycled water. Equation (17) suggests that at the boundary, $\hat{X}$, there is a wedge between the two rents equal to the marginal cost of pumping the residual amount of water $z(\hat{X})$ one more unit of distance (see Figure 3). At $\hat{X}$, the rent from recycled water denoted by R_c is higher than the rent from fresh water denoted by R_f. Thus the boundary is closer to the city than at the location where the two rents are equal. This makes sense intuitively because the cost of transporting residual fresh water to the city must be considered. Farmers are allocated fresh water up to a distance of $\hat{X}$ even though the rents from fresh water are lower than from recycled water.

Several additional insights can be derived from the above solution.

- If urban demand is strictly positive at all prices, then there will always be a positive amount of water available for recycling, so $\hat{X} < X$, i.e., a portion of the agricultural land will always be supplied with recycled water.
- If recycling of urban wastewater has environmental benefits, then $\hat{X}$ will be smaller and thus more land will be supplied by treated wastewater than is shown in Figure 3.
- If pumping costs are high compared to the costs of water generation, i.e., m >> C'(z0), then z(X) may be close to zero, and $\hat{X}$ may be close to zero, in which case most of the land will be supplied by wastewater irrigation. On the other hand, if, for example, pumping costs are low,

then the residual flow, z(X), may be large; the amount of desalination, S, may be low; and $\hat{X}$ may move to the right. Pumping costs may be low due to low fuel costs or a high cost of generating fresh water supplies.

- Although we have assumed only one type of crop is produced, the model can be extended for multiple cropping. In this context, one can consider two types of crops, described as tolerant and sensitive to salinity. Then the crop that is sensitive to water quality could be grown under fresh water agriculture in the north, while salinity-resistant crops may be grown with wastewater in the south. Thus, our model predicts that higher-valued crops may be grown in the north while water-intensive crops should be grown in the south, close to the Negev. This also suggests a possible extension of the model, with quality differences reflecting fresh water, desalinated water and recycled water. For example, if the quality of the recycled water is lower than that of fresh water, then $\hat{X}$ may move to move to the right because rents to production from recycled water will be lower than shown in Figure 3.
- Although this model is deterministic, the effects of weather and supply variability can be observed directly. For example, in a drought year, fresh water availability may decrease, in which case the cost of generating fresh water will go up. This would lead to a lower $\hat{X}$, i.e., less fresh water agriculture; lower z(X), i.e., less residual flow; and an increase in desalination for urban and agricultural use. On the other hand, in a year with abundant rainfall, less water may be desalinated.

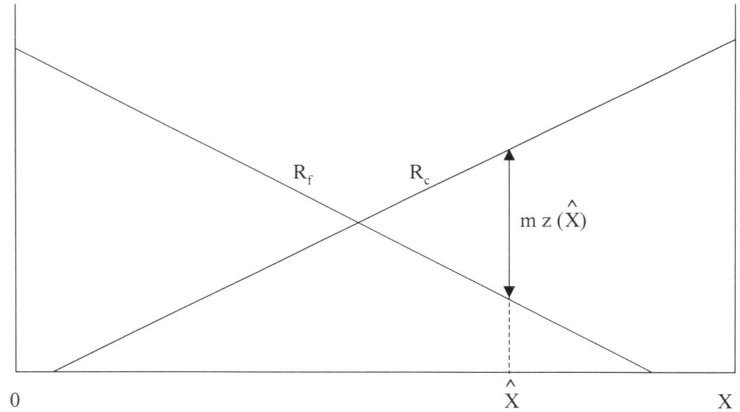

Figure 3. Distribution of Rent Over Space

4. FEASIBILITY AND STABILITY OF THE OPTIMAL SOLUTION

The previous section introduced the conditions for optimal allocation of water between the urban and the agricultural sector along a canal. However, the framework has several flaws.

The first flaw of the model, in which the optimal control solution differs from conditions in reality, is the assumption of the existence of a central planner who is able to calculate the optimal solution and impose it on participants. In a trans-boundary allocation problem, such a planner likely does not exist. Hence, optimal allocation if achieved has to be based on economic incentives for both sides, i.e., both participants must have substantial gains from cooperation. It is important, therefore, to identify the conditions under which both participants have maximum incentives for cooperation.

Second, inter-temporal allocation of water must take into consideration the natural recharge of water at the head of the canal. The cost function $C(Z_0)$, which represents the costs of generating water at the head of the canal, impacts optimal allocation along the canal but is myopic with respect to intertemporal water allocation. A key element is whether cooperation will increase or decrease the quantity of fresh water, Z_0, generated at the head of the canal. If cooperation results in an increase in Z_0, hydrological constraints imposed by natural recharge might turn the optimal solution into a non-feasible one. On the other hand, if cooperation results in a decrease in Z_0, it may improve the water balance over time and, therefore, contribute to the stability of the solution.

A third element that influences the stability of the optimal solution, which is not included in the framework presented in Section 3, is the level of trust between both sides. Palestinians may reject dependency on outside supply of household fresh water because in times of crisis Israel may use it as a threat. However, the costs of desalinating water might be beyond the ability of the average household in the Gaza Strip to pay. Thus, only very large gains from cooperation for the Palestinians may outweigh the uncertainty of out-of-region water supply. Cooperation depends also on Israeli trust in the quality of recycled water from the Gaza. Low quality recycled water from the Palestinians may force Israel to maintain its own recycling facilities and result in high costs and low gains from cooperation.

Several components may affect the motivation of both sides to cooperate. First is the cost of desalination relative to the cost of conveying fresh water to the urban area. In our case, desalination is considered as the backstop technology for fresh water. It is preferred by the urban area if $c'(z_0) + mx > a$. However, if the cost of conveying fresh water to the urban area is smaller than the cost of desalination, i.e., $c'(z_0) + mx < a$, then trading in fresh water will result in surplus for both sides. In such a case, a bargaining process between the parties may determine the distribution of this surplus. The bargaining issue may be the price of fresh water paid by the urban sector (henceforth, w_0), or the price of recycled water paid by the agricultural sector (henceforth, w_1).

The second component, which may influence farmers' motivation for cooperation, is changes in income distribution resulting from cooperation. As will be described in the next section under several different scenarios, the location of and the values of w_0 will affect income distribution and may strengthen or weaken the motivation for cooperation.

5. SIMULATION MODEL

In what follows a simulation model is used to assess the stability and feasibility of the optimal solution. Although simulation is used, some of the data are based on real-world estimation and were designed to reflect reality. The data used for the simulation are summarized in Table 1. The distance between the Sea of Galilee and the Gaza Strip is about 150 kilometers. According to Israeli calculations (see Mosenzon 1986), the cost of conveying water along the Israeli national aqueduct over such a distance is 30-35 cents (m = \$0.35). The cost of generating water at the head of the canal (the Sea of Galilee) depends on the shadow price of water. During wet years with no scarcity of water, when the water level of the Sea of Galilee is above the long-run steady state level, the shadow price is nearly zero. During dry years, when scarcity of water forces reduction in irrigation water for Israeli agriculture, its shadow price is estimated at \$0.30 cents (see Yaron et al. 1996).[2] If the scarcity of water is severe enough to influence household fresh water supply, its shadow price might be even higher. An important question is what can guarantee $c'(Z_0) = 0$ during dry years also. Just et al. (1997) discuss a possible increase in the amount of water available in the Sea of Galilee. Two possibilities are the construction of the Mukheiba dam and, alternatively, storing Yarmouk River water in the Sea of Galilee. Both projects could increase the amount of water in the Sea of Galilee by 180-240 MCM and decrease the costs of generating water at the head of the NWC to zero.

Table 1. A Summary of Cost Data for the Simulation

Item	Model Notation	Value \$
Cost of generating water at the Kinneret	$C'(z_0)$	0-0.30
Transportation costs: Kinneret-Gaza Strip	M	0.30-0.35
Desalination of sea water	A	0.70
Desalination of brackish water	A	0.43
Land-intensive recycling costs	B	0.10
Capital-intensive recycling costs	B	0.30
Shaf-Dan conveyance cost to the Negev	B	0.40

Desalination costs depend on the source of water being desalinated and the desalination technology. According to Abu Mayla and Abu Jaball (1995), the annual cost of desalination plants to supply household water demand (55 MCM) for the year 2000 is estimated as \$24 million. The average cost of \$0.43 per cubic meters is more than twice the highest current household water price in the Gaza

[2] The shadow price is estimated as the average productivity of water in agricultural production.

Strip (hence, a = \$0.43). If seawater is used for desalination, fresh water costs might rise to \$0.70-0.80 per cubic meters.

The cost of recycling water depends on the technology and the distance the water must be conveyed from its source at the urban area to its potential agricultural consumers. Two major types of recycling technologies are being used in Israel: capital-intensive and land-intensive. According to Ben-Meir (1996), the cost of recycling water under the land-intensive technology to a quality safe for agricultural usage is estimated as \$0.06-0.10 per cubic meter, while the cost under the capital-intensive technology (such as in the Shaf-Dan or the Kishon plants) is \$0.30-0.35 per cubic meters. Considering the cost of conveying water from the Shaf-Dan to the Negev, the cost of Tel-Aviv's recycled water used for irrigation in the Negev is \$0.40-0.50 per cubic meters. Alternatively, the cost of Palestinian recycled water used by Israeli agriculture depends on the location of the recycling plants. If Palestinians are responsible for second- and third-stage recycling, land-intensive technology cannot be utilized (because of scarcity of land in the Gaza Strip). In this case, the recycled water price is assumed to be higher than in the case where the Palestinian authority is responsible only for the first stage of recycling. Under such a solution, Israel (utilizing land-intensive technology in the North of the Negev, where land is not scarce) would be responsible for the second and third stages.

5.1 The Simulation Model

In order to keep the algorithm as simple as possible, a discrete model was simulated assuming 15 homogenous regions located along the canal, each of equal width. In the simulation, the continuous variable x is replaced by the index i. Each region is assumed to have a quadratic production function represented by

$$f(q) = q - 0.03q^2 . \tag{18}$$

The derived inverse demand function for water in each region is given by

$$D_a^{-1} = 1 - 0.06q . \tag{19}$$

The price for agricultural products, P, varied from \$0.30 to \$0.50.

The cost function for generating water at the head of the fresh water canal is assumed to be

$$C(z) = 0.01 + 0.0005z^2 , \tag{20}$$

where z is the aggregate quantity of fresh water produced at point A. The cost of desalination was set to \$0.43 and the cost of recycling was set to \$0.10. The transportation cost between regions is assumed to be \$0.015, which yields a cost of

\$0.30 over the entire canal. This corresponds to a transportation cost along the canal of around \$0.30, which is similar to the cost of transportation from the Sea of Galilee to the Negev.

5.2 Urban Water Demand

The urban inverse demand for water is denoted by

$$D_u^{-1} = 500 - 10Q\,, \tag{21}$$

where Q denotes the aggregate quantity of water used by the city. The inverse demand function is highly inelastic and expresses the ultimate demand for drinking water.

5.3 Simulation Results

The simulation model examines three different scenarios as listed in Table 2. The scenarios differ according to the values of the product produced. The first one is with relative low value (P = 0.30), the second with high value (P=0.50), and the third with an intermediate value (p = 0.40). The resulting solutions to these three scenarios are characterized by the marginal cost of transferring fresh water to the urban area, which is less than, greater than, or equal to the costs of desalination, respectively. Detailed information on the scenarios can be found in Appendix A. In each scenario, two solutions are compared. The first is based on the solution as calculated by optimal control where cooperation is examined. In the second, the absence of cooperation results in a solution in which fresh water is generated for irrigation only, while household supply in the urban area is based on desalination only. The rows in Table 2 depict the resulting water quantities of fresh water generated at the head (x = 0), fresh water used for irrigation along $[0, \hat{X}]$, recycled water used for irrigation along $[\hat{X}, X]$, fresh water used at the urban center, and desalinated water used at the urban center.

Table 2. Simulation Results According to the Value of Agricultural Production

Scenario	Scenario 1 P = \$0.30 $c'(z_0) + mx < a$		Scenario 2 P = \$0.50 $c'(z_0) + mx > a$		Scenario 3 P = \$0.40 $c'(z_0) + mx = a$	
Water Quantities	No Cooperation	Optimal Solution	No Cooperation	Optimal Solution	No Cooperation	Optimal Solution
Z_0	88.64	116.98	141.53	134.01	100.00	129.96
Q_a (Fresh)	88.64	67.02	141.53	134.01	100.00	100.65
Q_a (Recycled)	0	24.98	0	24.98	0	24.98
Q_u	0	49.96	0	0	0	29.31
S	49.96	0	49.96	49.96	100.00	20.66

The remainder of this section examines the three scenarios according to their effects on Z_0 and the rent distribution along the conveyer. Under the noncooperative solution, the city always uses desalinated water, while the cooperative solution allows the city to use fresh water and the agriculture sector recycled water.

5.4 Scenario 1

This scenario is characterized by a low value of the agricultural product. The noncooperative solution results in 25 percent of the regions having no irrigation. With the cooperative solution, the aggregate amount of water generated at the head of the canal, z_0, increases from 88.64 to 116.98. This increase occurs because the cost of transferring water to the urban area is less than the desalination cost [$c'(z0) + mx = 0.37 < a = 0.43$; see Appendix A, Table A1]. Hence, 49.96 units of water consumed by the urban center is furnished from the NWC instead of using desalination. The total amount of water used for irrigation is increased from 88.64 to 92.0, out of which 24.98 is recycled water. Note that the amount of fresh water used for irrigation is decreased. As a result, the flow and transportation costs are greater at each point along the canal. Farmers in the regions near the urban area are better off as a result of cooperation because they are able to buy water at a price that is lower than under the noncooperative scenario. However, farmers in the regions near the source of fresh water are worse off as a result of cooperation because they buy less water at a higher price. Note that an increase in z_0 can violate hydrological constraints on the aquifer level dictated by water conservation goals. Also, farmers at the head of the conveyer may form a lobby group opposing cooperation.

5.5 Scenario 2

This scenario is characterized by a high value market for agricultural products. Accordingly, the noncooperative solution is characterized by utilization of fresh water for irrigation. The aggregate quantity of fresh water generated at the head is increased under the noncooperative solution of Scenario 2. Hence, the cost of fresh water along the conveyer is relatively high, and greater than desalination costs at the boundary point of the urban center [$c'(z_0) + mx > 0.43 = a$; see Appendix A, Table A2]. In this case, cooperation results in a decrease in the aggregate amount of water generated at the head of the canal, from 141.53 to 134.01. This decrease is a result of using recycled water for irrigation without using fresh water for urban supply. The cost of water at each point along the canal is lower under cooperation relative to the noncooperative case, and as a result all farmers are better off. Note that the urban sector furnishes its supply of water from desalination under both alternatives. However, it can benefit from selling its recycled water to farmers.

5.6 Scenario 3

The third scenario is a combination of scenarios 1 and 2. Thus, under cooperation, water supply to the urban area is based on conjunctive use of fresh and desalinated water [S = 20.66 and z(X) = 29.31; see Appendix, Table A3]. In this case, the effect on z_0 and the rent distribution may work in both directions.

As can be seen from the above scenarios, farmers may be highly motivated for cooperation under Scenario 2, in which their rent distribution is preferred, and may reject cooperation under Scenario 1, in which their rent is worse for some farmers. The urban sector may be highly motivated for cooperation under Scenario 1, in which fresh water is supplied by the agricultural sector at a cost less than desalination. Note that these results depend heavily on the price w_0 charged by the agricultural sector for fresh water supplied to the urban sector, and on w_1 charged by the urban sector for recycled water supplied to the agricultural sector.

6. CONCLUSIONS

This chapter discusses the problem of trans-boundary allocation of water between Israel and the Gaza Strip. A stylized model of allocation over space is used in order to formalize the problem. The solution is obtained by solving the problem of allocation over space by using an optimal control method. Finally, using the optimal trajectory of the shadow prices, a yardstick of Pareto efficiency for the agreement between the two parties is given.

The main contribution of the chapter is twofold:

- It develops a model of intersectoral water allocation that is appropriate for regions with limited fresh water supplies, rising water demand in urban areas, and environmental problems of wastewater disposal.
- It applies the model to water sharing between Israel and the Gaza Strip.

Further work should be done in developing a bargaining solution in a more precise fashion. However, the framework developed here can assist in guiding necessary estimation.

APPENDIX: SIMULATION DETAILED RESULTS

This appendix contains detailed results about the three scenarios discussed in Section 5. For all three scenarios, a = 0.43, b = 0.1, and m = 0.015.

	Non-Cooperative Solution		Cooperative Solution				
	$z_0 = 88.64$ $c'(z_0) = 0.09$	$w(X) = 0$ $S = 49.96$ $z(X) = 0$		$z_0 = 116.98$ $c'(z_0) = 0.12$	$w(X) = 24.9$ $S = 0$ $z(X) = 49.9$		Cost Along the Canal
Region	Fresh Water	z(x)	Fresh Water	Recycled Water	z(x)	w(x)	$c'(z_0) + mx$
1	11.74	76.89	10.17		106.81		0.12
2	10.91	65.98	9.33		97.48		0.13
3	10.08	55.91	8.50		88.97		0.15
4	9.24	46.67	7.67		81.31		0.16
5	8.41	38.26	6.83		74.47		0.18
6	7.58	30.68	6.00		68.47		0.19
7	6.74	23.94	5.17		63.30		0.21
8	5.91	18.03	4.33		58.97		0.22
9	5.08	12.95	3.50		55.47		0.24
10	4.24	8.71	2.67		52.80		0.25
11	3.41	5.30	1.83		50.96		0.27
12	2.58	2.73	1.00		49.96		0.28
13	1.74	0.98		0.21	49.96	0.00	0.30
14	0.91	0.08		1.04	49.96	0.21	0.31
15	0.08	0.00		1.87	49.96	1.25	0.33
16				2.71	49.96	3.12	0.34
17				3.54	49.96	5.82	0.36
18				4.37	49.96	9.36	0.37
19				5.21	49.96	13.74	0.39
20				6.04	49.96	18.94	0.40

Table A1. Scenario 1

	Non-Cooperative Solution		Cooperative Solution				Cost Along the Canal
	$z_0 = 141.53$ $c'(z_0) = 0.14$	$w(X) = 0$ $S = 49.96$ $z(X) = 0$		$z_0 = 134.01$ $c'(z_0) = 0.13$	$w(X) = 24.9$ $S = 49.97$ $z(X) = 49.9$		
Region	Fresh Water	z(x)	Fresh Water	Recycled Water	z(x)	w(x)	$c'(z_0) + mx$
1	11.95	129.58	12.20		121.81		0.13
2	11.45	118.13	11.70		110.11		0.15
3	10.95	107.18	11.20		98.91		0.16
4	10.45	96.73	10.70		88.21		0.18
5	9.95	86.79	10.20		78.01		0.19
6	9.45	77.34	9.70		68.31		0.21
7	8.95	68.39	9.20		59.11		0.22
8	8.45	59.94	8.70		50.41		0.24
9	7.95	51.99	8.20		42.21		0.25
10	7.45	44.54	7.70		34.51		0.27
11	6.95	37.59	7.20		27.31		0.28
12	6.45	31.14	6.70		20.61		0.30
13	5.95	25.19	6.20		14.41		0.31
14	5.45	19.74	5.70		8.71		0.33
15	4.95	14.80	5.20		3.51		0.34
16	4.45	10.35	3.51	1.18			0.36
17	3.95	6.40		5.20		1.18	0.37
18	3.45	2.95		5.70		6.38	0.39
19	2.95			6.20		12.08	0.41
20				6.70		18.28	0.43

Table A2. Scenario 2

	Non-Cooperative Solution		Cooperative Solution				Cost Along the Canal
	$z_0 = 100$, $c'(z_0) = 0.10$	$w(X) = 0$, $S = 49.96$, $z(X) = 0$		$z_0 = 129.96$, $c'(z_0) = 0.13$	$w(X) = 24.9$, $S = 20.66$, $z(X) = 29.3$		
Region	Fresh Water	z(x)	Fresh Water	Recycled Water	z(x)	w(x)	$c'(z_0) + mx$
1	12.50	87.50	11.25		118.71		0.13
2	11.67	75.83	10.63		108.08		0.15
3	10.83	65.00	10.00		98.08		0.16
4	10.00	55.00	9.38		88.70		0.18
5	9.17	45.83	8.75		79.95		0.19
6	8.33	37.50	8.13		71.82		0.21
7	7.50	30.00	7.50		64.32		0.22
8	6.67	23.33	6.88		57.44		0.24
9	5.83	17.50	6.25		51.19		0.25
10	5.00	12.50	5.63		45.56		0.27
11	4.17	8.33	5.00		40.56		0.28
12	3.33	5.00	4.38		36.19		0.30
13	2.50	2.50	3.75		32.43		0.31
14	1.67	0.83	3.13		29.31		0.33
15	0.83	0.00		2.60	29.31		0.34
16				3.23	29.31	2.60	0.36
17				3.85	29.31	5.83	0.37
18				4.48	29.31	9.68	0.39
19				5.10	29.31	14.15	0.41
20				5.73	29.31	19.26	0.43

Table A3. Scenario 3

REFERENCES

Abu Mayla, Y., and M. Abu Jaball. 1995. Desalination for the Gaza Strip. Paper presented at the meeting of the European Network to integrate renewable energy into water production in Pikermi, Greece, January 30-31, 1995.

Awartani, H. 1990. A Projection for the Demand for Water in the West Bank and Gaza Strip. Unpublished manuscript. The World Bank, Washington, D.C.

Ben-Meir, M., and T. Shofer. 1996. Yezreel Valley Sewage Project, *Water and Irrigation* 353. In Hebrew.

Chakravorty, U., E. Hochman, and D. Zilberman. 1995. A Spatial Model of Optimal Water Conveyance, *Journal of Environmental Economics and Management* 28.

Cohen, I. 1996. Water—A Crucial Question in the Israeli-Palestinian Peace Process, *JIPF New Bulletin* 95: 2. Stockholm.

Goldberger, S. 1992. The Potential of Brackish Water in Israel. Committee on Brackish Water, Israel Water Commission.

Jerusalem Post. 1997. The Price of Water. January 15, p. 6.

Just R.E., J.K. Horowitz, and S. Netanyahu. 1997. Peace and Prospects for International Water Projects in the Jordan-Yarmouk River Basin. In D. Parker and Y. Tzur, eds., *Decentralization and Coordination of Water Resource Management.* Boston: Kluwer Academic Publishers.

Mosenzon, R. Water Industry—Economic Aspects of Regulation. 1990. In Y. Kislev, ed., *Technology, Cooperation, Growth and Policy—Researches in Agricultural Economics.* Hebrew University: MAGNES Publishers. In Hebrew.

Nevo, N. 1992. Comments. Paper presented at the Seventh Continuing Workshop on Israel Water Issues, Center for Agricultural Economics, Rehovot. In Hebrew.

Tahal. 1988. Water Master Plan. Interim Report, Tahal Consulting Co., Tel Aviv (in Hebrew).

Yaron, D. 1994. An Approach to the Problem of Water Allocation to Israel and the Palestinian Entity, *Resource and Energy Economics* 16: 271-286.

Yaron, D., R. Spektor, and F. Mayers. 1996. Future Utilization of Water in Irrigated Agriculture in Israel 2010. Rehovot: The Center of Agricultural Research, Hebrew University.

Zusman, P., Y. Kislev, and Z. Lerman. 1989. Credit in the Moshav—Experience and Lessons, *The Economic Quarterly* 141.

PART V

POTENTIAL CONTRIBUTIONS OF ECONOMIC ANALYSIS TO TRANS-BOUNDARY WATER COOPERATION

23 RESOLVING TRANS-BOUNDARY WATER DISPUTES: ECONOMISTS' INFLUENCE ON POLICY CHOICES IN THE UNITED STATES

David Sunding
President's Council of Economic Advisers, USA

Economic analysis of trans-boundary water allocation has yielded some interesting insights. This chapter reviews trans-boundary water allocation issues that are currently of interest to the federal government of the United States, and identifies some of the major decision-makers. Then, the chapter highlights the actual role of economic analysis in the resolution of these disputes. I conclude that while economists have influenced a number of decisions, often by stopping projects with low political support, economists could clearly increase their influence by reorienting their research. The final section of the chapter identifies some promising avenues for further work.

1. INTRODUCTION

The issue of trans-boundary water allocation is an interesting economic problem, and also one with significant real world impacts. A number of chapters in this volume focus on positive analysis of trans-boundary disputes using techniques of cooperative and non-cooperative game theory. Predicting outcomes of bargaining processes is important to the extent that it suggests potential improvements to the bargaining process itself, and shows where outside intervention is welfare improving.

This chapter has three basic objectives. Section 2 reviews trans-boundary water allocation issues that are currently of interest to the federal government, and identify some of the major decision-makers and agencies responsible for resolving these disputes. Section 3 highlights the role of economic analysis in the resolution of these disputes, and present a few success stories. Section 4 suggests how economists might increase their influence in the domestic policy process. In this section, I identify some aspects of trans-boundary water problems that are in need of more intensive research.

2. UNITED STATES GOVERNMENT EFFORTS TO RESOLVE TRANS-BOUNDARY WATER DISPUTES

"Boundary" is of course a relative term, and there are at least two types of boundaries that are of interest with regard to water disputes handled by the United States government: interstate and international boundaries. Both types of boundaries involve jurisdictional issues, and also entail divergent preferences and constraints on transfers of water rights. In the case of interstate boundaries, however, the problem is simplified somewhat by the fact that federal law often provides a single, overriding environmental quality objective as required by, say, the Clean Water Act or the Endangered Species Act.

Managing domestic interstate conflicts over water allocation is a task the federal government takes seriously, and there are several efforts underway to manage specific conflicts. Perhaps the best example of a domestic trans-boundary allocation problem is the Colorado River, where several states are vying to divert water for municipal and agricultural consumption. The river is currently over-allocated, partly as a result of base allocations being set in an above average water year. The various states along the River have reached an interstate agreement setting consumption levels, but also placing strict barriers on transfers among some regions. A federal role is justified here, not just because the river is so dramatically affected by the operation of federal power and diversion facilities, but also because of the national interest in maintaining the quality of the Colorado River. The river provides important recreation opportunities for all citizens, and also provides habitat for endangered and threatened species.

Federal flood control projects involve many of these same interstate issues, and the Clinton Administration has been involved in finding solutions to these problems as well. Recent floods in the Mississippi, Missouri and Red River valleys, and in central California, have focused attention on this important problem. The nature of the interstate conflict is somewhat different for flood control than it is for the allocation of diversion rights because flood control involves different river characteristics. Consider a typical flood control facility, which is a dam with large excess storage capacity even in an average year. Upstream users derive certain benefits from the project, usually diverted surface water supplies and flat-water recreation (the latter being an aspect of project management whose political significance is frequently underestimated by economists). Downstream users have very different preferences, namely reduction in the variability of river flows. All users may also derive some hydropower benefits because electricity is cheaper to convey over long distances than water.

Different regions have different preferences for various aspects of the project. Downstream users wish to reduce the risk of flooding, which entails increasing the variability of the lake level to enhance storage capacity during high-runoff months. Upstream users, who have the smallest travel costs for recreation, and thus the highest amounts demanded, wish to minimize the variation in lake levels to preserve the quantity and quality of boating trips. Recent conflicts over operation of some Pick-Sloan facilities on the Missouri River and construction of the Auburn

Dam on California's American River are examples of these types of inter-regional conflicts between upstream and downstream users.

Another type of domestic trans-boundary water dispute where the federal government is playing an active role involves Native American tribes. These tribes are increasingly asserting their rights to surface water, particularly in the western United States. Tribes wish to use these rights for consumption and economic development, and also to protect aquatic species with cultural significance by improving instream water quality. A good example of the latter phenomenon is the efforts of the Pyramid Lake Paiute Tribe in western Nevada to protect the cui-ui, an endangered species that resides at the terminus of the Truckee River. This fish has cultural significance to the Paiute Tribe, and federal legislation and court decisions have devoted water rights to ensure its protection.

Within this array of domestic trans-boundary water conflicts, there are several characteristics of federal-state interaction. First, while the federal role is certainly important, it also tends to be passive—more characterized by oversight than initiative. States have generally wide latitude in controlling their water resources, even when there are cross-border externalities. The federal role results from its usual function of adjudicating interstate conflicts, but, more importantly, from its statutory responsibility to enforce environmental laws. Federal government agencies do not have the resources to be proactive.

Second, the federal government prefers, whenever possible, to rely on stakeholder "consensus" processes. A good example of this deference occurs with respect to the San Francisco Bay/Delta estuary. There are several reasons for this phenomenon. Obviously, a consensus decision is easier to sell to budget committees and other parties in Washington, making it more likely that agencies can obtain funding for water projects. Further, there is a sense, perhaps exaggerated, that local residents have better information about water supplies and demands than federal regulators. Thus, locally negotiated outcomes have a greater aura of legitimacy than those that originate with the federal government do.

One major problem with local consensus processes is that there are frequently too many issues on the table to expect that the final outcome will be efficient. In the western United States, water has been used as a tool for social planning, and this tendency continues today. Farm labor groups are increasingly injecting themselves into water negotiations in an attempt to stabilize rural employment. Practically, this means that labor unions are voicing opposition to water transfers and marketing because this would entail a reduction, or at least a reallocation of, rural employment. Organic farmers are demanding water subsidies in recognition of the beneficial environmental aspects of limiting pesticide use. Traditional commercial farmers prefer water markets and other tools that minimize dislocations and allow them to obtain alternative supplies. In this sense, it is difficult to limit water negotiations to the topic of water allocation. Rather, water is seen as a means to accomplish other objectives.

The third main feature of the federal-state relationship is that while the federal government plays a relatively small role in making decisions about construction and operation of facilities, it continues to pay most project costs, both for capital investments and for operations and maintenance. Federal projects constructed and managed by the Bureau of Reclamation and the Army Corps of Engineers have three general sources of financing: the federal government, the state government and project beneficiaries. Project beneficiaries receive large subsidies from most federal projects. For example, projects constructed by the Bureau of Reclamation embody a number of different subsidies. These subsidies include interest-free repayment of capital invested in irrigation facilities, a limitation on repayment associated with "ability to pay" guidelines that do not necessarily reflect changing economic or market conditions or individuals' income, and the repayment of costs above an irrigator's estimated ability to pay by using hydropower revenues far into the future. The length of the repayment period is also important in determining the overall magnitude of the subsidy. It is important to note that recreation use of federal water projects is also subsidized, and is often free.

From an economic perspective, these are hardly characteristics of an ideal federal-state relationship. By taking a limited role in deciding project configuration and operation, the federal government is encouraging selection and construction of sub-optimal projects. The federal government is uniquely positioned to speak for the interests of citizens outside the immediate project area whose utility is nonetheless affected by project operation. Relying on local decision making processes tends to result in projects that provide a low level of benefits to people outside the project area, an inefficiency that can be remedied by more aggressive federal action.

The cost share arrangement is also problematic. By ceding much decision making authority to local entities while committing to pay for a majority of project costs, the federal government is creating a moral hazard problem: local agencies have an incentive to select a larger project than is socially optimal. The Water Resources Development Act of 1996 made a small step in the right direction, increasing the local government cost allocation from 25 to 30 percent. However, the cost share problem remains a serious distortion, and federal taxpayers are not being served by current policies.

Even aside from externalities and moral hazard, there are good reasons to believe that negotiations among various state entities, or among interest groups from the affected region, may not result in an efficient solution. As is the case with international disputes, problems of coordination and incomplete information plague these domestic negotiations. Further, environmental quality often plays a more central role in domestic disputes than in international negotiations, where consumption is the key issue. There is thus the possibility that preferences will be more diverse among the parties at the table in domestic negotiations, making it even less likely that an agreement will be reached.

Federal intervention has its own problems, of course. First, local residents may have more accurate information about project effects. More important, however, is the fact that rent seeking distorts federal decision-making, just as it distorts state decisions. In this sense, the possibility of federal "government failure" may outweigh the benefits of more active federal intervention to remedy the problems caused by local decision making. It may also suggest that direct bargaining among interest groups may be superior to decision making by elected officials. These are larger questions, however, and are not limited to water allocation problems. This is an important area for further economic research.

Turning to international issues, the most serious international water conflict facing the United States concerns the Colorado River, which drains (or, more accurately, formerly drained) into the Gulf of California. The United States has recently agreed that it will reduce its diversion of Colorado River water and will even construct a desalination facility at the border to improve water quality.

Freshwater allocation conflicts in the United States are generally localized and require less international coordination than other international environmental problems such as climate change. Most of the environmental problems that are dealt with on a high level within the Administration are true common property resource conflicts; ocean fisheries and the global climate are two good examples. International conflicts over water allocation usually can be dealt with in bilateral or regional negotiations.

There is reason to believe that regional negotiations should often work more effectively in the case of water than in more complex problems like climate change. Water disputes typically involve a smaller number of nations than other environmental problems, and there is generally much better information about the welfare impacts of various policies. Further, there is often the possibility of real welfare gains for all sides upon resolution of water conflicts. In the Middle East, for example, more efficient management of the Jordan River may provide significant benefits to both Israel and Jordan. In the case of climate change, cutting U.S. carbon emissions entails a large and somewhat uncertain redistribution of wealth from the United States to future generations in developing countries.

3. ASSESSING ECONOMISTS' CONTRIBUTIONS TO U.S. POLICY

Economists have had an influence on the resolution of trans-boundary water conflicts, and also on federal efforts to deal with these conflicts. This is not surprising in one sense because federal intervention often occurs in response to conditions of scarcity, an environment where economists feel most comfortable, and where economic tools are most relevant. As usual, economists' comparative advantage is to conduct welfare analysis of various policies, and to focus on incentives.

Ex ante economic analysis has helped end consideration of some inefficient water projects, particularly projects that had weak political support to begin with. For example, the Water Resources Development Act of 1996 contained authorization for 37 water projects. It is unlikely that even a small number of these projects will actually receive appropriations, primarily because the majority of them do not make economic sense; they were included in the bill simply because some representative or senator wanted to send a signal to his or her constituents. Agency review of these projects using economic principles is sufficient to end consideration.

Ex post analysis of failed projects also helps in this regard. Federal water projects have become such an integral part of the western landscape that it is easy to forget how recently most of them were constructed. Inefficient projects are thus resulting in current welfare losses, sometimes of significant magnitude. Pointing out the continuing harm resulting from these projects helps ensure that fewer of them are built in the future. The Central Arizona Project is a case where *ex post* analysis would prove interesting and useful. Economists predicted that the project water would be too expensive to be profitable, and this prediction has largely come true. Pointing out the continuing losses from project operation would help guide future debate and would vindicate the numerous economists who predicted the project's failure long before it was even constructed.

Fortunately, federal law provides for economic analysis of water projects designed to resolve conflicts over allocation. Executive Order 12866 requires cost-benefit analysis of significant federal projects. More important, though, is the requirement that economic analysis of projects affecting water quality should be conducted according to the method described in the federal *Principles and Guidelines* (P&G). The P&G are used by various agencies, including the Bureau of Reclamation and the Army Corps of Engineers, to select water projects according to a net social benefit criterion. Recently, a team of economists analyzed the P&G method, and considered its impacts in a number of specific cases (Zilberman et al. 1996). This study found that, while the basic idea of selecting water projects to maximize net benefit is appealing, there are significant problems with the P&G, and problems with its implementation as well.

This issue is significant because the P&G lays out the basic ground rules for economic input into analysis of potential water projects. The Zilberman study identified several serious flaws with the P&G method:

- Asymmetric treatment of environmental costs and benefits. While environmental benefits assessment is required for project evaluation, these methods are not used to assess the costs of new projects.
- Inappropriate choice of interest and discount rates, and inadequate recognition of uncertainty.
- Little attention to non-structural alternatives such as water marketing or conservation.
- No consideration of project timing, and no recognition of the value of waiting to construct a project.
- No recognition of the opportunity cost of federal funds, either for water projects or for other uses. These considerations suggest that it may also be appropriate to use rates of return in selecting projects, in addition to net benefits.

The Galloway Commission that reviewed federal flood control policy also wrote of the need to reform the P&G. While their analysis was broad and did not identify many specific changes, the efforts of Zilberman and others are important in this regard. The methodological flaws inherent in the current P&G are serious, and affect the quality of economic advice given in the design of capital projects that resolve interstate and intrastate water conflicts.

Rent seeking being a powerful force in American government, however, also implies that many inefficient projects may be built nonetheless. Even if the P&G

were ideal, it is unreasonable to expect that some inefficient projects would still go forward. Perhaps the worst example of an inefficient project that is still receiving serious consideration, and funding, is the Animas-La Plata Project in southern Colorado.

An interesting question is whether it is most useful for economists to identify the first-best outcome (that is, show politicians the way), or take political constraints into account and describe the best project that is politically feasible. My own opinion is that it is better for economists to place a benchmark on the discussion and focus on describing the first-best resource allocation. These analyses raise the cost of rent seeking. Further, economists are often not skilled at political analysis, especially in the context of an individual project. We are not trained in vote counting, and individual political agendas may be hard to discern. The danger in focusing on the first best solution, however, is that economists will be treated as if they are "out of the loop."

4. INCREASING ECONOMISTS' INFLUENCE IN U.S. WATER POLICY

Economists' greatest potential contribution to the resolution of trans-boundary water disputes is accurate and timely welfare analysis. Laying out benefits and costs to all groups is the best way to have an impact on policy choices, and can even change the political dynamic. Political choices about water allocation are, like many other collective decisions, motivated at least as much by the desire to redistribute assets as by efficiency considerations. Identifying losers and winners can significantly influence political outcomes by informing decision-makers and stakeholders about the consequences of various policies.

On the international level, economic analysis can play a large role in increasing visibility within the United States government of trans-boundary water disputes. Economists can help other policy analysts and decision-makers understand the true costs of action and inaction. Political analysts and diplomats are adept at understanding both the political situation in foreign countries, and also understanding how foreign conflicts affect the national security of the United States. What is most often missing from the equation is an accurate understanding of how water disputes affect the well being of various parties abroad. This is the proper role of economic analysis.

There are other, more specific, ways in which economists could increase their marginal productivity with regard to resolving trans-boundary water conflicts, both domestically and internationally. For example, water economists should pay more attention to flood control; often, this is the aspect of water resource management that dominates both policy choices and social welfare.

The economic impacts of hydropower generation are significant, and have been relatively under-researched by water economists. Understanding hydropower in relation to irrigation, flood control and instream quality is an interesting and important problem both in the United States and internationally. Unfortunately, most analysis of hydropower is conducted by regulatory economists, who are more interested in the impact of hydropower regulation on utility markets. An important current problem in the United States, for example, is the impact of electricity market deregulation on the operation of hydropower facilities.

A major problem concerning federal water projects is *ex post* renegotiation of repayment obligations. Often, projects were constructed with the expectation that water and other services would be provided to various user groups; costs were then allocated based on relative benefits. However, some project benefits never materialized, perhaps because of changes in the local economy, environmental restrictions or lack of federal funding. In many important cases, costs shares are only now being changed to reflect actual, rather than expected, relative benefits, and to stop continuing taxpayer losses from project operations. Not surprisingly, such renegotiation has potentially large impacts on various user groups and regions. Economic analysis of cost reallocation is needed to more fully understand how changes in cost allocation affect demand for project services, and also to measure relative costs and benefits. This is, of course, just another role for disaggregated welfare analysis of water project impacts.

Economists generally have a keener understanding of the importance of institutions than do engineers and other professionals, and water economists would do well to maintain their focus on non-structural alternatives to conflict resolution. Water marketing, for example, is an enduring contribution that economists have made to trans-boundary water conflicts in the United States, and is a good example of the social value of improving institutions for resource management. Helping water users create maximal value from given water supplies is potentially Pareto-improving. Identifying such "win-win" solutions, of which water markets and other financial incentives are a good example, increases the likelihood that stakeholders can find lasting solutions to trans-boundary conflicts.

REFERENCES

Goulder, L. 1995. Effects of Carbon Taxes in an Economy with Prior Tax Distortions: An Intertemporal General Equilibrium Analysis, *Journal of Environmental Economics and Management* 22: 271-297.

World Resources Institute. 1997. *The Costs of Climate Protection*. Washington, D.C.

Zilberman, D., R. Griffin, D. Parker, M. Mazzotta, J. Opaluch, W. Hyde, D. Larson, and A. Fisher. 1996. Review of Federal Principles and Guidelines. Report prepared for the U.S. Environmental Protection Agency, University of California at Berkeley.

24

USING BARGAINING THEORY AND ECONOMIC ANALYSIS AS AN AID TO TRANS-BOUNDARY WATER COOPERATION*

Richard E. Just
George Frisvold
Verna Harrison
Joe Oppenheimer
David Zilberman

Economists must consider the practical applicability of their analyses to problems of trans-boundary water problems. Mathematical tools and a specialized jargon complicate communication. In addition, policy makers often have interests dictated by political interests that are inconsistent with economists' objective concept of economic efficiency. Departures from competitive efficiency may be undertaken to meet legitimate social objectives of equity that cannot be reached by other means. Trans-boundary water problems where sovereignty, lack of infrastructure, or weak enforcement mechanisms cause market imperfections are examples. Economic analysis must be conditioned on realities of the policy process to find practical applicability. Perhaps individual economic analyses must serve selfish individual

---

* This chapter is added largely as a summary of the panel discussion which was held in conjunction with the Fifth Meeting of the International Water and Resource Economics Consortium entitled "Conflict and Cooperation on Trans-Boundary Water Resources" co-sponsored by the University of Maryland, Annapolis, Maryland, April 13-16, 1997. The panel consisted of George Frisvold, Chief of the Resource and Environmental Policy Branch, Economic Research Service, USDA, and former Staff Economist, U.S. President's Council of Economic Advisers; Verna Harrison, Assistant Secretary of the Chesapeake Bay and Watershed Program, Maryland Department of Natural Resources; William Matuszeski, Chairman of the Implementation Committee and Director, Chesapeake Bay Program, U.S. Environmental Protection Agency; Joe Oppenheimer, Professor and Associate Chair of Government and Politics at the University of Maryland; and David Zilberman, Professor and Chair of Agricultural and Resource Economics at the University of California, Berkeley. Richard E. Just, Distinguished University Professor, Department of Agricultural and Resource Economics at the University of Maryland, College Park, served as moderator and principal author.

policy maker or agency goals to find usefulness by them. Debate among policy makers where each is informed according to their own selfish interests may reach informed policy choice. Alternatively, if each policy maker or agency rejects available economic analyses because they try to serve comprehensive and objective criteria, then policy choice is economically uninformed.

1. INTRODUCTION

While the studies comprising this book demonstrate a variety of economic theories and empirical approaches to water cooperation problems, the purpose of this chapter is to consider the *practical* relevance of economic analysis to solving trans-boundary water conflicts in reality. Some key questions that must be addressed include the following. Do the optimality tools of economic analysis carry practical potential for facilitating trans-boundary water cooperation or do they simply represent an ideal which is not of practical relevance for one reason or another? What realistic relevance does economics have in solving trans-boundary water conflicts? Are the solutions currently provided by economic analysis useful to policymakers and negotiators? Could they be more useful? How practical are the traditional concepts of economic efficiency and competitive markets? How important or practical are the solutions provided by more modern and sophisticated tools such as control theory and game theory? How much can the solutions reached by policy makers be improved by considering the complexities of uncertainty, information, and spatial variation? Can economists communicate such complexities of their analyses to policymakers sufficiently well to make the analyses useful? Are the results of economic analysis sufficiently reliable to be useful? Are economic analyses too efficiency-oriented to be of use to policymakers who are responding to political pressures determined by distributional outcomes?

Because economic theory has much to say about efficiency and because answers are less clear regarding distribution and equity, economic analysis has accordingly focused relatively more on efficiency. This focus has led some consumers of economic analysis to conclude that economics is too narrow to provide useful answers. Some perceive economists as able to agree on only a few general principles such as "preferences lead to choices" and "Pareto optimality is preferred." These perceptions raise some important questions. First, are such perceptions true? Second, if such perceptions are not true, how do they limit the contribution of economics to solution of practical policy questions? Third, how can these perceptions be changed? This chapter attempts to address these questions in the context of trans-boundary water problems.

2. LACK OF AGREEMENT AMONG ECONOMISTS

A common observation is that economists tend to disagree. They disagree with respect to which data to analyze, how to analyze a given set of data, and the theories to use as a basis for data analysis. Understandably, economists tend to analyze problems for which they have tools, which by consequence leaves aside the problems for which tools are not yet available. This approach may be somewhat like looking for a lost wallet under a lamp post because the light is better (McCloskey, 1985). Alternatively, the tools of economics may represent the best approaches to problems of historical relevance while new tools of relevance to current problems yet await crafting.

In one sense, the professional economics literature represents the debate centered around which tools, concepts, issues and views are of current relevance. The expanding nature of this debate is represented by the evolving scope of trans-boundary water problems into interdisciplinary and spatially diverse settings (Burke et al. 1997), airshed management (Austin et al. 1997), endogenizing ecological linkages (Milon et al. 1997), considering regeneration potential (Lichtenberg and Olson 1997), consideration of multi-quality water systems and recycling (Brill et al. 1997b), and of integrating political-economic aspects of water policy (Tsur and Zemel 1997). Because the political, economic, social, and physical landscape are constantly changing, the tools of relevance to current problems should also be expected to change. Because economics is not a perfect science, it is understandable that several potential answers may emerge to each new question. The associated debate and professional disagreement that has come to be characteristic of economics is likely a healthy approach to sorting out the answers to these questions. Perhaps, economists tend to become comfortable with the associated uncertainty in the midst of debate about which answers will ultimately prove to apply. On the other hand, it is not surprising that users of economic analysis are uncomfortable with this level of uncertainty, or even find the cost of this uncertainty to outweigh the benefits of economic analysis on new issues requiring immediate action.

3. EFFICIENCY VERSUS EQUITY IN WATER POLICY IMPLEMENTATION

With little doubt, policymakers and government agencies are concerned with more dimensions of the problems they face than neoclassical economics addresses. For example, agencies are concerned with fairness, implementability, and ecological matters. In reality, one of the biggest challenges of government is to coordinate social planning. Almost every single issue involves alternatives that have conflicting interests. Few political issues can be solved by individual projects that make everyone better off. Projects typically make some groups better off at the expense of making other groups worse off.

Traditionally, the approach of neoclassical economics has been to assume an ability to redistribute income with lump sum transfers. (Lump sum transfers permit redistribution of income without affecting anyone's marginal behavior.) If lump sum transfers are possible, economists can focus solely on maximizing the size of the social pie (economic efficiency), claim the moral high ground of perfect objectivity, and leave the choice of distribution of the pie solely in the hands of politicians.

Perhaps because economists tend to evaluate all actions from the standpoint of efficiency implications, many policymakers have come to perceive economists as having a disdain for anything other than efficiency. If policymakers perceive economists as having the same answer to every question—namely, efficiency—then the advice of economists tends to be viewed as narrow and, accordingly, is undervalued.

Even on reasonably sophisticated economic grounds, however, economic efficiency is viewed as a limited economic concept that must be supplemented with equity considerations (Just 1988). The approach of studying efficiency issues in isolation from equity concerns has long been criticized because of the economic infeasibility of defining lump sum transfers that truly do not affect the marginal behavior of economic agents as well as because of the political and administrative infeasibility of lump sum transfer schemes (Layard and Walters 1978). When transfers are made, they can affect market equilibrium simply by changing expectations about future transfers related to other policy issues even in the rare case where they do not directly effect intensity of efforts to maximize profits or satisfy consumptive wants.

The two economic justifications usually given for policy intervention are (1) to improve Pareto efficiency by correcting distortions and (2) to redistribute income (Rausser 1982). Only a very naive view of the policy process can ignore the intent of policies to redistribute income. But if policies have specific goals of income redistribution and if equity and efficiency are not separable issues, as they are not when lump sum transfers are not possible, then competitive efficiency analyses address the wrong question for those policy purposes. Policies that improve efficiency according to the competitive norm may indeed be inferior according to the preferences of policymakers.

Some have argued that efficiency analysis is still important in measuring the social cost of attaining certain distributional results. However, this argument also fails when efficiency and distribution are not separable issues. The competitive efficiency norm is only relevant in comparing alternative distributions if lump sum transfers are possible so that any desired distribution can be achieved once the efficiency norm is reached. Otherwise, alternatives must be evaluated according to the preferences that matter.

Given that policymakers are interested in both efficiency and distribution, economists must entertain the hypothesis that policies currently in effect which are economically inefficient achieve their intended effects better than the policies that enforce the competitive efficiency norm. It is naive and unrealistic for economists to expect economic analysis to make an important contribution in policy formulation until these considerations are made. Otherwise, the scope of applicability will likely be limited to cases where either (1) distributional considerations are not important, (2) important policy preferences happen to be consistent with equal welfare weightings, (3) the separation of equity and efficiency happens to be small so that lump sum redistribution is approximately feasible, or (4) the results of economic efficiency analysis happen to support the policy decisions that would have been made in absence of economic analysis.

In perhaps slow but clear progress, the economic literature has been making progress towards addressing these problems. Increasingly, economic analysis has focused on the distributional consequences of policies. And as economic models have focused on the consequences of projects for individual groups, increased attention has been drawn to understanding the political economy and both the role and consequences of rent-seeking behavior.

4. THE POLITICAL ECONOMY OF POLICY FORMATION

Politicians face a balancing act whereby a limited government budget must be allocated among competing projects so as to preserve political support. Distribution is typically more important than efficiency in two senses. First, a politically unacceptable distributional consequence can often block acceptance of policies that have clear and large efficiency gains. Second, policies are often implemented to achieve a certain distributional goal even though efficiency consequences are adverse. Water policies that allocate water to the lower-valued uses of agriculture while high prices are imposed on domestic users are clear examples. Certainly, one role of economics is to evaluate the efficiency losses associated with individual policies as a means of putting a price tag on the distributional consequences. But is this the only role that efficiency-oriented economic analysis can play in a practical world? Both politicians and special interest groups may have little interest in determining the price tags they are imposing on the general public.

Typically, when policy alternatives arise, the groups that will be affected thereby evaluate the consequences they will enjoy or suffer. Then they lobby for or against various alternatives with the implied threat of switching support among political candidates and voting accordingly. In this process, policymakers become keenly aware of who is affected by their actions. With the high cost of political campaigns, politicians are well aware of the need for financial as well as voter support. Political survival thus dictates an overriding concern for distributional consequences.

In this political environment, a distinct possibility is that perceived consequences at the time of policy formation, rather than actual ultimate consequences, are more important in affecting policy outcomes. This raises important questions about whether policies and projects are being selected that will really make society better off or simply appease contemporary concerns. Certainly, economics has a useful role to play in educating players about the actual effects of policies (to the extent that outcomes can be accurately predicted) so that perceived and actual consequences coincide so far as available cost-effective information permits. Then politicians responding to political pressures are more likely to make the choices that truly improve social well-being.

Increasingly, economists have attempted to understand and model rent-seeking behavior and political power. For example, some economists have attempted to estimate the weights assigned by the political process to the interests of individual groups (e.g., Rausser and Freebairn 1974). Statistical tests have shown that weights are not equal as they would be if economic efficiency were the political criterion. Others have attempted to explain choices of inefficient policies by a preference for equality. For example, adoption of welfare programs for the poor may be explained by such criteria. Some economists have, in fact, argued that social welfare functions should be adopted that value equality at the expense of some inefficiency (Blackorby and Donaldson 1990). Still others such as Becker (1983), Peltzman (1976), and Zusman (1976) have developed more sophisticated models of social power which reflect the influence of individual groups on political actions. Political influence is reflected by economic resources that can be used to influence policy makers and the concentration of individual sectors that reflects the ability to organize forces without heavy transaction costs.

These models have given many insights that seem to explain the selection of policies that do not maximize economic efficiency. However, the underlying mechanism by which power is exerted is usually implicit and unclear. Furthermore, policies can often be 'explained' by more than one model or mechanism. As a result, economic modeling of political power at this point is more at a level of verifying that it exists rather than explaining how it works. Therefore, economists may not be in a position to determine a unique course of action that is preferred. Alternatively, a realistic role for economics for many problems may of necessity be constrained to illustrating how alternative courses of action affect distributional outcomes. But the value of this information should not be overlooked. With this information, the choice of distributional outcome can be more effectively made by those who are elected to work in the political process for the very purpose of making such judgements. A major product of economic analysis, if adapted to the preferences of those who use it, will be improved information.

5. THE COLLECTIVE NATURE OF GOVERNMENT

One difficulty economists face is their textbook view of government. Economists usually model government as a single entity that chooses and enforces all policies, levies all taxes, and makes all government expenditures. Economic analysis often approaches policy analysis from the standpoint of what makes sense for the economy as a whole as if one central decision maker were in control. In contrast, realities of policy formation are that actions are decided by collective bodies of Congress and are often negotiated outcomes acceptable to the balance of powers in executive, legislative, and even judicial roles. Once written into law, the implementation of broad policies is often left in the hands of a vast array of separate agencies at different levels of government. Individual agencies compete for budget dollars. Most agencies focus on the needs of a limited component of society. Federal, state, and local agencies often compete over control of the same problems.

In contrast to the comprehensive nature of their efficiency criteria, individual economists usually interact with only one or two law makers or one or two agencies, if at all. Thus, their criteria may not be well suited to the interests of their clientele. An alternative would be for economists to build models that maximize the role of individual agencies or maximize the benefits to individual sectors or constituencies of society. Such approaches may well find more direct acceptance of economic analyses by specific clientele. But such approaches would also make economists vulnerable to criticism for loss of objectivity.

On a broader scale, however, if all agencies and groups involved in the policy process were using such an approach, and economic analysis is sufficiently precise, the result should be more informed negotiation. In other words, by atomistically inducing the use of economic analysis through direct pursuit of individual interests, a kind of invisible hand could lead to a better-informed negotiated outcome of the policy process. In contrast, continued insistence on objectivity may prevent economists from being valued by any individual group, law maker, or agency so that the net outcome is non-use or limited-use of their work. With this perspective, the cost to economists of continuing to cling to objectivity is high and the cost of relinquishing objectivity is ultimately small. On the other hand, economists have a rich area of research associated with recognition that Adam Smith's invisible hand may fail when decisions are made by collective action rather than according to individual preferences.

6. CONFLICTING TIME HORIZONS OF ECONOMISTS AND POLITICIANS

Economists' and politicians' time horizons tend to differ in two major respects. First, one problem in acceptance of economic analysis is that economists and politicians tend to have different time horizons of concern for outcomes. Politicians tend to have a short time horizon associated with the need to survive the next election. They may be out of office by the time consequences of their actions are realized. The performance evaluations that count are based on expected rather than actual consequences. By comparison, once economists go beyond simple static analysis, they rarely stop short of infinite time horizons.

Second, economists and politicians tend to have different horizons for determining answers to problems. Intense media coverage of certain issues can impose significant pressures on politicians to act quickly. By comparison, because of the time required to develop theories, collect data, analyze data, and write up results, economists often cannot respond with complete analyses within a time frame required for political action by media coverage. Furthermore, many economists are not retained by employers who require immediate analysis. Many economists are employed in academia and by research foundations which permit slow and well-reasoned analyses of government action. Relatively few economists are employed in direct support of legislative or executive action and many of those are consumed with responsibilities to simply forecasting budget implications of alternative actions. Most government agency economists are consumed with day-to-day operations as opposed to analysis of alternative policy proposals.

It seems unrealistic for economists to expect politicians to change in response to these issues. Rather, economists need to adapt to the realities dictating political time horizons if economic analyses are to be used in public decision making.

7. STRATEGIC USE OF ECONOMIC ANALYSIS IN POLICY FORMATION

Another reality of the policy arena has to do with managing the flow of information. Of course, special interest groups benefit from revealing only information that tends to induce policymakers toward their desired outcome and withholding information that would induce policymakers away from their desired outcome. However, policymakers also benefit from revealing only that information that tends to keep powerful interest groups satisfied and withholding information that tends to raise unrest and dissatisfaction. Understanding this political reality is important for economists. Often policymakers are reluctant to discuss costs of programs explicitly because price tags are thereby placed on them. Price tags tend to serve as a basis for opposition and comparison. Cost evaluation also tends to call for assessing benefits and then comparing costs and benefits. However, benefit evaluations usually rest on more debatable conditions and contingencies that exacerbate controversies.

Economists must recognize the incentives of various groups in marketing the products of economic analysis. If a major value of economic analysis is improved information, then economists should seek to provide each component of information to the respective group who benefits from it. Historically, economists have likely been rather naive in trying to provide a broad information product to all groups, most of whom are adversely affected by part of it. For example, a complete economic efficiency analysis of replacing government-administered and agriculturally favorable water quotas with water markets may have large efficiency gains resulting from large gains for domestic households at the expense of losses to agricultural concerns and reduced need for government administration. Such analysis is not likely to be appreciated by farmers who currently enjoy favorable access to water. Moreover, the administering agency may feel a threat of downsizing from such a switch and may also have incentive to suppress such results.

8. HETEROGENEITY, RISK AVERSION, AND INFORMATION

One of the fundamental characteristics of trans-boundary water problems is the heterogeneity among players. Countries as well as local jurisdictions differ in many important characteristics some of which are observable and some of which are not readily observable. When characteristics differ in unobservable ways, a number of other difficulties with the concept of economic efficiency have been identified. Stiglitz (1982) among others has demonstrated that lump sum tranfers necessarily affect efficiency considerations when risk markets are incomplete, preferences are heterogeneous, or information is imperfect. Stiglitz (1985) finds in the case of imperfect information and incomplete risk markets that "the separation between equity and efficiency considerations is no longer generally valid" (p. 31). Of course, water problems are heavily plagued by risk associated with weather and resulting river flows and the players in many trans-boundary problems have distinctly different preferences. A fruitful area for economists to contribute lies in structuring agreements with self-enforcing mechanisms that provide sustainability (e.g., Frisvold and Schimmelpfennig 1997). A key step in doing so is to recognize the potential of unforseen contingencies and structure agreements accordingly (e.g., Barrett 1997; Netanyahu 1997).

Because players benefitting from certain actions may not want to have other players recognize how favorably they are affected, there is potential benefit to withholding information about individual characteristics. When players differ in ways that are not perceptible by other players, one of the most challenging aspects of cooperation has to do with structuring agreements so that such information will be revealed and incorporated into outcomes either explicitly or implicitly. This is an area in which modern economic analysis can offer important insights. The studies of principal-agent relationships and mechanism design are relatively new areas of study that address such issues (Arrow 1985). With such economic analysis, agreements can be structured so as to induce individual players to act in ways that takes private information into account. Such information-enhancing approaches to policy can be an important contribution of economics.

9. TRIBUTARY VERSUS WATERSHED MANAGEMENT

Historically, one of the errors of big government has been that all problems were treated the same regardless of their individual characteristics. As a result of the political process including media and lobbying efforts, politicians have become acutely aware of local problems caused by insensitive nationwide policies. In a few instances, these problems have been addressed by a federal-state partnership with measured success (e.g., Cairo 1997; Pigram and Musgrave 1997). In other cases, federal policy making has been avoided or relaxed in favor of state and local policy making. This seems to be particularly true of water problems requiring interstate cooperation. Policy for many environmental problems has begun to be implemented in ways that give considerable flexibility for local implementation (Wayland and Lewicki 1997). In some areas of water policy, laws have been adopted which impose legislative initiatives but leave implementation in the hands of agencies such as the U.S. Environmental Protection Agency (EPA). In this framework, the EPA and various state agencies try to work at the lowest feasible level where feasibility is determined by the complexity of community-level problems and available information. The outcome is that water quality strategies are different in each tributary. The goal is to implement local policies adapted to considerations of (1) cost effectiveness, (2) fairness, (3) implementability, and (4) the impact on living resources.

In this policy framework, balancing the marginal costs and benefits to national goals across individual watersheds becomes an interesting issue (McConnell and Strand 1997). Seemingly, leaving watershed policy entirely in the hands of local tributaries would achieve this result only by chance. Economists have long recognized the importance of heterogeneity and the local nature of interactions they imply (Hochman and Zilberman 1978; Just and Antle 1990). Many economic analyses have argued that increased efficiency was possible by tailoring the parameters of policies to local circumstances. However, broad empirical application has been difficult because of limited data availability.

With the age of information, facilitated by satellite imaging and increased computer capacity and accessability, new horizons are now available to economists. The development of these horizons holds significant prospects in an era where agencies are seeking to adapt policies to local circumstances. Clearly, economic efficiency can be achieved only by adapting policies to local circumstances in a systematic way. If data are available, economists' tools can be readily adapted to determine how policy instruments should be tailored to local circumstances to equate marginal costs and marginal benefits thus achieving the traditional concept of efficiency.

However, tailoring policy instruments to local circumstances to address equity issues and other political concerns is a major challenge. If unmet, this challenge can render economists' advice irrelevant. Economists have made great strides in including local environmental and economic conditions in models to determine how policies can be adjusted to local circumstances to improve national efficiency. Yet, if equity or other political concerns are of overriding importance to policy makers, the results of such complex modeling exercises will be of limited use.

10. BUILDING BRIDGES TO ECONOMIC ANALYSIS

Economists are frequently criticized for being too detached from real life and the political process, and for lacking understanding of ecological and hydrological processes. Undoubtedly, there is much truth behind these criticisms. But much progress is being made by incorporating more sophisticated science models representing ecological and hydrological processes. The papers by Milon et al. (1997) and Burke et al. (1997) in this volume are examples. Perhaps, economists can help to address some of the other concerns by spending more time in practical settings such as could be provided by exchange programs among academia, government agencies, and private organizations. Perhaps the interaction facilitated by such exchanges could help economists become involved with agencies at the early conceptual stages of implementation where their input is proactive and thus more useful. However, such exchanges are not without precedent and have addressed these problems in only limited ways.

Apparently, more radical steps are also required. One of the greatest difficulties is building a bridge from the simple abstract models and concepts of economics to the practical problems faced by public and private decision makers. This is primarily a problem of communication. The simple principles of economics are powerful but must be adapted to application in reality where a host of practical complexities cannot be swept under the rug with simplifying assumptions. This means that theoretical models, which are so useful to economists in identifying and defining economic concepts and principles, likely have no role in presenting results to lay decision makers. Rather, economic concepts and principles must be presented intuitively using little more than bar charts and spreadsheets, a considerable constraint relative to typical graduate student training in economics.

Historically, economists' approaches to considering practical complexities have been to create large and unwieldy empirical models that are often perceived as black boxes by lay decision makers. If the intuition of the results of such models cannot be explained in simple layman's terms, they are also not likely to be used except as justification for decisions that would have been made anyway. A growing availability of Geographic Information System (GIS) data and the increasing ability to present results of such models pictorially promises to address this problem at least partially. For example, the results presented by Bockstael and Bell (1997) exemplify how large amounts of information can be presented in maps comprehensible to lay audiences.

Two additional difficulties with such models have also limited the role of economics. First, forecasting error has typically been understated. As a result, the performance of empirical models in assessing effects of policy changes has been worse than lay audiences have been led to expect. A negative view of such exercises has ensued accordingly. Second, the complexity of such models is often so great that economists have been led to reduce the dimension of policy criteria to simple economic efficiency in order to understand or present results. As the above discussion suggests, economists cannot always lead with efficiency if distribution is more important to the policymaker. Alternatively, economists must become more effective in explaining how to make the pie bigger in ways that benefit all concerned parties. This requires going beyond identifying increments in net surplus. Practical schemes that make everyone better off without relying on lump sum transfers must be identified. If all players can see themselves gaining individually, political support is easier to achieve. Practical possibilities are provided by new approaches to linked games (e.g., Bennett et al. 1997, in this volume) in which several policies are tied together to attain a package that benefits all parties. They must also become instrumental in demonstrating how the structure of agreements can be improved. If agreements can be structured as self-enforcing mechanisms, they can be sustainable and are more likely to be adaptable to unanticipated circumstances (e.g., Frisvold and Schimmelpfennig 1997).

Another approach to increasing the direct usefulness of economic models in policy formation is to consider the political process as a constraint in maximizing social welfare. If this can be done, then economists can help society to find the best outcomes that are politically feasible. This information is of much greater practical use than knowing the best possible outcome regardless of political feasibility or the economic efficiency loss associated with a particular policy selection. Perhaps economists have a long way to go before this approach can be implemented. As mentioned above, economic models are weak in explaining how the political process works. Nevertheless, this is exactly the information that currently defines the gap of communication between economists and policy makers. Thus, further work is called for.

Finally, improving timeliness of economic analysis appears to hold significant potential in providing policy makers a bridge to use of economic analysis. No one wants to be told that something is a bad idea after they have made an irreversible decision to adopt it. Yet this is typically the role that economics has pursued. That is, because economists are often reluctant to respond to immediate policy needs until solutions are well thought out and carefully completed, politicians responding to media pressures often become wedded to specific proposals before careful economic analysis.

The dilemma of economists is whether (i) to provide feasible advice to the political process which may be imperfectly reasoned, based on limited data, and possibly later proven faulty, or (ii) to withdraw from providing relevant advice to the political process and only later provide complete analyses that provide ex post criticism of political choices. Unfortunately, most economists tend to choose the latter option and criticize the few who choose the former option. A third option is to teach correct economic principles which politicians can apply given their own perceptions. This approach may be more realistic for issues that require immediate action but is not very pleasing to either economists or politicians. Politicians want answers rather than more things to think about. And economists, because of their technical jargon, often find articulating their principles to lay audiences difficult. A willingness to invest in overcoming the latter obstacle may be crucial in making economics a more integral part of negotiating teams and the nuts-and-bolts process that actually determines political outcomes.

11. CONCLUSIONS

Perhaps this chapter and book can be best concluded by a brief review of where we are, where we need to go from here, and how we get there.

11.1 Assessing What Economics Has to Offer

The most common tools of economics stem largely from neoclassical economic roots and place heavy emphasis on growing the size of the economic pie. In this context, economists have tended to focus on policies that can potentially make everyone better off. Two primary approaches have potential to reach this end: (i) development of new technology (e.g., Brill et al. 1997b; Zilberman and Parker 1997), and (ii) facilitating competitive market behavior including eliminating market distortions and establishing market-enabling property rights (e.g., Bjornlund and McKay 1997; Brill et al. 1997a; Easter et al. 1997; Moncur et al. 1997). Policy makers largely appreciate the role of technology in making the pie bigger but the role of markets in doing so has been largely down-played by government. Whether policy makers simply have other agendas of overriding importance or economists have simply been ineffective in selling their message is perhaps unclear.

Perhaps the optimal role for economists is not understood. Some economists complain that they are perceived by the world simply as glorified accountants who are used only to evaluate budget costs of alternative actions. In contrast, economists feel they have the tools to examine distributional issues and trade-offs, to suggest improved policies and to be forward looking, and to suggest improved rules and remedies that will make private pursuits tend toward overall social well-being. While economists have continued for many decades in this vein arguing for property rights and free markets, seemingly without response, some areas of policy have begun to reflect such thinking (e.g., U.S. agricultural policy). Perhaps these messages will eventually have an equally large impact on water allocation policy once sufficient public education occurs regarding the adverse implications of current water policy.

Clearly, the social sciences are weak and imprecise compared to the natural sciences. Indeed, economics is the only social science with a theory. However, considerable evidence has been developed both supporting and not supporting economic theory.

Moreover, the neoclassical theory is not designed to apply nor does experimental testing tend to support the theory outside of market situations. Neoclassical economic theory is based on market participation by individual decision makers.

In contrast, in the context of trans-boundary water problems, much of the decision making is collective or negotiated action. Markets often do not exist and property rights are often contested. Neoclassical economic theory based on efficiency alone has little to say about how property rights should be allocated—only that they should be allocated. In the real world, agencies, governments, and organizations cannot be expected to act like competitive market participants. Even if property rights are allocated, it is not clear that efficiency is the ultimate outcome when the players make decisions by collective action.

Clearly, assessments based on neoclassical theory paint a bleak potential for economics to address important trans-boundary water problems. However, many tools beyond neoclassical concepts of efficiency have been developed and applied recently. This volume attempts to illustrate some of these approaches. Concepts of bargaining theory and the tools of game theory present a host of opportunities for modeling and predicting the outcomes of collective action and negotiation. Economic models can be enhanced to consider complex ecological and scientific processes and linkages. Models can be enhanced to consider spatially and characteristically diverse populations. Even the surplus tools of neoclassical economics can be applied to focus on the distributional differences implied by alternative policies among competing groups. Perhaps a major problem is simply that economists are still perceived to be prevented by their neoclassical tools from studying collective action and distributional consequences in heterogeneous environments.

11.2 Making Economics Useful

A widespread view by non-economists is that economists are too self-absorbed and too technical. The profession is weak empirically (in spite of the usefulness of econometrics in examining historical data), and lacks simple visual aids for communicating ideas and results to non-economists. If this is the prevailing assessment, then economists must learn how to better tell their stories, how to talk convincingly about institutional design and change, and how to use raw data to explain their answers convincingly with simple charts and tables. They must allow themselves to be more inspired by the real world, to interact more effectively, to identify problems that attract broad attention. Economists must become more proficient at justifying their results rather than just communicating their implications.

Economists must learn how to communicate. Perhaps one approach is to increase exchange programs between agencies and academia. But, in any case, economists must learn to communicate outside the circles of economics. For example, externalities and market failure are concepts largely unknown to non-economists. Economists must develop a more popular vocabulary if they expect to be understood. They need to incorporate ever broadening accuracy in representing scientific and engineering relationships depicting physical and ecological phenomena of interest to decision makers. In short, economists must become responsive to the consumers of economics.

Apart from communication, economists must continue generalizing their models and theories toward the real world. For certain problems, which apparently include trans-boundary water problems, they need to consider non-unitary actors as addressed by game theory. If economists can identify an opportunity to make everyone better off (a non-empty core in economic jargon) clearly and convincingly for policy makers, then acceptance should be enhanced compared to the case where economists can simply say that everyone taken as a whole is better off. Possibly, one of the greatest contributions by economists can be identifying combinations of policies that make this possible such as through the linked game approach illustrated by Bennett et al. (1997).

11.3 Acknowledging the Constraints of Communication

Perhaps the communication problem of economics is best illustrated by using the example of control theory. Optimal management of water resources is frequently studied using control theory because it permits careful and complete consideration of the dynamic nature of the problem. Yet the policies suggested by optimal control theory involve instantaneous adjustment of policies over time according to complicated differential equations that are typically not understood by policy makers. If policy makers do not understand such equations, there is little hope they will be incorporated into policies even it they are accurate. Clearly, economists face constraints of communication.

Considering communication constraints, one possibility is to try to communicate only the optimal steady state, which is hopefully more understandable, and let policy makers find a way to get to it. The problem with this approach is that much of the efficiency of optimal control may be lost if the optimal approach is not taken. Should economists accept that they cannot communicate the right path to the right steady state? Should they simply communicate the right steady state? Does the right steady state differ depending on how it is approached? Should constraints of communication be built into economic models? Should economists do more simulation to see how close benefits of an optimal control process can be imitated by a realistic policy? Clearly, there is much to be done to bridge the gap of communication to the policy process. And given the dynamic nature of water problems, this gap is probably greater than in many other areas of policy making. Economists simply must face up to the problem of how to find simple solutions to complex problems that approximate the potential of complex solutions.

11.4 Post Script

At the conclusion to this closing chapter, we come to the question of whether the economist's cup is half empty or half full. As in most other contexts, the answer is one of perception. Many problems exist with adapting economic analysis to the policy process but many problems have been solved or the tools to address them have been developed. Many feel that economics is not relevant but yet it has had a profound impact on the approach to many policy problems. Perhaps patience is a great virtue affecting this perception. The process of bringing economic principles to bear on important problems is a lengthy one. Education is a slow process. For example, economists have long propounded the virtues of the free market for distributing water efficiently. While these principles seemed to be ignored for years and even decades, market principles have been increasingly adopted and implemented recently (see Zilberman and Parker 1997).

Perhaps a useful gauge of the value of economic analysis is given by the payment it commands in private markets. Economic consulting has been growing rapidly in the last few decades. The premium paid for it is quite high compared to other professions. But the highly paid work is for private industry rather than for government. Economic consulting rates for government are comparatively low. The highly paid work is often for litigation support. In this context it is interesting to consider whether consultants are employed simply to defend a position rather than determine the right decision. The answer is not clear because litigation can require both types of services. However, it is noteworthy that the outcome of litigation is often determined by which economist's arguments capture the decision maker be it judge or jury. This outcome typically depends on the economist's ability to communicate economic concepts with simplicity. Perhaps if this art were better perfected for government and policy-related work, it would be more highly valued there as well.

REFERENCES

Arrow, K.J. 1985. The Economics of Agency. In J. Pratt and R. Zeckhauser, eds., *Principals and Agents: The Structure of Business*. Boston: Harvard Business School Press.

Austin, D.H., A.J. Krupnick, and V.D. McConnell. 1997. Trans-Boundary Airshed Management as an Approach to Trans-Boundary Water Cooperation: The Case of the Chesapeake Bay, in this volume.

Barrett, R. 1997. The Efficient Sharing of an Uncertain Natural Resource: A Contract Theory Approach, in this volume.

Becker, G.S. 1983. A Theory of Competition Among Pressure Groups for Political Influence, *Quarterly Journal of Economics* 97: 371-400.

Bennett,L, S. Ragland, and P. Yolles. 1997. Facilitating International Agreements Through An Interconnected Game Approach: The Case of River Basins, in this volume.

Bjornlund, H., and J. McKay. 1997. Overcoming the Introspective Legacy of Tradeable Water Entitlement Policies in South Eastern Australia, in this volume.

Blackorby, C., and D. Donaldson, 1990. A Review Article: The Case Against the Use of the Sum of Compensating Variations in Cost Benefit Analysis, *Canadian Journal of Economics* 23: 471-494

Bockstael, N., and K. Bell. 1997. Land-Use Patterns and Water Quality: The Effect of Differential Land Management Controls, in this volume.

Brill, E., U. Chakravorty, and E. Hochman. 1997a. Trans-Boundary Water Allocation Between Israel and the Gaza Strip: Desalination, Recycling and Fresh Water, in this volume.

Brill, E., G. Hochman, and E. Hochman. 1997b. Privitization and Regulation of Multi-Source Water Usage, in this volume.

Burke, S., R.M. Adams, and A. Draper. 1997. Estimating Economic and Environmental Benefits of Water Markets in a Spatially Diverse Setting, in this volume.

Cairo, R. 1997. Dealing with Interstate Water Issues: The Federal Interstate Compact Experience, in this volume.

Easter, K.W., A. Dinar, and M. Rosegrant. 1997. The Performance of Water Markets: Transaction Costs,

Interjurisdictional Barriers and Institutional Options, in this volume.

Frisvold, G., and D. Schimmelpfennig. 1997. Potential for Sustainability and Self-Enforcement of Trans-Boundary Water Agreements, in this volume.

Hochman, E., and D. Zilberman. 1978. Examination of Environmental Policies Using Production and Pollution Microparameter Distributions *Econometrica* 46: 739-760.

Lichtenberg, E., and L. Olson. 1997. Noncooperative and Cooperative Management of an Accumulative Water Pollutant, in this volume.

Just, R.E. 1988. Making Economic Welfare Analysis Useful in the Policy Process: Implications of the Public Choice Literature, *American Journal of Agricultural Economics* 70: 448-453.

Just, R.E., and J.M. Antle. 1990. Interactions Between Agricultural and Environmental Policies: A Conceptual Framework, *American Economic Review* 80: 197-202.

Layard, P.R.G., and A.A. Walters. 1978. *Microeconomic Theory.* New York: McGraw-Hill.

McCloskey, D.N. 1985. *The Rhetoric of Economics.* Madison: University of Wisconsin Press.

McConnell, K.E., and I.E. Strand. 1997. Inter-Jurisdictional Water Quality Management: Experiences From Chesapeake Bay Nutrient Management, in this volume.

Milon, W., C. Kyker, and D. Lee. 1997. Ecosystems and Social Conflict: Lessons from the Florida Everglades, in this volume.

Moncur, J., J. Roumasset, and R. Smith. 1997. Optimal Allocation of Ground and Surface Water in Oahu: Water Wars in Paradise, in this volume.

Netanyahu, S. 1997. Bilateral Water Policy Coordination Under Uncertainty, in this volume.

Peltzman, S. 1976. Toward a More General Theory of Regulation. *Journal of Law and Economics* 19: 211-40.

Pigram, J.J., and W.F. Musgrave. 1997. Sharing the Waters of the Murray-Darling Basin: Cooperative Federalism Under Test in Australia, in this volume.

Rausser, G.C. 1982. Political Economic Markets: PERTs and PESTs in Food and Agriculture, *American Journal of Agricultural Economics* 64: 821-833.

Rausser, G.C., and J.W. Freebairn. 1974. Estimation of Policy Preference Functions: An Application to U.S. Beef Import Quotas, *Review of Economics and Statistics* 56: 437-449

Stiglitz, J.E. 1982. The Inefficiency of the Stock Market Equilibrium, *Review of Economic Studies* 149: 241-61.

Stiglitz, J.E. 1985. Information and Economic Analysis: A Perspective. *Economic Journal* 95 (Conference Papers Supplement): 21-41.

Wayland, R., and C. Lewicki. 1997. Evolving Legal and Institutional Responses to Inter-Jurisdictional Water Management Issues in the U.S., in this volume.

Zilberman, D., and D. Parker. 1997. Internal Water Disputes: Causes and Solutions, in this volume.

Zusman, P. 1976. The Incorporation and Measurement of Social Power in Economic Models, *International Economic Review*, 17: 447-462.

INDEX